COMPUTATIONAL MODELING OF MULTIPHASE GEOMATERIALS

COMPUTATIONAL MODELING OF MULTIPHASE GEOMATERIALS

FUSAO OKA
SAYURI KIMOTO

CRC Press
Taylor & Francis Group
Boca Raton London New York

CRC Press is an imprint of the
Taylor & Francis Group, an **informa** business
A SPON PRESS BOOK

CRC Press
Taylor & Francis Group
6000 Broken Sound Parkway NW, Suite 300
Boca Raton, FL 33487-2742

Printed in the United States of America on acid-free paper
Version Date: 20120501

International Standard Book Number: 978-0-415-80927-6 (Hardback)

Library of Congress Cataloging-in-Publication Data

Oka, Fusao.
Computational modeling of multi-phase geomaterials / Fusao Oka, Sayuri Kimoto.
p. cm.
Includes bibliographical references and index.
ISBN 978-0-415-80927-6 (hardback)
1. Earthwork--Materials--Mathematical models. 2. Engineering geology--Mathematics. 3. Soil physics--Mathematics. 4. Plasticity--Mathematical models. I. Kimoto, Sayuri. II. Title.

TA721.O43 2012
624.1'8--dc23 2012015481

Visit the Taylor & Francis Web site at
http://www.taylorandfrancis.com

and the CRC Press Web site at
http://www.crcpress.com

Contents

Preface

Over the last three decades, studies on constitutive models and numerical analysis methods have been well developed. Nowadays, numerical methods play a very important role in geotechnical engineering and in a related activity called computational geotechnics. This book deals with the constitutive modeling of multiphase geomaterials and numerical methods for predicting the behavior of geomaterials such as soil and rock. The book provides fundamental knowledge of continuum mechanics, constitutive modeling, numerical methods for multiphase geomaterials, and their applications. In addition, the monograph includes recent advances in this area, namely, the constitutive modeling of soils for rate-dependent behavior, strain localization, the multiphase theory, and their applications in the context of large deformations. The presentation is self-contained. Much attention has been paid to viscoplasticity, water–soil coupling, and strain localization.

Chapter 1 presents the fundamental concept and results in continuum mechanics, such as motion, deformation, and stress, which are necessary for understanding the following chapters. This chapter helps readers make a self-consistent study of the contents of this book.

Chapter 2 deals with the governing equations for multiphase geomaterials based on the theory of porous media, such as water-saturated and air–water–soil multiphase soils including soil–water characteristic curves. This chapter is essential for the study of computational geomechanics.

Chapter 3 starts with the elastic constitutive model and reviews the fundamental constitutive models including plasticity and viscoplasticity. For the plasticity theory, the stability concept in the sense of Lyapunov is discussed. At the end of the chapter, cyclic plasticity and viscoplasticity models are presented with kinematical hardening rules.

In Chapter 4, failure criteria and the Cam-clay model are reviewed. For the failure criteria, many well-known criteria have been proposed in this chapter, from Coulomb's criterion to Matsuoka–Nakai's criterion. Then, the Cam-clay model is reviewed since the model includes a description of the basic properties of soil behavior such as dilatancy and the critical state concept.

Chapter 5 is devoted to the rate- and time-dependent behavior and modeling of soils. At first, typical rate- and time-dependent behaviors of soils are reviewed based on the experimental measurements. Several rate-dependent models are discussed and elastoviscoplastic models based on the Cam-clay model and Perzyna's viscoplasticity theory are presented. Adachi and Oka's model is first described and then an elastoviscoplastic model considering structural degradation is introduced. The chapter ends with the calibration of these models using the experimental results.

In Chapter 6, the virtual work theorem is presented and then the finite element method for two-phase materials is described for quasi-static and dynamic problems within the framework of the infinitesimal strain theory.

Chapter 7 deals with a typical multiphase phenomenon of soils; namely, the consolidation problem. In particular, the effects of sample thickness on consolidation, using Aboshi's well-known data, and the anomalous behavior of pore water development in the clay foundation beneath the embankment, during loading and after the end of construction embankment, are numerically analyzed.

Chapter 8 starts with a review of the study on the strain localization behavior of soils. Several issues related to the strain localization are then discussed for rate-independent and rate-dependent models. Finally, a numerical analysis of the strain localization of water-saturated clay is presented for triaxial tests and practical problems.

In Chapter 9, a liquefaction analysis method is presented with a cyclic elastoplastic model using the two-phase theory presented in Chapter 2 for water-saturated soils. Applications of the liquefaction behavior to a man-made island during an earthquake and of the soil–pile–structure interaction are shown.

Chapter 10 deals with recent advances in geomechanics. It includes the temperature-dependent behavior of soils such as consolidation due to the change in temperature, and the numerical analysis of air–water–soil coupled problems; namely, the deformation–seepage flow coupled analysis of an unsaturated river embankment is presented.

Acknowledgments

During the writing and preparation of this book, the authors became indebted to many researchers and students. In particular, we express our sincere thanks to Dr. K. Akai and Dr. T. Adachi, Emeritus Professors of Kyoto University; Dr. H. Aboshi, Emeritus Professor of Hiroshima University, for giving us data on consolidation; Dr. S. Leroueil, Professor of Laval University; Dr. A. Yashima of Gifu University; Dr. T. Kodaka of Meijo University; Dr. R. Uzuoka of Tokushima University; Dr. Y. Higo of Kyoto University; Dr. F. Zhang of Nagoya Institute of Technology; Dr. K. Sekiguchi; Dr. A. Tateishi; Dr. Y. Taguchi; Dr. S. Sunami; Dr. M. Kato; Dr. M. J. Jiang, Dr. C.-W. Lu; Y.-S. Kim; Dr. Garcia; Dr. Mojtaba Mirjalili; Dr. R. Kato; Dr. Young Seok Kim; Dr. A. W. Karnawardena; Dr. H. Feng; Dr. Md. R. Karim; Dr. B. Siribumrungwong; Mr. T. Takyu; Mr. T. Satomura; Mr. N. Nishimatsu; Ms. T. Ichinose; Mr. Takada; and the graduate students of the Geomechanics Laboratory of Kyoto University for their contributions and discussions. We thank Ms. Chikako Itoh for her daily assistance; Mr. Shahbodagh Khan Babak, a PhD student of Kyoto University, for his assistance in preparing the figures; and Ms. H. Griswold for her English corrections. Finally, we dedicate this book to our families, in particular, O. Keiko and K. Keiko.

Many thanks are also due to the following organizations and the researchers for permission for use the indicated figures: Professor H. Aboshi, Figure 5.8; Professor T. Adachi, Figure 5.7a,b; Professor Liam Finn, Figure 5.3a,b and Figure 5.5; Gihodo Syuppan Co. Ltd. (Dr. M. Saito), Figure 5.6; Professor G. Sällfore, Figure 5.9; American Society of Civil Engineers (ASCE), Figure 10.1 and Figure 10.3; ASTM, Figure 5.2 and Figure 5.12a,b; Institution of Civil Engineers (*Géotechnique*), Figure 5.1a,b and Figure 5.11; and National Research Council of Canada (NRC) (*Canadian Geotechnical Journal*), Figure 5.10.

Chapter 1

Fundamentals in continuum mechanics

In this book, we use vectors and tensors in components, and the direct notations for these vectors and tensors are given without further explanation. A dot denotes a contraction of the inner indices, for example, $a_i b_i \equiv a \cdot b$ so that $A_{ij} B_{ij} \equiv A{:}B$.

1.1 MOTION

The position of the material point $X_i (i = 1,2,3)$ of a body at time t is expressed by

$$x_i = \hat{x}_i(X_j, t) \tag{1.1}$$

Material point X_i can be given by the position of x_i at a time $t = 0$. Equation (1.1) expresses the motion of the material point of the body. The rectangular Cartesian coordinates used in this book are described by $(o, \tilde{e}_1, \tilde{e}_2, \tilde{e}_3)$ with origin o and unit base vector $\tilde{e}_i$.

There are two methods for describing the motion of a particle. One is the material description, in which the motion is expressed by material point X_i, and the other is the spatial description, in which the motion is expressed by spatial coordinates x_i. The material description is called the Lagrangian description and the spatial description is called the Eulerian description.

The velocity vector of a particle is given by

$$v_i = \frac{\partial x_i(X_j, t)}{\partial t} \tag{1.2}$$

In the material description, the acceleration of a particle in a body is expressed by

$$a_i = \frac{\partial v_i(X_j, t)}{\partial t} \tag{1.3}$$

In the spatial description, on the other hand, the acceleration of a particle is given by

$$a_i = \frac{\partial v_i(x_j,t)}{\partial t} + v_k \frac{\partial v_i(x_j,t)}{\partial x_k} \tag{1.4}$$

1.2 STRAIN AND STRAIN RATE

1.2.1 Strain tensor

Strain is the change in shape or the change in volume of a body during the application of force to the body. We need an objective measure of strain that can be derived through changes in the variation of the line element.

Let us consider the motion of the body shown in Figure 1.1. Material points P and Q have moved to points P′ and Q′ after the deformation. Points Q and Q′ are the points located in the vicinity of points P and P′.

Distance, dS, between points P and Q, is given by

$$dS^2 = dX_a dX_a \tag{1.5}$$

and the distance between points P′ and Q′ after the deformation, ds, is given by

$$ds^2 = dx_b dx_b \tag{1.6}$$

where the summation convention is used for $a,b = 1,2,3$.

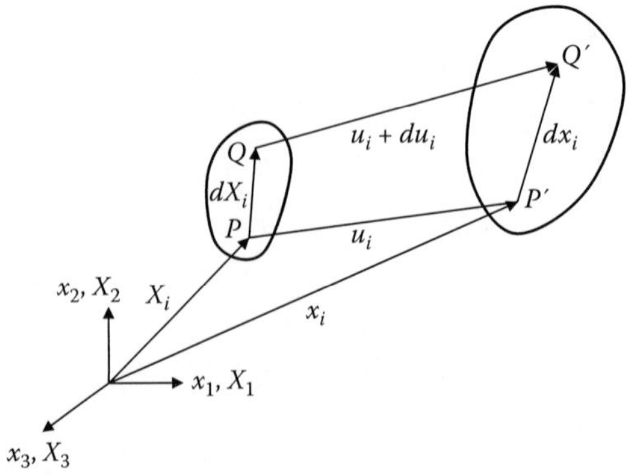

Figure 1.1 Motion.

Displacement vector $u_i (i = 1,2,3)$ is given by

$$x_i = X_i + u_i \tag{1.7}$$

Taking the difference between Equations (1.5) and (1.6), we have

$$\begin{aligned} ds^2 - dS^2 &= dx_k dx_k - dX_k dX_k = (F_{ki}F_{kj} - \delta_{ij})dX_i dX_j \\ &= \left(\frac{\partial x_k}{\partial X_i}\frac{\partial x_k}{\partial X_j} - \delta_{ij} \right) dX_i dX_j = 2E_{ij} dX_i dX_j \end{aligned} \tag{1.8}$$

where $F_{ij} = \frac{\partial x_i}{\partial X_j}$ is the deformation gradient and $\delta_{ij} = (1, i = j; 0, i \neq j)$ is Kronecker's delta. E_{ij} in Equation (1.8) is called the Green strain tensor.

Substituting Equation (1.7) into Equation (1.8), we get

$$E_{ij} = \frac{1}{2}\left(\frac{\partial u_i}{\partial X_j} + \frac{\partial u_j}{\partial X_i} + \frac{\partial u_k}{\partial X_i}\frac{\partial u_k}{\partial X_j} \right) \tag{1.9}$$

For the case of infinitesimal strain, that is, $\left| \frac{\partial u_i}{\partial X_j} \right| << 1$, the strain tensor is called an infinitesimal strain tensor. It is expressed as

$$\varepsilon_{ij} = \frac{1}{2}\left(\frac{\partial u_i}{\partial x_j} + \frac{\partial u_j}{\partial x_i} \right) \tag{1.10}$$

Engineering strain, γ_{ij}, is defined for the shear components as

$$\begin{aligned} \gamma_{ij} &= 2\varepsilon_{ij} (i \neq j) \\ \gamma_{ij} &= \varepsilon_{ij} (i = j) \end{aligned} \tag{1.11}$$

It is worth noting that the engineering strain is not a tensor.

Relative displacement, du_i, is determined by the gradient of the displacement, $\frac{\partial u_i}{\partial x_j}$, as

$$du_i = \frac{\partial u_i}{\partial x_j} dx_j = \varepsilon_{ij} dx_j + \omega_{ij} dx_j \tag{1.12}$$

where ω_{ij} is given by

$$\omega_{ij} = \frac{1}{2}\left(\frac{\partial u_i}{\partial x_j} - \frac{\partial u_j}{\partial x_i}\right) \tag{1.13}$$

and represents the rotation of a small element, for example, the rotation is not zero for the rigid body motion.

1.2.2 Compatibility relation of strain

The strain tensor has six components, although the displacement vector has only three components. This indicates that we need three independent equations to obtain the displacement vector from the strain tensor. However, six compatibility equations exist among the strain components. As for the compatibility equations, three of them are independent, and compatibility equations are necessary and provide sufficient conditions for single-value displacements in a simple connected body (see Malvern 1969).

As for the differentiation of the displacement–strain relations with respect to coordinates, we obtain the compatibility equations as

$$\frac{\partial^2 \varepsilon_{xx}}{\partial y^2} + \frac{\partial^2 \varepsilon_{yy}}{\partial x^2} = \frac{\partial^2 \gamma_{xy}}{\partial x \partial y} \tag{1.14}$$

$$\frac{\partial^2 \varepsilon_{yy}}{\partial z^2} + \frac{\partial^2 \varepsilon_{zz}}{\partial y^2} = \frac{\partial^2 \gamma_{yz}}{\partial y \partial z} \tag{1.15}$$

$$\frac{\partial^2 \varepsilon_{zz}}{\partial x^2} + \frac{\partial^2 \varepsilon_{xx}}{\partial z^2} = \frac{\partial^2 \gamma_{zx}}{\partial z \partial x} \tag{1.16}$$

$$2\frac{\partial^2 \varepsilon_{xx}}{\partial y \partial z} = \frac{\partial}{\partial x}\left(-\frac{\partial \gamma_{yz}}{\partial x} + \frac{\partial \gamma_{xz}}{\partial y} + \frac{\partial \gamma_{xy}}{\partial z}\right) \tag{1.17}$$

$$2\frac{\partial^2 \varepsilon_{yy}}{\partial x \partial z} = \frac{\partial}{\partial y}\left(\frac{\partial \gamma_{yz}}{\partial x} - \frac{\partial \gamma_{xz}}{\partial y} + \frac{\partial \gamma_{xy}}{\partial z}\right) \tag{1.18}$$

$$2\frac{\partial^2 \varepsilon_{zz}}{\partial x \partial y} = \frac{\partial}{\partial z}\left(\frac{\partial \gamma_{yz}}{\partial x} + \frac{\partial \gamma_{xz}}{\partial y} - \frac{\partial \gamma_{xy}}{\partial z}\right) \tag{1.19}$$

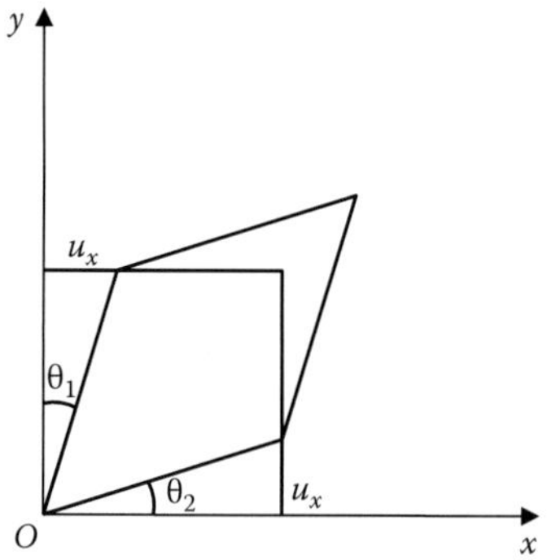

Figure 1.2 Shear deformation.

Equations (1.14) to (1.19) can be expressed by a tensor notation as

$$\varepsilon_{ij,kl} + \varepsilon_{kl,ij} - \varepsilon_{ik,jl} - \varepsilon_{jl,ik} = 0 \tag{1.20}$$

where $i,j,k,l = 1,2,3$.

1.2.3 Shear strain and deviatoric strain

Let us consider the deformation shown in Figure 1.2. The displacement vector is given by $u_x = c_1 y,\ u_y = c_2 x, c_1, c_2 > 0$.

Then,

$$\gamma_{xy} = 2\varepsilon_{xy} = \left(\frac{\partial u_x}{\partial y} + \frac{\partial u_y}{\partial x}\right) = (c_1 + c_2) \tag{1.21}$$

and the strain components, $\varepsilon_{xx}, \varepsilon_{yy}$, are zero.

Assuming a small deformation gradient, $\frac{\partial u_x}{\partial y}$ is given by θ_1 and $\frac{\partial u_y}{\partial x}$ is given by θ_2, namely,

$$\frac{\partial u_x}{\partial y} + \frac{\partial u_y}{\partial x} = \theta_1 + \theta_2 \tag{1.22}$$

This indicates that γ_{xy} expresses changes in the angle, in other words, changes in the shape, that is, shearing deformation.

$$e_{ij} = \varepsilon_{ij} - \frac{1}{3}\delta_{ij}\varepsilon_{kk} \tag{1.23}$$

is defined as the deviatoric strain tensor.

1.2.4 Volumetric strain

Setting V as the volume after the deformation and V_0 as the volume before the deformation, volumetric strain ε_v is expressed by $\varepsilon_v = \varepsilon_{kk} = (V - V_0)/V_0$.

If we express the volume before the deformation by $V_0 = dX_1 dX_2 dX_3$,

$$\begin{aligned}\varepsilon_v &= [(1+\varepsilon_{11})(1+\varepsilon_{22})(1+\varepsilon_{33})dX_1dX_2dX_3 - dX_1dX_2dX_3]/\,dX_1dX_2dX_3 \\ &= \varepsilon_{11} + \varepsilon_{22} + \varepsilon_{33} + o(\cdot)\end{aligned} \tag{1.24}$$

where $o(\cdot)$ is the higher-order small term.

We can disregard the higher-order term for the small deformation case. Next, we will consider the changes in volume for the finite deformation case.

The volume of the small hexahedron after the deformation, dV, is given by

$$dV = (d\mathbf{x} \times d\mathbf{x}) \cdot d\mathbf{x} = \varepsilon_{ijk} dx_i dx_j dx_k \tag{1.25}$$

where ε_{ijk} is a permutation (or alternating) symbol.

The volume of the small hexahedron before the deformation, dV_0, is given by

$$dV_0 = (d\mathbf{X} \times d\mathbf{X}) \cdot d\mathbf{X} = \varepsilon_{pqr} dX_p dX_q dX_r \tag{1.26}$$

Using the deformation gradient, $F_{ij} = \frac{\partial x_i}{\partial X_j}$, we get

$$J \equiv det(F_{mn}) = \frac{1}{6} \varepsilon_{ijk} \varepsilon_{pqr} F_{ip} F_{jq} F_{kr} \tag{1.27}$$

Using the following relation:

$$\varepsilon_{ijk}\varepsilon_{pqr} = \begin{vmatrix} \delta_{ip} & \delta_{iq} & \delta_{ir} \\ \delta_{jp} & \delta_{jq} & \delta_{jr} \\ \delta_{kp} & \delta_{kq} & \delta_{kr} \end{vmatrix} \tag{1.28}$$

we obtain

$$\varepsilon_{pqr} det(F_{mn}) = \varepsilon_{ijk} F_{ip} F_{jq} F_{kr} \tag{1.29}$$

As for Equation (1.29), it is worth noting that $det(F_{mn}) = \varepsilon_{ijk} F_{i1} F_{j2} F_{k3}$ following the expansion of the determinant $det(F_{mn})$.

If pqr is an even permutation of 1,2,3, we have the determinant and if pqr is odd we have the negative one.

Consequently, we have

$$dV = \varepsilon_{ijk} F_{ip} F_{jq} F_{kr} dX_p dX_q dX_r$$
$$= \varepsilon_{pqr} det(F_{mn}) dX_p dX_q dX_r = J dV_0 = \frac{\rho_0}{\rho} dV_0 \tag{1.30}$$

where ρ_0 and ρ are the initial mass density and the current mass density, respectively.

Disregarding the higher-order term leads to

$$J = det(F_{ij}) \approx 1 + \frac{\partial u_i}{\partial X_i} \tag{1.31}$$

Therefore, we obtain the following relation consistent to Equation (1.24) as

$$\frac{dV - dV_0}{dV_0} = \varepsilon_{ii} = \varepsilon_v \tag{1.32}$$

1.3 CHANGES IN AREA

The changes in area have been estimated by Nanson's formula in continuum mechanics (Malvern 1969) and are given by

$$\mathbf{n} ds = J \mathbf{F}^{-T} \mathbf{N} dS_0 \tag{1.33}$$

where n is the unit normal to area ds in the current configuration, ds is an area in the current configuration, $\mathbf{N}$ is the unit normal to the initial configuration, $\mathbf{F}^{-1}$ is the inverse of the deformation gradient, and dS_0 is an area in the initial configuration.

Surface vector $\mathbf{N}dS_0$ at point X in the referential configuration is expressed by

$$\mathbf{N} dS_0 = d\mathbf{X} \times d\mathbf{X} \tag{1.34}$$

where $d\mathbf{X}$ is an infinitesimal vector at point $\mathbf{X}$.

Surface vector $\mathbf{n}ds$ at point x in the current configuration is expressed by

$$\mathbf{n} ds = d\mathbf{x} \times d\mathbf{x}, \ (n_s ds = \varepsilon_{spq} dx_p dx_q) \tag{1.35}$$

where $d\mathbf{x}$ is an infinitesimal vector at point $\mathbf{x}$.

From Equation (1.34), we obtain

$$N_i dS_0 = \varepsilon_{ijk} \frac{\partial X_j}{\partial x_p} \frac{\partial X_k}{\partial x_q} dx_p dx_q \tag{1.36}$$

Then, multiplying both sides of Equation (1.36) by $\frac{\partial X_i}{\partial x_s}$ and using Equations (1.28) and (1.35), we get

$$J \frac{\partial X_i}{\partial x_s} N_i dS_0 = n_s ds \tag{1.37}$$

in which the relation $\varepsilon_{spq} J^{-1} = \varepsilon_{ijk} \frac{\partial X_i}{\partial x_s} \frac{\partial X_j}{\partial x_p} \frac{\partial X_k}{\partial x_q}$

Consequently, we get Equation (1.33), called Nanson's theorem, since components of $\mathbf{F}^{-1}$ and $\mathbf{F}^{-T}$ are

$$F_{ij}^{-1} = \frac{\partial X_i}{\partial x_j}, \; F_{ij}^{-T} = \frac{\partial X_j}{\partial x_i}$$

1.4 DEFORMATION RATE TENSOR

When we deal with the large deformation of a body, the material configuration changes each time, and the deformation rate tensor is useful for the analysis.

Taking a time derivative of Equation (1.8) leads to

$$\frac{d}{dt}(ds^2 - dS^2) = 2dx_k \frac{d}{dt}(dx_k) \tag{1.38}$$

$$\frac{d}{dt}(dx_k) = \frac{d}{dt}\frac{\partial x_k}{\partial X_m} dX_m + \frac{\partial x_k}{\partial X_m}\frac{d}{dt} dX_m$$

$$= \frac{\partial v_k}{\partial X_m} dX_m = dv_k = L_{km} dx_m \tag{1.39}$$

$$L_{ij} = \frac{\partial v_i}{\partial x_j} \tag{1.40}$$

where L_{ij} is called the velocity gradient tensor.

The velocity gradient tensor can be separated into symmetric part D_{ij} and antisymmetric part W_{ij} as

$$L_{ij} = D_{ij} + W_{ij} \tag{1.41}$$

$$D_{ij} = \frac{1}{2}\left(\frac{\partial v_i}{\partial x_j} + \frac{\partial v_j}{\partial x_i}\right) \tag{1.42}$$

$$W_{ij} = \frac{1}{2}\left(\frac{\partial v_i}{\partial x_j} - \frac{\partial v_j}{\partial x_i}\right) \tag{1.43}$$

Substituting the preceding equations into Equation (1.39) gives

$$\begin{aligned}\frac{d}{dt}(ds^2 - dS^2) &= 2dx_i \frac{d}{dt}(dx_i) = 2dx_i v_{i,m} dx_m = 2dx_i L_{im} dx_m \\ &= 2dx_i D_{im} dx_m + 2dx_i W_{im} dx_m\end{aligned} \tag{1.44}$$

Since W_{ij} is skew symmetric, $2dx_i W_{im} dx_m = 0$.
Hence, we have

$$\frac{d}{dt}(ds^2 - dS^2) = \frac{d}{dt}(ds^2) = 2dx_i D_{im} dx_m \tag{1.45}$$

Subsequently, D_{ij} is used to express the measure of the deformation rate at the current configuration, which is called the rate-of-deformation tensor or the stretching tensor. In contrast, W_{ij} denotes the rate of rotation and is called the spin.

In a small deformation field, we do not distinguish the deformation between the current and the reference configurations. Hence, we can use strain rate $\dot{\varepsilon}_{ij}$ instead of the deformation rate tensor as:

$$\dot{\varepsilon}_{ij} = \frac{1}{2}\left(\frac{\partial \dot{u}_i}{\partial x_j} + \frac{\partial \dot{u}_j}{\partial x_i}\right) \tag{1.46}$$

1.5 STRESS AND STRESS RATE

1.5.1 Stress tensor

The forces acting on a body can be classified into two forces: the body force and the surface force. The body force is the force acting on the body remotely, such as the gravitational force, which is proportional to the mass volume. The surface force is the force acting on the body through the surface, which is proportional to the area of the surface and is called the stress vector.

Let us consider the surface force $\mathbf{t}(\mathbf{x},t,\mathbf{n})ds$, shown in Figure 1.3(a), acting on the small surface element ds of the cross section of the body at a position $\mathbf{x}$. $\mathbf{t}$ is a surface traction vector per unit area acting on side II from side I.

In contrast, the force acting on side I from side II has the same magnitude of force as that from side I to side II, only in an opposite direction.

$$\mathbf{t}(\mathbf{x},t,-\mathbf{n})ds = -\mathbf{t}(\mathbf{x},t,\mathbf{n})ds \tag{1.47}$$

in which $\mathbf{n}$ is the unit normal vector to the surface. The surface force per unit area $\mathbf{t}$ is called the "stress vector" or the "traction vector."

Consider the stress state of a tetrahedron, which is in equilibrium under the surface, and the body forces shown in Figure 1.3(b).

The area of ΔABC is ΔS, the area of ΔOBC is ΔS_1, the area of ΔOCA is ΔS_2, and the area of ΔOAB is ΔS_3. Then,

$$\Delta S_i = \Delta S n_i, \quad \mathbf{n} = (n_1, n_2, n_3) \tag{1.48}$$

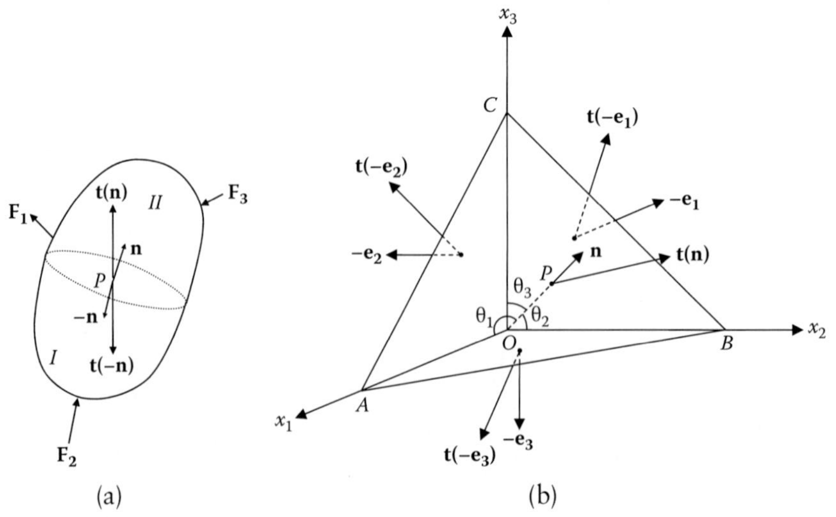

Figure 1.3 (a) Force balance. (b) Traction vectors in equilibrium.

The reason is as follows: When we set the intersection point between the perpendicular line from point O and ΔABC as point P, and $OP = h$, the volume of the tetrahedron is $\Delta Sh/3$.

Hence,

$$\Delta Sh = \Delta S_1 \overline{AO} = \Delta S_2 \overline{BO} = \Delta S_3 \overline{CO} \tag{1.49}$$

Then,

$$\Delta S_1 / \Delta S = h / \overline{AO} = \cos\theta_1 \equiv n_1 \tag{1.50}$$

where n_i is the direction cosine $cos\theta_i$.

From the equilibrium of the forces acting on the tetrahedron, we have

$$\mathbf{t}(\mathbf{n})\Delta S + \mathbf{t}(-\mathbf{e}_1)\Delta S_1 + \mathbf{t}(-\mathbf{e}_2)\Delta S_2 + \mathbf{t}(-\mathbf{e}_3)\Delta S_3 + \Delta Sh(\mathbf{F} - \mathbf{a}) / 3 = 0 \tag{1.51}$$

where F is the gravitational force and a is the inertial force.

As $h \to 0$, Equation (1.51) becomes

$$\mathbf{t}(\mathbf{n}) + \mathbf{t}(-\mathbf{e}_1)n_1 + \mathbf{t}(-\mathbf{e}_2)n_2 + \mathbf{t}(-\mathbf{e}_3)n_3 = 0 \tag{1.52}$$

Equation (1.52) can be rewritten as

$$\mathbf{t}(\mathbf{n}) + \sum_{i=1}^{3} \mathbf{t}(-\mathbf{e}_i)n_i = 0 \tag{1.53}$$

Then, if we use the following expression

$$\mathbf{n} = \sum_{k=1}^{3} n_k \mathbf{e}_k \tag{1.54}$$

the traction vector becomes

$$\mathbf{t}(n_k \mathbf{e}_k) = -n_k \mathbf{t}(-\mathbf{e}_k) = n_k \mathbf{t}(\mathbf{e}_k) \tag{1.55}$$

where Einstein's summation convention $\sum_{k=1}^{3} n_k \mathbf{e}_k \equiv n_k \mathbf{e}_k$ is used.

The stress vector can be defined as

$$\mathbf{t}(\mathbf{e}_m) \equiv \sum_{k=1}^{3} \sigma_{mk}\mathbf{e}_k \tag{1.56}$$

From Equations (1.55) and (1.56), we get Cauchy's fundamental theorem of stress vector as

$$\mathbf{t}(\mathbf{n}) = \sum_{m,k=1}^{3} \sigma_{mk}\mathbf{e}_k n_m \tag{1.57}$$

where σ_{mk} is called the stress tensor.

The kth component of t is given by $t_k = \sigma_{mk}n_m$. σ_{mk} denotes the component of the stress vector in the x_k direction acting on the perpendicular plane to the x_m axis.

When we disregard the couple stress (see Section 1.8), the stress tensor becomes symmetric from the equilibrium of the moment as

$$\sigma_{ij} = \sigma_{ji} \tag{1.58}$$

The Cauchy stress tensor is expressed by both σ_{ij} and T_{ij} in this book.

1.5.2 Principal stresses and the invariants of the stress tensor

In general, the stress vector is not parallel to the normal vector of the section, as shown in Figure 1.3. In a certain direction, however, the stress vector is parallel to the direction of normal vector, n_i, in which direction stress vector, t_i, can be expressed by

$$t_i = \sigma_{ji}n_j = \sigma n_i \tag{1.59}$$

where σ is the magnitude of the stress vector.

Since δ_{ij} is symmetric, we have

$$(\sigma_{ji} - \sigma\delta_{ij})n_j = (\sigma_{ij} - \sigma\delta_{ij})n_i = 0 \tag{1.60}$$

Equation (1.60) is a set of linear homogeneous equations for n_i and has a nontrivial solution, that is, $n_i \neq 0$ if and only if the following relation holds:

$$det\,|\,\sigma_{ij} - \sigma\delta_{ij}\,| = 0 \tag{1.61}$$

Equation (1.61) is an eigenvalue equation. When the stress tensor is symmetric, Equation (1.61) has three real roots (real eigenvalues). These three eigenvalues, $\sigma_1, \sigma_2, \sigma_3$, are called principal stresses. The direction of n_i, which satisfies Equation (1.60), is called the principal stress direction.

Equation (1.61) can be written as

$$\sigma^3 - I_1\sigma^2 + I_2\sigma - I_3 = 0 \tag{1.62}$$

$$I_1 = \sigma_{11} + \sigma_{22} + \sigma_{33} \tag{1.63}$$

$$I_2 = \sigma_{11}\sigma_{22} + \sigma_{22}\sigma_{33} + \sigma_{33}\sigma_{11} - (\sigma_{12}^2 + \sigma_{23}^2 + \sigma_{31}^2) \tag{1.64}$$

$$I_3 = \sigma_{11}\sigma_{22}\sigma_{33} + 2\sigma_{12}\sigma_{23}\sigma_{31} - (\sigma_{11}\sigma_{23}^2 + \sigma_{22}\sigma_{31}^2 + \sigma_{33}\sigma_{12}^2) \tag{1.65}$$

Since Equation (1.61) holds for the principal stress conditions, we have

$$\sigma^3 - I_1\sigma^2 + I_2\sigma - I_3 = (\sigma - \sigma_1)(\sigma - \sigma_2)(\sigma - \sigma_3) = 0 \tag{1.66}$$

Consequently, from the relation between roots and coefficients, I_1, I_2, I_3 can be expressed as

$$I_1 = \sigma_1 + \sigma_2 + \sigma_3 \tag{1.67}$$

$$I_2 = \sigma_1\sigma_2 + \sigma_2\sigma_3 + \sigma_3\sigma_1 \tag{1.68}$$

$$I_3 = \sigma_1\sigma_2\sigma_3 \tag{1.69}$$

Since I_1, I_2, I_3 are invariants under the rotation of the coordinates, they are called the first, the second, and the third invariants, respectively.

Alternatively, these three invariants can be expressed by I_1', I_2', I_3' as

$$I_1' = \sigma_{ii} \tag{1.70}$$

$$I_2' = \frac{1}{2}\sigma_{ij}\sigma_{ij} \tag{1.71}$$

$$I_3' = \frac{1}{3}\sigma_{ij}\sigma_{jk}\sigma_{ki} \tag{1.72}$$

The difference between the stress tensor and the mean value of stress tensor, σ_m, is called the deviatoric stress tensor, s_{ij}, as

$$s_{ij} = \sigma_{ij} - \sigma_m \delta_{ij} \tag{1.73}$$

$$\sigma_m = \frac{1}{3}(\sigma_{11} + \sigma_{22} + \sigma_{33}) \tag{1.74}$$

For the deviatoric stress tensor, three stress invariants exist, J_1, J_2, J_3, as

$$J_1 = 0, \ J_2 = \frac{1}{2} s_{ij} s_{ij}, \ J_3 = \frac{1}{3} s_{ij} s_{jk} s_{ki} \tag{1.75}$$

The angle of the coordinates for specifying the principal stresses is obtained by setting $\sigma'_{xy} = 0$ in Equation (1.77) as

$$\tan 2\theta = \frac{2\sigma_{xy}}{\sigma_{xx} - \sigma_{yy}} \tag{1.76}$$

$$\sigma'_{xy} = \sigma_{xy} \cos 2\theta - \frac{(\sigma_{xx} - \sigma_{yy})}{2} \sin 2\theta \tag{1.77}$$

where σ'_{xy} is a component of the stress tensor, which is transformed with respect to the rotation of the coordinates, and θ is the angle between the reference coordinates and corresponds to the principal stress directions.

PROBLEM

Show the principal stresses and their directions for the following stress tensor:

$$[\sigma_{ij}] = \begin{pmatrix} 4 & 1 & 1 \\ 1 & 2 & 1 \\ 1 & 1 & 2 \end{pmatrix}$$

Answer: If we set σ as the principal stress,

$$det \begin{bmatrix} 4-\sigma & 1 & 1 \\ 1 & 2-\sigma & 1 \\ 1 & 1 & 2-\sigma \end{bmatrix} = 0$$

This yields $(2-\sigma)^2(4-\sigma)+2-(4-\sigma)-2(2-\sigma)=0$. Then, the principal stresses are $\sigma_1=5$, $\sigma_2=2$, and $\sigma_3=1$.

From Equation (1.60), the direction corresponding to $\sigma_1=5$ is given as $-n_1+n_2+n_3=0$, $n_1-3n_2+n_3=0$, $n_1+n_2-3n_3=0$. Then, $n_1:n_2:n_3=2:1:1$, where $n_i=(n_1,n_2,n_3)$ are the components of the unit principal direction vector. Similarly, the principal direction for $\sigma_2=2$ is obtained as $2n_1+n_2+n_3=0$, $n_1+n_3=0$, $n_1+n_2=0$. Then, $n_1:n_2:n_3=-1:1:1$. The principal direction for $\sigma_3=1$ is $3n_1+n_2+n_3=0$, $n_1+n_2+n_3=0$, and then $n_1:n_2:n_3=0:1:-1$.

Let us consider the current force vector by the nominal stress vector that is the stress vector with respect to surface area dS_0 in the reference configuration as

$$t_i=T_{ji}n_j \tag{1.78}$$

$$s_i=\Pi_{ji}N_j \tag{1.79}$$

where Π_{ij} is the nominal stress tensor or the first Piola–Kirchhoff stress tensor.

Since the force is in equilibrium,

$$t_i ds=s_i dS_0 \tag{1.80}$$

Nanson's theorem, Equation (1.37), gives

$$JN_j dS_0=\frac{\partial x_s}{\partial X_j}n_s ds \tag{1.81}$$

By substituting Equations (1.78) and (1.79) for Equation (1.80) and using Equation (1.81), we obtain

$$JT_{ki}=\frac{\partial x_k}{\partial X_j}\Pi_{ji} \tag{1.82}$$

Then,

$$\Pi=J\mathbf{F}^{-1}\mathbf{T} \quad \text{or} \quad \Pi_{ij}=J\frac{\partial X_i}{\partial x_k}T_{kj} \tag{1.83}$$

1.5.3 Stress rate tensor and objectivity

When we consider the stress rate, we have to examine the objectivity of it. The objectivity is defined as the independence of the motion with respect to the observers. Since the physical law has to be objective, the physical law including the constitutive equation of materials should satisfy the objectivity. Here, we will discuss the objectivity of the stress and the stress rate tensor. The objectivity for the constitutive equations will be discussed in the next section. Herein, the objectivity follows mostly that by Malvern (1969).

The observer for an event is called a reference frame. For the different observers, there exists a transformation among them, which is expressed by a Euclid transformation.

The Euclid transformation between two frames, $(\mathbf{x},t)$ and $(\mathbf{x}^*,t^*)$, is given by

$$\mathbf{x}^* = \mathrm{Q}(t)\mathbf{x} + \mathbf{c}(t) \tag{1.84}$$

$$t^* = t - t_0 \tag{1.85}$$

where $\mathrm{Q}(t)$ denotes an orthogonal tensor that expresses the rotation between two frames, $\mathbf{c}(t)$ is the relative motion of the origin, and t_0 expresses the time difference.

For the change in reference frame from $\mathbf{x}$ to $\mathbf{x}^*$ by Equations (1.84) and (1.85), scalar C, vector $\mathbf{u}$, and tensor E are transformed as

$$\mathrm{C}^* = C \tag{1.86}$$

$$\mathbf{u}^* = \mathrm{Q}\mathbf{u} \tag{1.87}$$

$$\mathrm{E}^* = \mathrm{QEQ}^{\mathrm{T}} \tag{1.88}$$

For physical laws to be objective, they have to be described by tensor quantities. The differentiation of Equation (1.84), with respect to time, provides

$$\begin{aligned} \mathbf{v}^* &= \dot{\mathbf{c}}(t) + \mathrm{Q}(t)\mathbf{v} + \dot{\mathrm{Q}}(t)\mathbf{x}(t) \\ &= \dot{\mathbf{c}}(t) + \mathrm{Q}(t)\mathbf{v} + \dot{\mathrm{Q}}(t)\mathrm{Q}^T(t)(\mathbf{x}^* - \mathbf{c}(t)) \end{aligned} \tag{1.89}$$

where the superimposed dot $(\,\cdot\,)$ indicates time differentiation.

If we set

$$\mathbf{A} = \dot{\mathbf{Q}}\mathbf{Q}^T \tag{1.90}$$

A is the angular velocity tensor of the unstarred frame to the starred frame.

In the following, we will examine the deformation gradient ($F_{iJ} = \partial x_i / \partial X_J$), velocity gradient ($L_{ij} = \partial v_i / \partial x_j, \mathbf{L} = \dot{\mathbf{F}}\mathbf{F}^{-1}$), rate of deformation (stretching) tensor **D**, spin tensor **W**, and Cauchy stress tensor **T**.

If both the new and the old frames in the reference state are the same, from $d\mathbf{x}^* = \mathbf{Q}\mathbf{dx} = \mathbf{QF}\mathbf{dX}$, we obtain

$$\mathbf{F}^* = \mathbf{QF} \tag{1.91}$$

Differentiating the preceding equation,

$$\dot{\mathbf{F}}^* = \mathbf{Q}\dot{\mathbf{F}} + \dot{\mathbf{Q}}\mathbf{F} \tag{1.92}$$

$$\dot{\mathbf{F}}^*\mathbf{F}^{*-1} = \dot{\mathbf{F}}^*\mathbf{F}^{-1}\mathbf{Q}^T = \mathbf{Q}\dot{\mathbf{F}}\mathbf{F}^{-1}\mathbf{Q}^T + \mathbf{A} \tag{1.93}$$

Then, by setting $\mathbf{L}^* = \dot{\mathbf{F}}^*\mathbf{F}^{*-1}$, velocity gradient tensor L becomes

$$\mathbf{L}^* = \mathbf{QLQ}^T + \mathbf{A} \tag{1.94}$$

Consequently, **L** is not objective.

Since the stretching tensor (or the deformation rate tensor), $\mathbf{D} = \frac{1}{2}(\mathbf{L} + \mathbf{L}^T)$, satisfies the transformation as

$$\mathbf{D}^* = \mathbf{QDQ}^T \tag{1.95}$$

D is objective.

Moreover, the spin tensor, $\mathbf{W} = \frac{1}{2}(\mathbf{L} - \mathbf{L}^T)$, transforms as

$$\mathbf{W}^* = \mathbf{QWQ}^T + \mathbf{A} \tag{1.96}$$

Hence, **W** is not objective. **W** is a continuum spin tensor and does not represent rigid body motion.

The Cauchy stress tensor follows the transformation as

$$\mathbf{T}^* = \mathbf{QTQ}^T \tag{1.97}$$

Taking a time derivative of Equation (1.97) gives

$$\dot{\mathbf{T}}^* = \dot{\mathbf{Q}}\mathbf{T}\mathbf{Q}^{\mathrm{T}} + \mathbf{Q}\dot{\mathbf{T}}\mathbf{Q}^{\mathrm{T}} + \mathbf{Q}\mathbf{T}\dot{\mathbf{Q}}^{\mathrm{T}} \tag{1.98}$$

Then, the time derivative of the stress tensor is not objective. Let us consider the quantity as $\dot{\mathbf{T}} - \mathbf{W}\mathbf{T} + \mathbf{T}\mathbf{W}$. Since

$$\dot{\mathbf{Q}} = \mathbf{A}\mathbf{Q} \tag{1.99}$$

$$\mathbf{Q}\mathbf{Q}^{\mathrm{T}} = \mathbf{I} \tag{1.100}$$

$$\dot{\mathbf{Q}}^{T} = -\mathbf{Q}^{\mathrm{T}}\mathbf{A} \tag{1.101}$$

we have

$$\dot{\mathbf{T}}^* - \mathbf{W}^*\mathbf{T}^* + \mathbf{T}^*\mathbf{W}^* = \mathbf{Q}[\dot{\mathbf{T}} - \mathbf{W}\mathbf{T} + \mathbf{T}\mathbf{W}]\mathbf{Q}^{\mathrm{T}} \tag{1.102}$$

Hence, $\hat{\mathbf{T}}$ is objective.

$$\hat{\mathbf{T}} = \dot{\mathbf{T}} - \mathbf{W}\mathbf{T} + \mathbf{T}\mathbf{W} \tag{1.103}$$

where $\hat{\mathbf{T}}$ is called the Jaumann stress rate tensor.

The other stress rate, the Jaumann derivative of Kirchhoff stress ($J\mathbf{T}$), $\overset{\hat{o}}{\mathbf{T}}$, is objective.

$$\overset{\hat{o}}{\mathbf{T}} = \hat{\mathbf{T}} + \mathbf{T}tr(\mathbf{L}), (tr\mathbf{L} = L_{ii}) \tag{1.104}$$

Differentiating Equation (1.82) with respect to time,

$$\frac{\partial v_k}{\partial X_j}\Pi_{ji} + \frac{\partial x_k}{\partial X_j}\dot{\Pi}_{ji} = \dot{J}T_{ki} + J\dot{T}_{ki} \tag{1.105}$$

Hence,

$$\begin{aligned}\frac{\partial x_k}{\partial X_j}\dot{\Pi}_{ji} &= \dot{J}T_{ki} + J\dot{T}_{ki} - \frac{\partial v_k}{\partial X_j}\Pi_{ji} \\ &= J\left(\dot{T}_{ki} + \frac{\dot{J}}{J}T_{ki} - \frac{\partial v_k}{\partial x_p}T_{pi}\right) \\ &= J\left(\dot{T}_{ki} + L_{pp}T_{ki} - L_{kp}T_{pi}\right)\end{aligned} \tag{1.106}$$

Multiplying Equation (1.106) by $\frac{\partial X_q}{\partial x_k}$, we have

$$\frac{\partial X_q}{\partial x_k}\frac{\partial x_k}{\partial X_j}\dot{\Pi}_{ji} = J\frac{\partial X_q}{\partial x_k}\left(\dot{T}_{ki} + L_{pp}T_{ki} - L_{kp}T_{pi}\right) \tag{1.107}$$

Subsequently,

$$\dot{\Pi}_{ji} = J\frac{\partial X_j}{\partial x_k}\dot{S}_{ki} \tag{1.108}$$

$$\dot{S}_{ki} \equiv \dot{T}_{ki} + L_{pp}T_{ki} - L_{kp}T_{pi} \tag{1.109}$$

where $\dot{S}_{ij}$ is the nominal stress rate with respect to the current configuration.

1.6 CONSERVATION OF MASS

The conservation of mass continuous medium V is expressed as "the balance of mass of volume V holds if there is no mass inflow into volume V and no mass is produced in V." In this case, the balance of mass for an arbitrary volume V is expressed by

$$\frac{D}{Dt}\int_V \rho dv = 0 \tag{1.110}$$

where D/Dt denotes the material time derivative.

Equation (1.110) gives

$$\int_V \left[\frac{D\rho}{Dt} + \rho\frac{\partial v_i}{\partial x_i}\right] dv = 0 \tag{1.111}$$

When the integrand is a continuous function, the local form for the balance of mass is

$$\frac{D\rho}{Dt} + \rho\frac{\partial v_i}{\partial x_i} = 0 \tag{1.112}$$

In contrast, using Equation (1.29), the mass conservation law is expressed by

$$\rho dV = \rho_0 dV_0 \quad \text{or} \quad \rho_0 = J\rho \tag{1.113}$$

where ρ is the mass density after the deformation and ρ_0 is the density before the deformation.

Equation (1.112) is an Eulerian description and Equation (1.113) is a Lagrangian description of the mass conservation law.

1.7 BALANCE OF LINEAR MOMENTUM

The balance of linear momentum is given by the statement "the change in the linear momentum of the body occupying region R (= volume V + the boundary S) is proportional to the force acting on the body." The balance of momentum is expressed as

$$\frac{D}{Dt}\int_V \rho v_i dv = \int_S t_i ds + \int_V \rho b_i dv \tag{1.114}$$

where $\frac{D}{Dt}$ is the material time derivative, v_i is the velocity vector, t_i is the stress vector, b_i is the body force, and ρ is the mass density.

The left-hand side of Equation (1.114) indicates the time change of the linear momentum, and the first and second terms on the right-hand side express the surface force and the body force, respectively.

Using Cauchy's theorem (Equation 1.57, $t_i = \sigma_{ji} n_j$) and the Gauss theorem, and considering the balance of mass, we have

$$\int_V \left(\rho a_i - \rho b_i - \frac{\partial \sigma_{ji}}{\partial x_j} \right) dv = 0 \tag{1.115}$$

where $a_i = Dv_i / Dt$ is the acceleration term.

If Equation (1.115) holds locally,

$$\frac{\partial \sigma_{ji}}{\partial x_j} + \rho b_i = \rho a_i \tag{1.116}$$

When disregarding the acceleration term, that is, for the quasi-static case, Equation (1.116) is called the equilibrium equation.

Equation (1.116) can be expressed in component form for a two-dimensional problem as $x_1 \to x, x_2 \to y, x_3 \to z$, x,y components.

$$\frac{\partial \sigma_{xx}}{\partial x} + \frac{\partial \sigma_{yx}}{\partial y} + \frac{\partial \sigma_{zx}}{\partial z} + \rho b_x = 0 \tag{1.117}$$

$$\frac{\partial \sigma_{xy}}{\partial x}+\frac{\partial \sigma_{yy}}{\partial y}+\frac{\partial \sigma_{zy}}{\partial z}+\rho b_y=0 \tag{1.118}$$

$$\frac{\partial \sigma_{xz}}{\partial x}+\frac{\partial \sigma_{yz}}{\partial y}+\frac{\partial \sigma_{zz}}{\partial z}+\rho b_z=0 \tag{1.119}$$

were b_x, b_y, b_z are components of the body force vector.

The balance of linear momentum in the reference configuration is given by

$$\int_V \rho_0 a_i \, dV_0 = \int_V \Pi_{ji,j} dV_0 + \int_V \rho_0 b_i dV_0 \tag{1.120}$$

where a_i is the acceleration vector and Π_{ij} is the nominal stress tensor in Equation (1.83).

Taking a time derivative of the first term on the right-hand side of Equation (1.120) and using Nanson's theorem and $\rho J = \rho_0$, we find

$$\begin{aligned}\frac{D}{Dt}\int_S \Pi_{ji} N_j dS_0 &= \int_S J\dot{S}_{ki}\frac{\partial X_j}{\partial x_k} N_j dS_0 \\ &= \int_S J\dot{S}_{ki}\frac{\partial X_j}{\partial x_k}\frac{1}{J}\frac{\partial x_p}{\partial X_j} n_p ds \\ &= \int_S \dot{S}_{ki} n_k ds\end{aligned} \tag{1.121}$$

Hence, a rate type of balance of linear momentum, with respect to the current configuration, is obtained as

$$\int_V \rho \, \dot{a}_i \, dv = \int_S \dot{S}_{ki} n_k \, ds + \int_V \rho \, \dot{b}_i dv \tag{1.122}$$

Under the static conditions with constant body force, the preceding equation becomes

$$\int_V \dot{S}_{ki,k} dv = 0 \tag{1.123}$$

The above rate equilibrium equation will be used for the updated Lagrangian formulation of the boundary value problem.

1.8 BALANCE OF ANGULAR MOMENTUM AND THE SYMMETRY OF THE STRESS TENSOR

From the balance of angular momentum, the sum of the momentum is zero in the case of a zero time rate for the angular momentum. Then,

$$\int_S \mathbf{t}(\mathbf{n})\times \mathbf{r}ds+\int_v \rho\mathbf{b}\times \mathbf{r}dv-\int_v \rho\mathbf{a}\times \mathbf{r}dv+\int_S \mathbf{M}ds=0 \tag{1.124}$$

where × denotes the vector product, **t**(**n**) is the stress vector, **b** is the body force vector, **a** is the inertia force vector, **r** is the position vector, and **M** is the couple stress vector. Couple stress is called moment stress and cannot be disregarded for materials with a significant rotation of the particles, such as granular materials.

Equation (1.124) can be written in component form as

$$\int_s \varepsilon_{ijk}t_k x_j ds+\int_V \varepsilon_{ijk}\rho b_k x_j dv-\int_V \varepsilon_{ijk}\rho a_k x_j dv+\int_s M_i ds=0 \tag{1.125}$$

where ε_{ijk} is the permutation symbol.

Using Cauchy's theorem and the divergence theorem, we have

$$\int_s \varepsilon_{ijk}t_k x_j ds=\int_s \varepsilon_{ijk}x_j\sigma_{mk}n_m ds=\int_v \varepsilon_{ijk}\left(x_j\frac{\partial\sigma_{mk}}{\partial x_m}+\sigma_{jk}\right)dv \tag{1.126}$$

Considering couple stress tensor μ_{ij}, we obtain $M_i=\mu_{ji}n_j$; and Equation (1.125) becomes

$$\int_V \varepsilon_{ijk}\left[x_j\left(\frac{\partial\sigma_{mk}}{\partial x_m}+\rho b_k-\rho a_k\right)\right]dv+\int_V\left(\varepsilon_{ijk}\sigma_{jk}+\frac{\partial\mu_{ji}}{\partial x_j}\right)dv=0 \tag{1.127}$$

Upon substituting Equation (1.116), the first term of Equation (1.127) becomes zero.

Hence,

$$\int_V\left(\varepsilon_{ijk}\sigma_{jk}+\frac{\partial\mu_{ji}}{\partial x_j}\right)dv=0 \tag{1.28}$$

When the stress distribution is continuous, the local form for Equation (1.128) is

$$\varepsilon_{ijk}\sigma_{jk} + \frac{\partial \mu_{ji}}{\partial x_j} = 0 \tag{1.129}$$

when couple stress tensor μ_{ji} is zero.

For $i = 1$, $\varepsilon_{123}\sigma_{23} + \varepsilon_{132}\sigma_{32} = \sigma_{23} - \sigma_{32} = 0$, $\sigma_{23} = \sigma_{32}$.

In general,

$$\sigma_{ij} = \sigma_{ji} \tag{1.130}$$

Consequently, the stress tensor is symmetric when couple stress tensor μ_{ji} is zero.

1.9 BALANCE OF ENERGY

The energy conservation law is called the first law of thermodynamics, and it is described as follows: "The time rate of total energy of the mass system is equal to the sum of the external mechanical work rate done by the body force and the surface force, heat inflow through the surface of the body and the other energy supply."

$$\dot{K} + \dot{E} = F + Q \tag{1.131}$$

$$\dot{K} = \frac{D}{Dt}\int_V \frac{1}{2}\rho v_i v_i dv \tag{1.132}$$

$$F = \int_V \rho b_i v_i dv + \int_S t_i v_i ds \tag{1.133}$$

$$\dot{E} = \int_V \rho \dot{e} dv \tag{1.134}$$

$$Q = \int_V \rho h dv - \int_S q_i n_i ds \tag{1.135}$$

where $\dot{K}$ is the rate of the mechanical energy, E is the internal energy, e is the internal energy density, F is the external work rate, Q is the heat inflow and other supplies of energy, h is the energy supply density such as radiation, and q_i is the heat flow vector.

From the conservation of linear momentum, we have

$$F = \frac{D}{Dt}\int_V \frac{1}{2}\rho v_i v_i dv + \int_V D_{ij}\sigma_{ij} dv \tag{1.136}$$

Then,

$$\dot{E} = \int_V D_{ij}\sigma_{ij} dv + \int_V \rho h dv - \int_S q_i n_i ds \tag{1.137}$$

The local form for Equation (1.131) becomes

$$\rho\dot{e} = D_{ij}\sigma_{ij} + \rho h - q_{i,i} \tag{1.138}$$

1.10 ENTROPY PRODUCTION AND CLAUSIUS–DUHEM INEQUALITY

The second law of thermodynamics is described as follows: "The time rate of the entropy of the body is not less than the change in entropy associated with the heat inflow and the other supplies of energy."

In other words, the entropy production during the motion of a body is not always negative.

$$\dot{N} \geq \dot{H} \tag{1.139}$$

$$\dot{N} = \frac{D}{Dt}\int_V \rho\eta dv \tag{1.140}$$

where η is the entropy density.

$$\dot{H} = \int_V \frac{h}{\theta}\rho dv + \int_S -\frac{q_i}{\theta} n_i ds \tag{1.141}$$

where θ is the temperature.

$$\rho\frac{D\eta}{Dt} \geq \frac{h}{\theta}\rho - \left(\frac{q_i}{\theta}\right)_{,i} \tag{1.142}$$

The local form for Equation (1.139), Clausius–Duhem Inequality, is

If we set $\dot{\eta} = D\eta / Dt$, using the first law of thermodynamics, the preceding equation becomes

$$\rho\dot{\eta}\theta + \sigma_{ij}D_{ij} - \rho\dot{e} - \frac{1}{\theta}q_i\frac{\partial\theta}{\partial x_i} \geq 0 \tag{1.143}$$

If we set Helmholtz's free energy function as $\Psi = e - \eta\theta$, Equation (1.143) becomes

$$-\rho(\dot{\Psi} + \eta\dot{\theta}) + \sigma_{ij}D_{ij} - \frac{1}{\theta}q_i\frac{\partial\theta}{\partial x_i} \geq 0 \tag{1.144}$$

Let us consider the case with a small change in density ρ and the following relations:

$$\Psi = \hat{\Psi}(\varepsilon_{ij}^e, \theta) \tag{1.145}$$

$$\dot{\varepsilon}_{ij} = \dot{\varepsilon}_{ij}^e + \dot{\varepsilon}_{ij}^{vp}, \quad D_{ij} = \dot{\varepsilon}_{ij} \tag{1.146}$$

where $\dot{\varepsilon}_{ij}^e$ and $\dot{\varepsilon}_{ij}^{vp}$ are the elastic strain rate and the inelastic strain rate, respectively.

From Equation (1.144), we have

$$-\rho\left(\frac{\partial\Psi}{\partial\theta} + \eta\right)\dot{\theta} - \rho\left(\frac{\partial\Psi}{\partial\varepsilon_{ij}^e}\right)\dot{\varepsilon}_{ij}^e + \sigma_{ij}\dot{\varepsilon}_{ij} - \frac{1}{\theta}q_i\frac{\partial\theta}{\partial x_i} \geq 0 \tag{1.147}$$

$$\eta = -\frac{\partial\Psi}{\partial\theta}, \quad \sigma_{ij} = \rho\frac{\partial\Psi}{\partial\varepsilon_{ij}^e} \tag{1.148}$$

Hence,

$$\sigma_{ij}\dot{\varepsilon}_{ij}^{vp} - \frac{1}{\theta}q_i\frac{\partial\theta}{\partial x_i} \geq 0 \tag{1.149}$$

Equation (1.149) indicates that the internal entropy production occurs due to the inelastic strain and the heat flow.

Truesdell and Noll (1965) defined internal entropy production δ as

$$\delta = \rho\dot{\eta} + \frac{1}{\theta}\sigma_{ij}D_{ij} - \frac{1}{\theta}\rho\dot{e} - \frac{1}{\theta^2}q_i\frac{\partial\theta}{\partial x_i} \geq 0 \quad (1.150)$$

The strong sufficient conditions for Equation (1.150) to be true are given by the following two inequalities:

$$\delta = \rho\dot{\eta} + \frac{1}{\theta}\sigma_{ij}D_{ij} - \frac{1}{\theta}\rho\dot{e} \geq 0 \quad (1.151)$$

$$-\frac{1}{\theta^2}q_i\frac{\partial\theta}{\partial x_i} \geq 0 \quad (1.152)$$

When Fourier's law of heat flow is expressed as

$$q_i = -\kappa\theta_{,i} \quad \kappa: \text{heat conduction coefficient} \quad (1.153)$$

Equation (1.152) becomes

$$\frac{1}{\rho\theta^2}\kappa\left(\frac{\partial\theta}{\partial x_i}\right)^2 \geq 0 \quad (1.154)$$

Consequently,

$$\kappa \geq 0 \quad (1.155)$$

Namely, κ is nonnegative.

An important thermodynamical framework for the plasticity theory has been studied by Collins and Houlsby (1997) based on the Ziegler's theory of dissipation function. The related results will be presented in Chapter 3.

1.11 CONSTITUTIVE EQUATION AND OBJECTIVITY

As was been mentioned earlier, there are nine fundamental laws in continuum mechanics, except for electromagnetic laws. These include the mass conservation law (1), the conservation laws of linear momentum (3), the conservation laws of angular momentum (3), the conservation of energy (1), and the entropy production inequality (1, constraint condition).

On the other hand, there are nineteen variables contained in the laws, namely, the mass density (1), velocity components (3), the components of the stress tensor (9), temperature (1), the components of the heat flow vector (3), internal energy (1), and entropy (1).

Hence, eleven more equations are required to describe the response of materials. These eleven equations are called constitutive equations in order to specify the response characteristics of materials. The number of equations is eleven, that is, six for stress–strain relations, three for heat flux, one for internal energy, and one for entropy.

Constitutive equations are not given a priori but are derived based on experiments satisfying the fundamental laws and objectivity. The well-known Hooke's law is a typical constitutive equation for elastic materials.

1.11.1 Principle of objectivity and constitutive model

The response of a material to external action is independent of the observer. It indicates that the constitutive equation should be indifferent to changes in the coordinate frame. In the following, we will discuss the objectivity of the constitutive equations and how the principle of objectivity prescribes constitutive equations.

Constitutive equations describe the material's inherent response to external action and are expressed by a functional of the history of motion and deformation. This functional is called the constitutive functional.

For example, when the stress tensor **T** at a material point X is determined by the motion of material point **X**′ in the vicinity of X, constitutive functional **G** is given by

$$\mathbf{T} = \mathbf{G}[\mathbf{x}(\mathbf{X}', t), \theta(\mathbf{X}', t), \mathbf{X}] \tag{1.156}$$

Constitutive tensor functional **G** has to be indifferent with respect to the rigid rotation, the translational motion, and the time shift so that constitutive functional **G** satisfies the principle of objectivity as

$$\mathbf{x}^*(\mathbf{X}', t^*) = \mathbf{Q}(t)\mathbf{x}(\mathbf{X}', t) + \mathbf{c}(t) \tag{1.157}$$

$$t^* = t - t_0 \tag{1.158}$$

$$\mathbf{G}[\mathbf{x}(\mathbf{X}', t), \theta(\mathbf{X}', t), \mathbf{X}, t] = \mathbf{G}[\mathbf{x}^*(\mathbf{X}', t^*), \theta(\mathbf{X}', t^*), \mathbf{X}, t - t_0] \tag{1.159}$$

where $\mathbf{X}' \in \beta$ is a point in the vicinity of point **X**.

In the above, we assume that functional G depends on the motion of the material points in the vicinity of point X and, in general, the constitutive functional depends on the past history ($-\infty < t' \le t$). Herein, however, we disregard it for the sake of simplicity.

1.11.2 Time shift

When we set $\mathbf{Q}(t) = I$, $\mathbf{c}(t) = 0$, $t_0 = t$, and $\mathbf{x}^*(\mathbf{X}', t^*) = \mathbf{x}(\mathbf{X}', t), t^* = 0$. Then, for example, the stress tensor T can be expressed by functional G as

$$\mathbf{T} = \mathbf{G}[\mathbf{x}(\mathbf{X}', 0), \theta(\mathbf{X}', 0), \mathbf{X}, 0] \tag{1.160}$$

This indicates that the functional does not explicitly depend on time. In other words, the response functional, which explicitly depends on time, is not objective.

1.11.3 Translational motion

When we set $\mathbf{Q}(t) = I$, $\mathbf{c}(t) = -\mathbf{x}(\mathbf{X}, t)$, and $t_0 = 0$,

$$\mathbf{x}^*(\mathbf{X}', t^*) = \mathbf{x}(\mathbf{X}', t) - \mathbf{x}(\mathbf{X}, t), \quad t^* = t \tag{1.161}$$

$$\mathbf{T} = \mathbf{G}[\mathbf{x}(\mathbf{X}', t) - \mathbf{x}(\mathbf{X}, t), \theta(\mathbf{X}', t), \mathbf{X}, t] \tag{1.162}$$

1.11.4 Rotational motion

When $\mathbf{Q}(t)$ is set to be arbitrary and $\mathbf{c}(t) = 0$ and $t_0 = 0$,

$$\mathbf{Q}(t)\mathbf{T}(\mathbf{X}, t)\mathbf{Q}^T(t) = \mathbf{G}[\mathbf{Q}(t)\mathbf{x}(\mathbf{X}', t), \theta(\mathbf{X}', t), \mathbf{X}, t] = \mathbf{T}^* \tag{1.163}$$

From Equation (1.162), we get

$$\mathbf{Q}(t)\mathbf{G}[\mathbf{x}(\mathbf{X}', t) - \mathbf{x}(\mathbf{X}, t), \theta(\mathbf{X}', t), \mathbf{X}]\mathbf{Q}(t)^T = \mathbf{G}[\mathbf{Q}(t)(\mathbf{x}(\mathbf{X}'), t) - \mathbf{x}(\mathbf{X}, t)), \theta, \mathbf{X}] \tag{1.164}$$

Taking the Taylor series of $\mathbf{x}(\mathbf{X}', t)$ around $\mathbf{x}(\mathbf{X}, t)$, assuming the continuity of the functional, we have

$$\mathbf{x}(\mathbf{X}', t) - \mathbf{x}(\mathbf{X}, t) = \mathbf{F}(\mathbf{X}, t)d\mathbf{X} \tag{1.165}$$

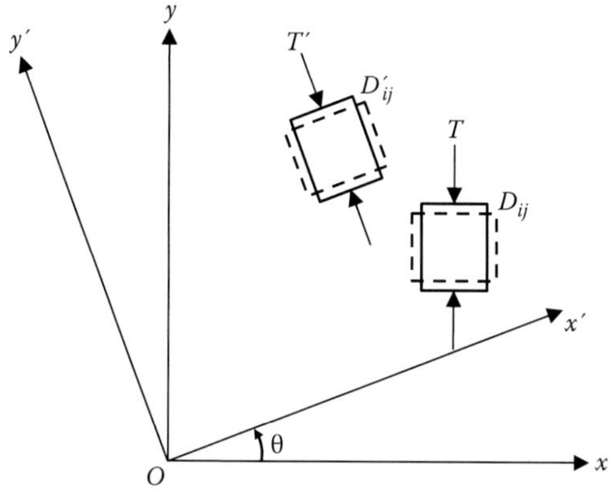

Figure 1.4 Change of reference frame.

Hence,

$$\mathbf{Q}(t)\mathbf{T}\mathbf{Q}^{T}(t) = \mathbf{G}[\mathbf{Q}\mathbf{F}(\mathbf{X},t),\theta(\mathbf{X}',t),\mathbf{X}] \tag{1.166}$$

The dependence of the functional on the relative position leads to the dependence on the deformation gradient, $F_{i,j} = \frac{\partial x_i}{\partial X_j}$, and the dependence on the strain.

For example, consider the case in which the deformation rate tensor, D, depends on the stress, namely,

$$\mathbf{D} = \mathbf{G}(\mathbf{T}) \tag{1.167}$$

The objective functional G satisfies

$$\mathbf{Q}\mathbf{D}\mathbf{Q}^{\mathrm{T}} = \mathbf{G}(\mathbf{Q}\mathbf{T}\mathbf{Q}^{\mathrm{T}}) \tag{1.168}$$

for a rotation Q.

From a physical point of view, it can be seen that the deformation rate tensor rotates according to the rotation of the loading system. Hence, the principle of objectivity can be viewed the principle of the space isotropy (Figure 1.4).

REFERENCES

Belytschko, T., Liu, W.K., and Moran, B. 2000. *Nonlinear Finite Elements for Continua and Structures*, John Wiley & Sons, New York.

Boehler, J.P. 1987. *Applications of Tensor Functions in Solid Mechanics*, CISM Courses and Lectures, No. 292, Springer-Verlag, New York.

Collins, I.F., and Houlsby, G.T. 1997. Application of thermomechanical principles to the modeling of geotechnical materials, *Proc. Roy. Soc. London A*, 453:1975–2001.

Eringen, A.C. 1967. *Mechanics of Continua*, New York: John Wiley & Sons.

Fung, Y.C., and Tong, P. 2001. *Classical and Computational Solid Mechanics*, World Scientific.

Gurtin, M.E. 1982. *An Introduction to Continuum Mechanics*, New York: Academic Press.

Malvern, L.E. 1969. *Introduction to the Mechanics of a Continuous Media*, New York: Prentice-Hall.

Maugin, G.A. 1992. *The Thermomechanics of Plasticity and Fracture*, Cambridge University Press.

Spencer, A.J.M. 1988. *Continuum Mechanics*, Longman Scientific and Technical, New York.

Truesdell, C., and Noll, W. 1965. The non-linear field theories of mechanics, In *Encyclopedia of Physics*, Vol. III, Part 3, ed. by S. Flügge, Berlin: Springer-Verlag.

Ziegler, H. 1983. *An Introduction to Thermomechanics*, 2nd ed., Elsevier Science, North-Holland, Amsterdam.

Chapter 2

Governing equations for multiphase geomaterials

One of the important characteristics of geomaterials is that the material is composed of solid, liquid, and gas in general. In this chapter, governing equations for the analysis including balance laws and constitutive equations are presented based on the theory of porous media, that is, an immiscible mixture of solid and fluids. First, governing equations for fluid–solid materials are shown, then in Section 2.2, the governing equations for gas–liquid–solid three-phase materials are formulated. In Section 2.3 the equations for the unsaturated saturated soils are presented.

2.1 GOVERNING EQUATIONS FOR FLUID–SOLID TWO-PHASE MATERIALS

2.1.1 Introduction

The governing equations for pore water–soil coupled problems can be derived from Biot's theory of water saturated porous media, which is based on continuum mechanics (Biot 1941, 1955, 1956, 1962; Atkin and Craine 1976; Bowen 1976). Before Biot's work, Fillunger (1913) proposed a theory of porous media filled with water. Historical development of the theory of porous media has been well documented by de Boer (2000a,b). Various methods are proposed for Biot's two-phase mixture theory depending on the method of approximation and the choice of unknown variables (Coussy 1995; Lewis and Schrefler 1998; Zienkiewicz et al. 1999; Ehlers, Graf, and Ammann 2004). In many computer programs for liquefaction and consolidation analyses, a u-p formulation is adopted in which the displacement (u) of the solid and pore water pressure (p) are used as the unknown variables, because we can reduce the degree of freedom although the u-p formulation provides a different solution in the high frequency range for the higher permeability

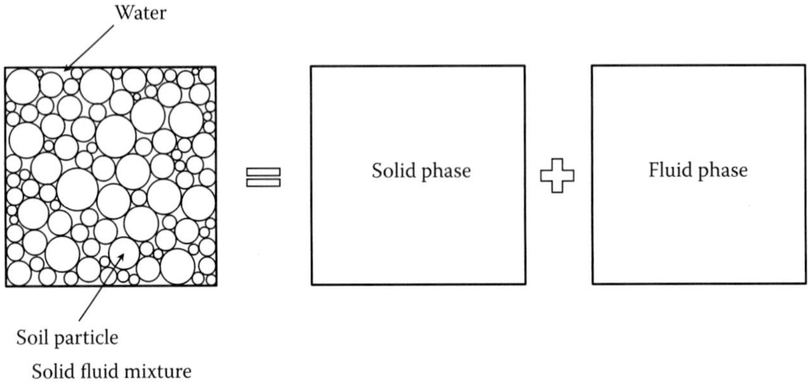

Figure 2.1 Superposition of solid and fluid phases.

(Zienkiewicz et al., 1980, LIQCA Res. Development Group, 2005). This u-p formulation can be easily applied to the consolidation problems since the displacement of pore water is explicitly introduced.

2.1.2 General setting

The following assumptions are adopted in the u-p formulation:

1. An infinitesimal strain is used.
2. The relative acceleration of the fluid phase to that of the solid phase is much smaller than the acceleration of the solid phase.
3. The grain particles in the soil are incompressible.
4. The effect of the temperature is disregarded.

The motion of the mixture for the multiphase medium is described by the superposition of multiphases in the context of continuum theory as shown in Figure 2.1. For the fluid–solid two-phase mixture, we assume that each point within the mixture is occupied simultaneously by two constituents and described by the rectangular Cartesian coordinates (Figure 2.2).

In the followings, the material time derivative is given by

$$\frac{D^{(a)}}{Dt} = \frac{\partial}{\partial t} + v_i^a \frac{\partial}{\partial x_i^a} \tag{2.1}$$

where superscript a indicates phase a and v_i^a is the velocity of the material in phase a, x_i^a is the position of particle of a phase.

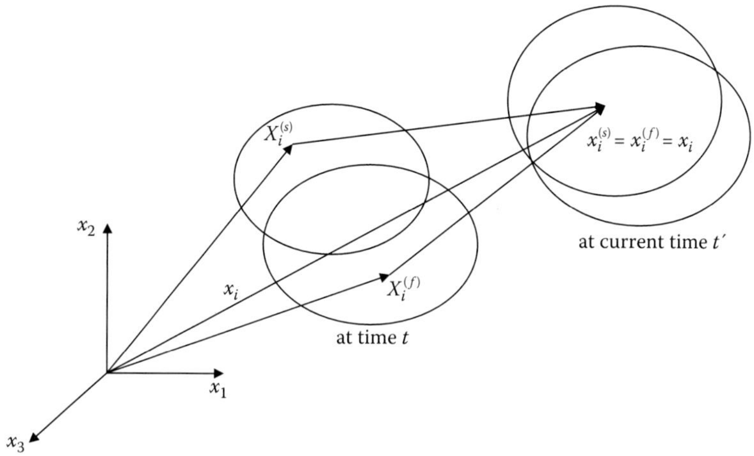

Figure 2.2 Geometric arrangement.

At current state at time t', $x_i^{(s)} = x_i^{(f)}$ for two phases; (s) stands for solid and (f) for fluid phases in Figure 2.2. Hence, material time derivative for the multiphases can be described in the spatial coordinate x_i and the superscripts can be neglected as:

$$\frac{D}{Dt} = \frac{\partial}{\partial t} + v_i^a \frac{\partial}{\partial x_i} \tag{2.2}$$

2.1.3 Density of mixture

The densities of the solid phase, $\bar{\rho}^s$, and the fluid phase, $\bar{\rho}^f$, are defined as

$$\bar{\rho}^s = (1-n)\rho^s, \bar{\rho}^f = n\rho^f \tag{2.3}$$

where n is the porosity, ρ^s is the density of the solid, and ρ^f is the density of the fluid.

The density of the mixture ρ is described using Equations (2.1) and (2.2) as

$$\rho = \bar{\rho}^s + \bar{\rho}^f = (1-n)\rho^s + n\rho^f \tag{2.4}$$

In Biot's theory, the water-saturated soil is described by the superposition of the solid phase and the fluid phase as shown in Figure 2.1.

2.1.4 Definition of the effective and partial stresses of the fluid–solid mixture theory

The total stress is given by the sum of the partial stresses acting on the phases as

$$\sigma_{ij} = \sigma_{ij}^s + \sigma_{ij}^f \tag{2.5}$$

where σ_{ij}^s is the partial stress tensor of the solid phase and σ_{ij}^f is the partial stress tensor of the fluid phase.

The partial stresses for the fluid phase and the solid phase are given by

$$\sigma_{ij}^f = -np\delta_{ij} \tag{2.6}$$

$$\sigma_{ij}^s = \sigma'_{ij} - (1-n)p\delta_{ij} \tag{2.7}$$

where σ_{ij} is the total stress tensor, σ'_{ij} is the effective stress tensor, p is the pore water pressure, n is the porosity, and δ_{ij} is the Kronecker's delta. In the derivation, tension is positive but the pore water pressure is positive for compression.

The total stress is described by the effective stress and the pore water pressure as

$$\sigma_{ij} = \sigma'_{ij} - p\delta_{ij} \tag{2.8}$$

2.1.5 Displacement–strain relation

From Assumption 1, the displacement–strain relations for the solid and the fluid phases are defined as

$$\varepsilon_{ij}^s = \frac{1}{2}\left(\frac{\partial u_i^s}{\partial x_j} + \frac{\partial u_j^s}{\partial x_i}\right), \ \varepsilon_{ij}^f = \frac{1}{2}\left(\frac{\partial u_i^f}{\partial x_j} + \frac{\partial u_j^f}{\partial x_i}\right) \tag{2.9}$$

where ε_{ij}^s is the strain tensor of the solid phase and u_i^s is the displacement vector of the solid phase, ε_{ij}^f is the strain tensor of the fluid phase, and u_i^f is the displacement vector of the fluid phase.

The strain rates are given by the time differentiation of strains as

$$\dot{\varepsilon}_{ij}^s = \frac{1}{2}\left(\frac{\partial \dot{u}_i^s}{\partial x_j} + \frac{\partial \dot{u}_j^s}{\partial x_i}\right), \ \dot{\varepsilon}_{ij}^f = \frac{1}{2}\left(\frac{\partial \dot{u}_i^f}{\partial x_j} + \frac{\partial \dot{u}_j^f}{\partial x_i}\right) \tag{2.10}$$

where $(\dot{\cdot})$ denotes the time differentiation.

2.1.6 Constitutive model

The constitutive relations of the solid phase are given by the relations between the incremental strains and effective stress increments as

$$\Delta\sigma'_{ij} = D_{ijkl}\Delta\varepsilon^s_{kl} \tag{2.11}$$

where $\Delta\sigma'_{ij}$ is the effective stress increment tensor, D_{ijkl} is the modulus tensor, and $\Delta\varepsilon^s_{kl}$ is the strain increment tensor of the solid phase.

In the case of the elastoplastic model, it becomes

$$D_{ijkl} = D^{ep}_{ijkl} \tag{2.12}$$

When disregarding the viscous resistance of fluid phase, the constitutive equation is given by

$$p = -K^f\varepsilon^f_{ii} \tag{2.13}$$

where K^f is the elastic volumetric modulus of the pore fluid.

2.1.7 Conservation of mass

The mass conservation laws for the solid and the fluid phases are given by

$$\frac{\partial\bar{\rho}^s}{\partial t} + \frac{\partial(\bar{\rho}^s\dot{u}^s_i)}{\partial x_i} = 0 \tag{2.14}$$

$$\frac{\partial\bar{\rho}^f}{\partial t} + \frac{\partial(\bar{\rho}^f\dot{u}^f_i)}{\partial x_i} = 0 \tag{2.15}$$

2.1.8 Balance of linear momentum

The linear momentum conservation laws for the two phases are given as

$$\bar{\rho}^s\ddot{u}^s_i - R_i = \frac{\partial\sigma^s_{ij}}{\partial x_j} + \bar{\rho}^s b_i \tag{2.16}$$

$$\bar{\rho}^f\ddot{u}^f_i + R_i = \frac{\partial\sigma^f_{ij}}{\partial x_j} + \bar{\rho}^f b_i \tag{2.17}$$

where b_i is the body force vector, and R_i is the term expressing the energy dissipation due to the relative motion between the solid and the fluid phases (Biot 1956).

$$R_i = n\frac{\gamma_w}{k}\dot{w}_i \tag{2.18}$$

$$\dot{w}_i = n(\dot{u}_i^f - \dot{u}_i^s) \tag{2.19}$$

where k is the permeability coefficient(assumed to be scalar because of isotropy), and $\dot{w}_i$ is the relative velocity vector of the fluid phase to the solid phase. $\gamma_w = \rho^f g$ is the unit weight of the pore water with the gravitational acceleration g.

When we describe R_i by Equation (2.18), it is easily shown that the equation of motion for the fluid phase is a general description of Darcy's law.

Considering Equation (2.19) and that the porosity is constant, we obtain the following equation from the equation of motion for the fluid phase Equation (2.17), after manipulation:

$$\bar{\rho}^f \ddot{u}_i^s + \frac{\bar{\rho}^f}{n}\ddot{w}_i + R_i = \frac{\partial \sigma_{ij}^f}{\partial x_j} + \bar{\rho}^f b_i \tag{2.20}$$

If the relative acceleration is almost zero,

$$\ddot{u}_i^s \gg \dot{w}_i \tag{2.21}$$

Equation (2.20) can be approximated considering Equation (2.21):

$$\bar{\rho}^f \ddot{u}_i^s + R_i = \frac{\partial \sigma_{ij}^f}{\partial x_j} + \bar{\rho}^f b_i \tag{2.22}$$

Substituting Equations (2.2), (2.5), and (2.18) into Equation (2.22), we obtain

$$n\rho^f \ddot{u}_i^s + n\frac{\gamma_w}{k}\dot{w}_i = -\frac{\partial np}{\partial x_i} + n\rho^f b_i \tag{2.23}$$

When we assume that the spatial gradient of the porosity is sufficiently small, the following equation holds:

$$\frac{\partial n}{\partial x_i} = 0 \tag{2.24}$$

Substituting Equation (2.24) into Equation (2.23), we have

$$\rho^f \ddot{u}_i^s + \frac{\gamma_w}{k}\dot{w}_i = -\frac{\partial p}{\partial x_i} + \rho^f b_i \tag{2.25}$$

Then, when the second term on the right-hand side of Equation (2.25) is a body force due to the gravitational force, and we disregard the dynamic term, the following equation holds:

$$-\frac{\partial p}{\partial x_i} + \rho^f b_i = 0 \tag{2.26}$$

Hence, $p = \rho^f g x_1$ in which x_1 is a coordinate in the direction of the gravitational force and g is the gravitational acceleration; then p is then called the hydrostatic pressure.

From Equation (2.25), we have the following equation:

$$\dot{w}_i = -\frac{k}{\gamma_w}\left(\rho^f \ddot{u}_i^s + \frac{\partial p}{\partial x_i} - \rho^f b_i\right) \tag{2.27}$$

Then, disregarding the acceleration term and setting the direction of x_1 for the direction of gravitational force g and $b_1 = g$, we have

$$\dot{w}_1 = -\frac{k}{\gamma_w}\left(\frac{\partial p}{\partial x_1} - \rho^f g\right) = -k\frac{\partial}{\partial x_1}\left(\frac{p}{\gamma_w} - x_1\right) = -k\frac{\partial h}{\partial x_1} \tag{2.28}$$

in which $\gamma_w = \rho^f g$ and h is the total head.

The total head is expressed by

$$h = \frac{p}{\gamma_w} - x_1 = \frac{p}{\gamma_w} + z \tag{2.29}$$

where $z = -x_1$, z is the elevation head, and $\frac{p}{\gamma_w}$ is the pressure head.

It is seen that Equation (2.28) is Darcy's law. It should be noted that the conservation law of the linear momentum of the fluid phase Equation (2.17) is a general description of Darcy's law. In the preceding, the fundamental equations of governing equations have been described. Next, we will derive the balance equation of the whole mixture and the continuity equation from the fundamental equations.

2.1.9 Balance equations for the mixture

By adding Equation (2.16) and Equation (2.17), we obtain the balance of linear momentum of the mixture as

$$\bar{\rho}^s \ddot{u}_i^s + \bar{\rho}^f \ddot{u}_i^f = \left(\frac{\partial \sigma_{ij}^s}{\partial x_j} + \frac{\partial \sigma_{ij}^f}{\partial x_j} \right) + \bar{\rho}^s b_i + \bar{\rho}^f b_i \tag{2.30}$$

Upon substitution of Equations (2.3), (2.4), and (2.19) into Equation (2.30), the following equation is derived:

$$\rho \ddot{u}_i^s + \rho^f \ddot{w}_i = \frac{\partial \sigma_{ij}}{\partial x_j} + \rho b_i \tag{2.31}$$

From Assumption 2, namely, in the case that the relative acceleration can be neglected as Equation (2.21), Equation (2.31) becomes the balance of linear momentum for the mixture as

$$\rho \ddot{u}_i^s = \frac{\partial \sigma_{ij}}{\partial x_j} + \rho b_i \tag{2.32}$$

For the balance of angular momentum of the multiphase materials, we assume the balance of angular momentum for the phases as well as for the whole mixture.

2.1.10 Continuity equation

Substitution of Equation (2.1) into the mass conservation equation of the solid phase Equation (2.14) leads to the following equation:

$$(1-n)\frac{\partial \rho^s}{\partial t} + \rho^s \frac{\partial (1-n)}{\partial t} + \rho^s \frac{\partial \{(1-n)\dot{u}_i^s\}}{\partial x_i} + (1-n)\dot{u}_i^s \frac{\partial \rho^s}{\partial x_i} = 0 \tag{2.33}$$

In a similar way, substituting Equation (2.2) into the mass balance equation of the fluid phase Equation (2.15) gives

$$n\frac{\partial \rho^f}{\partial t} + \rho^f \frac{\partial n}{\partial t} + \rho^f \frac{\partial (n\dot{u}_i^f)}{\partial x_i} + n\dot{u}_i^f \frac{\partial \rho^f}{\partial x_i} = 0 \tag{2.34}$$

Multiplying Equation (2.33) by ρ^f/ρ^s and adding the result and Equation (2.34), we have

$$\rho^f\left(\frac{\partial(1-n)}{\partial t}+\frac{\partial n}{\partial t}\right)+\rho^f\frac{\partial\{n(\dot{u}_i^f-\dot{u}_i^s)\}}{\partial x_i}+\rho^f\frac{\partial \dot{u}_i^s}{\partial x_i}$$
$$+n\left(\frac{\partial\rho^f}{\partial t}+\dot{u}_i^f\frac{\partial\rho^f}{\partial x_i}\right)+(1-n)\frac{\rho^f}{\rho^s}\left(\frac{\partial\rho^s}{\partial t}+\dot{u}_i^s\frac{\partial\rho^s}{\partial x_i}\right)=0 \tag{2.35}$$

In Equation (2.35), the first term is equal to zero. Taking into account Assumption 1 and substituting Equation (2.10) and Equation (2.19) into Equation (2.35), we have

$$\frac{\partial\dot{w}_i}{\partial x_i}+\dot{\varepsilon}_{ii}^s+\frac{n}{\rho^f}\left(\frac{\partial\rho^f}{\partial t}+\dot{u}_i^f\frac{\partial\rho^f}{\partial x_i}\right)+\frac{(1-n)}{\rho^s}\left(\frac{\partial\rho^s}{\partial t}+\dot{u}_i^s\frac{\partial\rho^s}{\partial x_i}\right)=0 \tag{2.36}$$

where $\dot{\varepsilon}_{ii}^s$ is the volumetric strain rate of the solid phase.

Considering the material time derivative, Equation (2.36) can be expressed as

$$\frac{\partial\dot{w}_i}{\partial x_i}+\dot{\varepsilon}_{ii}^s+\frac{n}{\rho^f}\dot{\rho}^f+\frac{(1-n)}{\rho^s}\dot{\rho}^s=0 \tag{2.37}$$

Herein, when we assume the incompressibility of the soil constituents such as soil particles (Assumption 3), the following equation holds:

$$\dot{\rho}^s=0 \tag{2.38}$$

Substitution of Equation (2.38) into Equation (2.37) leads to the following equation:

$$\frac{\partial\dot{w}_i}{\partial x_i}+\dot{\varepsilon}_{ii}^s+\frac{n}{\rho^f}\dot{\rho}^f=0 \tag{2.39}$$

The preceding equation is the continuity equation for the case of incompressibility of the solid constituent. The third term of the left-hand side of Equation (2.39) denotes a compressibility of the pore water.

Neglecting the effect of temperature, the following equation is given since the time derivative of mass of the pore fluid ($\rho^f V^f$) is zero:

$$\frac{D(\rho^f V^f)}{Dt} = 0 \tag{2.40}$$

After the manipulation of this equation we have

$$\frac{\dot{\rho}^f}{\rho^f} = -\frac{\dot{V}^f}{V^f} = -\dot{\varepsilon}_{ii}^f \tag{2.41}$$

Substituting the constitutive equation of the fluid, Equation (2.13), into Equation (2.41) gives

$$\frac{\dot{\rho}^f}{\rho^f} = \frac{\dot{p}}{K^f} \tag{2.42}$$

Upon substitution of Equation (2.42) into Equation (2.39), the continuity equation becomes

$$\frac{\partial \dot{w}_i}{\partial x_i} + \dot{\varepsilon}_{ii}^s + \frac{n}{K^f}\dot{p} = 0 \tag{2.43}$$

In Equation (2.43), this equation includes the relative velocity of the fluid phase to the solid phase.

If Assumption 2 of the u-p formulation is adopted, we can express w_i by the acceleration of the solid phase and the pore water pressure as has been shown in Equation (2.27) and can be eliminated in Equation (2.43).

Substituting Equation (2.27) into Equation (2.43), we get

$$-\frac{\partial}{\partial x_i}\left[\frac{k}{\gamma_w}\left(\rho^f \ddot{u}_i^s + \frac{\partial p}{\partial x_i} - \rho^f b_i\right)\right] + \dot{\varepsilon}_{ii}^s + \frac{n}{K^f}\dot{p} = 0 \tag{2.44}$$

If the body force b_i is constant, and the spatial gradients of permeability and the density of the fluid are sufficiently small, considering Equation (2.8), the final form of the continuity is given as

$$\frac{k}{\gamma_w}\left(-\rho^f \ddot{\varepsilon}_{ii}^s - \frac{\partial^2 p}{\partial x_i^2}\right) + \dot{\varepsilon}_{ii}^s + \frac{n}{K^f}\dot{p} = 0 \tag{2.45}$$

2.2 GOVERNING EQUATIONS FOR GAS–WATER–SOLID THREE-PHASE MATERIALS

2.2.1 Introduction

Geomaterials generally fall into the category of multiphase materials. They are basically composed of soil particles, water, and air. The behavior of multiphase materials can be described within the framework of a macroscopic continuum mechanical approach through the use of the theory of porous media (de Boer 2000b). The theory is considered to be a generalization of Biot's two-phase porous theory for saturated soil (Biot 1941, 1955, 1956).

Proceeding from the general geometrically nonlinear formulation, the governing balance relations for multiphase materials can be obtained (de Boer 2000b; Loret and Khalili 2000; Lewis and Schrefler 1998; Ehlers and Graf, 2003; Ehlers et al., 2004). Mass conservation laws for the gas phase as well as for the liquid phase are considered in those analyses. In the field of geotechnics, air pressure is assumed to be zero in many research works (Sheng et al. 2003), since geomaterials usually exist in an unsaturated state near the surface of the ground and we have not enough data on a development of air pressure. Considering gas hydrate dissociation in the sea bed ground, however, we have to deal with the high level of gas pressure that exists deep in the ground (Kimoto et al. 2007), this means that the mass balance for three phases must be considered. Oka et al. (2006) proposed an air–water–soil coupled finite element model in which the skeleton stress is used as a stress variable, and the suction effect is introduced in the constitutive equation for soil. Furthermore, the conservation of energy is required when there is a considerable change in temperature during the deformation process. Vardoulakis (2002) showed that the temperature of saturated clay rises with plastic deformation. Oka et al. (2004) and Kimoto et al. (2007) numerically simulated the thermal consolidation process, which will be shown in Chapter 10.

2.2.2 General setting

The material to be modeled is composed of three phases, namely, solid (S), water (W), and gas (G), which are continuously distributed throughout space. Total volume (V) is obtained from the sum of the partial volumes of the constituents, namely,

$$\sum_{\alpha} V^{\alpha} = V \quad (\alpha = S, W, G) \tag{2.46}$$

The volume of void, V^v, which is composed of water and gas, is given as follows:

$$\sum_{\beta} V^{\beta} = V^{v} \quad (\beta = W, G) \tag{2.47}$$

Volume fraction, n^{α}, is defined as the local ratio of the volume element with respect to the total volume, namely,

$$n^{\alpha} = \frac{V^{\alpha}}{V} \tag{2.48}$$

$$\sum_{\alpha} n^{\alpha} = 1 \quad (\alpha = S, W, G) \tag{2.49}$$

The volume fraction of the void, that is, porosity, n, is written as

$$n = \sum_{\beta} n^{\beta} = \frac{V^{v}}{V} = \frac{V - V^{S}}{V} = 1 - n^{S} \quad (\beta = W, G) \tag{2.50}$$

The volume fraction of the fluid, n^F, is given by

$$n^{F} = \sum n^{\gamma} \quad (\gamma = W, G) \tag{2.51}$$

The volume fraction concept has been adopted to construct the theory of mixture (Mills 1967; Morland 1972). The historical development of the volume fraction theory has been well discussed by de Boer (2000b).

In addition, the water saturation is required in the model, namely,

$$s_{r} = \frac{V^{W}}{V^{W} + V^{G}} = \frac{n^{W}}{n^{W} + n^{G}} = \frac{n^{W}}{n^{F}} \tag{2.52}$$

2.2.3 Partial stresses

By analogy to the water-saturated soil, we assume that

$$\sigma_{ij}^{S} = \sigma'_{ij} - n^{S} P^{F} \delta_{ij} \tag{2.53}$$

$$\sigma_{ij}^{W} = -n^{W} P^{W} \delta_{ij} \tag{2.54}$$

$$\sigma_{ij}^{G} = -n^{G} P^{G} \delta_{ij} \tag{2.55}$$

where P^F is the average pressure of the fluids surrounding the solid skeleton (Bolzon et al. 1996) given by

$$P^F = s_r P^W + (1 - s_r) P^G \tag{2.56}$$

and σ'_{ij} is a skeleton stress.

The skeleton stress, which will be explained in the following, is reasonable to describe the behavior of solid skeleton in the constitutive relation.

Total stress tensor, σ_{ij}, is obtained from the sum of the partial stresses, σ_{ij}^{α}, namely,

$$\sum_{\alpha} \sigma_{ij}^{\alpha} = \sigma_{ij} (\alpha = S, W, G) \tag{2.57}$$

and

$$\sigma'_{ij} = \sigma_{ij} + P^F \delta_{ij} \tag{2.58}$$

2.2.4 Conservation of mass

The conservation of mass for the solid, water, and gas phases, $\beta(= S,W,G)$, is given in the following equation:

$$\frac{\partial}{\partial t}\left(\rho^{\beta} n^{\beta}\right) = -q_{Mi,i}^{\beta} + \dot{m}^{\beta} \quad (\beta = S, W, G) \tag{2.59}$$

in which ρ^{β} is the material density, q_{Mi}^{β} is the flux vector, and $\dot{m}^{\beta}$ is the mass change rate of phase per unit volume. The flux vector is expressed in terms of the velocity of the flow as

$$q_{Mi}^{\beta} = n^{\beta} \rho^{\beta} v_i^{\beta} \quad (\beta = S, W, G) \tag{2.60}$$

where v_i^{β} is the velocity of phase β.

The relative velocity of the flow, V_i^{β}, with respect to the solid phase is

$$V_i^{\beta} = n^{\beta} (v_i^{\beta} - v_i^S) \quad (\beta = W, G) \tag{2.61}$$

The conservation laws in Equation (2.59) for the solid, water, and gas phases are expressed with water saturation, s_r, and the volume fraction of void, n^F, as

$$-\rho^s \dot{n}^F + (1-n^F)\rho^s v^s_{i,i} = \dot{m}^s \tag{2.62}$$

$$\dot{\rho}^W n^F s_r + \rho^{\overline{W}} \dot{s}_r n^F + \dot{n}^F s_r \rho^{\overline{W}} + n s_r \rho^W v^{\overline{W}}_{i,i} = \dot{m}^{\overline{W}} \tag{2.63}$$

$$\dot{n}^F(1-s_r)\rho^G - n^F \dot{s}_r \rho^G + n^F(1-s_r)\dot{\rho}^G + n^F(1-s_r)\rho^G v^G_{i,i} = \dot{m}^G \tag{2.64}$$

where we assume the incompressibility of soil particles, $\dot{\rho}^S = 0$, and $n^F = n$ is the porosity.

Assuming that the spatial gradient of the volume fractions are zero, we obtain following relations from Equations (2.62) to (2.64) as

$$s_r n^F \frac{\dot{\rho}^W}{\rho^W} + \dot{s}_r n^F + V^W_{i,i} + s_r v^S_{i,i} - s_r \frac{\dot{m}^s}{\rho^s} - \frac{\dot{m}^W}{\rho^W} = 0 \tag{2.65}$$

$$(1-s_r)n^F \frac{\dot{\rho}^G}{\rho^G} - \dot{s}_r n^F + (1-s_r)v^S_{i,i} + V^G_{i,i} - (1-s_r)\frac{\dot{m}^s}{\rho^s} - \frac{\dot{m}^G}{\rho^G} = 0 \tag{2.66}$$

As for describing changes in the gas density, the equation of ideal gases can be used, that is,

$$\rho^G = \frac{M^G P^G}{R\theta} \tag{2.67}$$

$$\dot{\rho}^G = \frac{M^G}{R}\left(\frac{\dot{P}^G}{\theta} - \frac{P^G\dot{\theta}}{\theta^2}\right) \tag{2.68}$$

in which M^G is the molecular weight of gas, R is the gas constant, θ is the temperature, and tension is positive in the equation.

Dividing Equation (2.68) by Equation (2.67) yields

$$\frac{\dot{\rho}^G}{\rho^G} = \frac{\dot{P}^G}{P^G} - \frac{\dot{\theta}}{\theta} \tag{2.69}$$

2.2.5 Balance of momentum

Momentum balance is required for each phase, namely,

$$n^{\alpha}\rho^{\alpha}\dot{v}_i^{\alpha} = \sigma_{ji,j}^{\alpha} + \rho^{\alpha}n^{\alpha}\bar{F}_i - \tilde{P}_i^{\alpha} \quad (\alpha = S, W, G) \tag{2.70}$$

in which $\bar{F}_i$ is the gravity force and $\tilde{P}_i^{\alpha}$ is related to the interaction term given in

$$\tilde{P}_i^{\alpha} = \sum_{\gamma} D^{\alpha\beta}(v_i^{\alpha} - v_i^{\gamma}), \quad D^{\alpha\gamma} = D^{\gamma\alpha} \quad (\alpha, \gamma = S, W, G) \tag{2.71}$$

where $D^{\alpha\beta}$ are parameters that describe the interaction with each phase. The momentum balance equation for each phase is obtained with the following equations when the acceleration is disregarded:

$$\sigma'_{ji,j} - (n^S P^F)_{,i} + \rho^S n^S \bar{F}_i - D^{SW}(v_i^S - v_i^W) - D^{SG}(v_i^S - v_i^G) = 0 \tag{2.72}$$

$$-(n^W P^W)_{,i} + \rho^W n^W \bar{F}_i - D^{WS}(v_i^W - v_i^S) - D^{WG}(v_i^W - v_i^G) = 0 \tag{2.73}$$

$$-(n^G P^G)_{,i} + \rho^G n^G \bar{F}_i - D^{GS}(v_i^G - v_i^S) - D^{GW}(v_i^G - v_i^W) = 0 \tag{2.74}$$

$D^{\alpha\beta}(\alpha, \beta = W, G)$ are given as

$$D^{WS} = -\frac{(n^W)^2 \rho^W g}{k^W}, \quad D^{GS} = -\frac{(n^G)^2 \rho^G g}{k^G} \tag{2.75}$$

in which k^W and k^G are the permeability coefficients for the water phase and the gas phase, respectively. We assume that the interaction between water and gas phases D^{GW} and D^{WG} is zero.

When the space derivative of volume fraction $n_{,i}^{\alpha}$ is negligible, Darcy's law for the water phase and the gas phase is obtained from Equations (2.73) and (2.74), respectively, as

$$V_i^W = n^W\left(v_i^W - v_i^S\right) = -\frac{k^W}{\rho^W g}\left(P_{,i}^W - \rho^W \bar{F}_i\right) \tag{2.76}$$

$$V_i^G = n^G\left(v_i^G - v_i^S\right) = -\frac{k^G}{\rho^G g}\left(P_{,i}^G - \rho^G \bar{F}_i\right) \tag{2.77}$$

The sum of Equations (2.72) to (2.74) leads to

$$\sigma_{ji,j} + \rho^E \bar{F}_i = 0, \rho^E = \sum_\alpha n^\alpha \rho^\alpha \quad (\alpha = S, W, G) \tag{2.78}$$

It is worth noting that Darcy-type laws such as Equations (2.76) and (2.77) are not objective since they include velocity but are a good approximation (Eringen 2003). In addition, as has been printed out in Section 2.2.1.9, it is assumed that the angular momentum is balanced for the phases for the whole mixture.

2.2.6 Balance of energy

The following energy conservation equation is applied in order to consider the heat conductivity:

$$(\rho c)^E \dot{\theta} = D_{ij}^{vp} \sigma'_{ij} - h_{i,i} + \dot{Q} \tag{2.79}$$

$$(\rho c)^E = \sum_\alpha n^\alpha \rho^\alpha c^\alpha \; (\alpha = S, W, G) \tag{2.80}$$

where c^α is the specific heat, θ is the temperature for all the phases, D_{ij}^{vp} is the viscoplastic stretching tensor, h_i is the heat flux vector, and $\dot{Q}$ is the heat source.

Heat flux, h_i, is given by

$$h_i = -\lambda^E \theta_{,i} \tag{2.81}$$

$$\lambda^E = \sum_\alpha n^\alpha \lambda^\alpha \qquad (\alpha = S,W,G) \tag{2.82}$$

in which λ^α is the thermal conductivity.

2.3 GOVERNING EQUATIONS FOR UNSATURATED SOIL

In the theory of porous media, the concept of the effective stress tensor is related to the deformation of the soil skeleton and plays an important role. The effective stress tensor has been defined by Terzaghi (1943) for water-saturated soil. However, the effective stress needs to be redefined if the fluid is made of compressible materials. In the present study, the skeleton stress tensor, σ'_{ij}, is defined and then used for the stress variable in the constitutive relation for the soil skeleton (Oka et al. 2006, 2008, 2010), which has been called "average skeleton stress" by Jommi (2000) and Gallipoli et al. (2003).

2.3.1 Partial stresses for the mixture

The total stress tensor is assumed to be composed of three partial stress values for each phase:

$$\sigma_{ij} = \sigma_{ij}^s + \sigma_{ij}^f + \sigma_{ij}^a \tag{2.83}$$

where σ_{ij} is the total stress tensor, and σ_{ij}^s, σ_{ij}^f, and σ_{ij}^a are the partial stress tensors for solid, liquid, and air, respectively.

Considering the volume fraction (Ehlers et al. 2004), the partial stress tensors for unsaturated soil can be given by

$$\sigma_{ij}^f = -nS_r p^f \delta_{ij} \tag{2.84}$$

$$\sigma_{ij}^a = -n(1 - S_r)p^a \delta_{ij} \tag{2.85}$$

$$\sigma_{ij}^s = \sigma'_{ij} - (1 - n)P^F \delta_{ij} \tag{2.86}$$

$$P^F = S_r p^f + (1 - S_r)p^a \tag{2.87}$$

where p^f and p^a are the pore water pressure and the pore air pressure, respectively, n is the porosity, S_r is the degree of saturation, σ'_{ij} is the skeleton stress, and P^F is the average pore pressure. The tension is positive in this chapter.

For the unsaturated soil, we will use the skeleton stress (Oka et al. 2006, 2008; Kimoto et al. 2010) as the basic stress variable in the model along with suction. The skeleton stress only applies to the soil skeleton and Equation (2.86) comes from the analogy to the effective stress for the water-saturated soil. By adding Equations (2.84), (2.85), and (2.86) we have skeleton stress as

$$\sigma'_{ij} = \sigma_{ij} + P^F \delta_{ij} \tag{2.88}$$

The skeleton stress was first advocated by Jommi (2000) as the average skeleton stress, which was defined as the difference between the total stress and the average fluid pressure. Ehlers et al. (2004) called it effective stress. However, herein we called it skeleton stress to avoid confusing it with mean value of the skeleton stress.

Adopting the skeleton stress provides a natural application of the mixture theory to unsaturated soil. The definition in Equation (2.88) is similar to Bishop's definition for the effective stress of unsaturated soil. In addition to Equation (2.88), the effect of suction on the constitutive model should always be taken into account. This assumption leads to a reasonable consideration of the collapse behavior of unsaturated soil, which has been known as a behavior that cannot be described by Bishop's definition for the effective stress of unsaturated soil. Introducing suction into the model, however, makes it possible to formulate a model for unsaturated soil, starting from a model for saturated soil, by using the skeleton stress instead of the effective stress.

2.3.2 Conservation of mass

The mass conservation law for the three phases is given by

$$\frac{\partial \bar{\rho}^J}{\partial t} + \frac{\partial (\bar{\rho}^J \dot{u}_i^J)}{\partial x_i} = 0 \tag{2.89}$$

where $\bar{\rho}^J$ is the average density for the J phase and $\dot{u}_i^J$ is the velocity vector for the J phase.

$$\bar{\rho}^s = (1-n)\rho^s \tag{2.90}$$

$$\bar{\rho}^f = nS_r\rho^f \tag{2.91}$$

$$\bar{\rho}^a = n(1-S_r)\rho^a \tag{2.92}$$

where J = s, f, and a, in which the superscripts s, f, and a indicate the solid, the liquid, and the air phases, respectively; n is the porosity; and S_r is the saturation. ρ^J is the mass bulk density of the solid, the liquid, and the gas.

2.3.3 Balance of linear momentum for the three phases

The conservation laws of linear momentum for the three phases are given by

$$\bar{\rho}^s \ddot{u}_i^s - Q_i - R_i = \frac{\partial \sigma_{ij}^s}{\partial x_j} + \bar{\rho}^s b_i \tag{2.93}$$

$$\bar{\rho}^f \ddot{u}_i^f + R_i = \frac{\partial \sigma_{ij}^f}{\partial x_j} + \bar{\rho}^f b_i \tag{2.94}$$

$$\bar{\rho}^a \ddot{u}_i^a + Q_i = \frac{\partial \sigma_{ij}^a}{\partial x_j} + \bar{\rho}^a b_i \tag{2.95}$$

where $\ddot{u}_i^J$ $(J = a, f, s)$ are the acceleration vectors for the three phases, b_i is the body force, Q_i denotes the interaction between the solid and the air phases, and R_i denotes the interaction between the solid and the liquid phases. These interaction terms, Q_i and R_i, can be described as

$$R_i = nS_r \frac{\gamma_w}{k^f} \dot{w}_i^f \tag{2.96}$$

$$Q_i = n(1 - S_r) \frac{\rho^a g}{k^a} \dot{w}_i^a \tag{2.97}$$

where k^f is the water permeability coefficient, k^a is the air permeability, $\dot{w}_i^f$ is the average relative velocity vector of water with respect to the solid skeleton, and $\dot{w}_i^a$ is the average relative velocity vector of air to the solid skeleton.

The relative velocity vectors are defined by

$$\dot{w}_i^f = nS_r \left(\dot{u}_i^f - \dot{u}_i^s \right) \tag{2.98}$$

$$\dot{w}_i^a = n(1 - S_r) \left(\dot{u}_i^a - \dot{u}_i^s \right) \tag{2.99}$$

Using Equation (2.98), Equation (2.94) becomes

$$\bar{\rho}^f\left(\ddot{u}_i^s + \frac{1}{nS_r}\ddot{w}_i^f\right) + R_i = \frac{\partial \sigma_{ij}^f}{\partial x_j} + \bar{\rho}^f b_i \tag{2.100}$$

We deal with the behavior of soil in which the difference between accelerations of the soil skeleton and pore fluid is sufficiently small. This assumption is reasonable except for the high frequency problem and very high permeability (Zienkiewicz et al. 1980). For this reason we assume that $\ddot{w}_i^f \cong 0$, in this case, using Equations (2.84), (2.91), and (2.96), Equation (2.100) becomes

$$nS_r\rho^f\ddot{u}_i^s + nS_r\frac{\gamma_w}{k^f}\dot{w}_i^f = -nS_r\frac{\partial p^f}{\partial x_i} + nS_r\rho^f b_i \tag{2.101}$$

in which we assume that the spatial gradients of porosity and saturation are sufficiently small. The same assumption will be taken in the following derivations of the governing equations.

After manipulation, the average relative velocity vector of water to the solid skeleton and the average relative velocity vector of air to the solid skeleton are shown as

$$\dot{w}_i^f = -\frac{k^f}{\gamma_w}\left(\frac{\partial p^f}{\partial x_i} + \rho^f\ddot{u}_i^s - \rho^f b_i\right) \tag{2.102}$$

$$\dot{w}_i^a = -\frac{k^a}{\rho^a g}\left(\frac{\partial p^a}{\partial x_i} + \rho^a\ddot{u}_i^s - \rho^a b_i\right) \tag{2.103}$$

in which $\ddot{w}_i^a \cong 0$ is assumed due to the reason mentioned earlier.

Based on the aforementioned fundamental conservation laws, we can derive equations of motion for the whole mixture. Substituting Equations (2.90), (2.91), and (2.92) into the given equation and adding Equations (2.93) to (2.95), we have

$$\rho\ddot{u}_i^s + nS_r\rho^f\left(\ddot{u}_i^f - \ddot{u}_i^s\right) + n(1-S_r)\rho^a\left(\ddot{u}_i^a - \ddot{u}_i^s\right) = \frac{\partial \sigma_{ij}}{\partial x_j} + \rho b_i \tag{2.104}$$

where ρ is the mass density of the mixture as $\rho = \bar{\rho}^f + \bar{\rho}^a + \bar{\rho}^s$, and σ_{ij} is the total stress tensor.

From the following assumptions,

$$\ddot{u}_i^s >> \left(\ddot{u}_i^f - \ddot{u}_i^s\right) \tag{2.105}$$

$$\ddot{u}_i^s >> \left(\ddot{u}_i^a - \ddot{u}_i^s\right) \tag{2.106}$$

equations of motion for the whole mixture are derived as

$$\rho\ddot{u}_i^s = \frac{\partial \sigma_{ij}}{\partial x_j} + \rho b_i \tag{2.107}$$

2.3.4 Continuity equations

Using the mass conservation law for the solid and the liquid phases, Equation (2.89) ($J = s,f$) and Equations (2.90) and (2.91), and assuming the incompressibility of soil particles, we obtain

$$\frac{\partial\left\{nS_r(\dot{u}_i^f - \dot{u}_i^s)\right\}}{\partial x_i} + S_r\dot{\varepsilon}_{ii}^s + nS_r\frac{\dot{\rho}^f}{\rho^f} + n\dot{S}_r = 0 \tag{2.108}$$

Incorporating Equation (2.102) and $p = -K^f\varepsilon_{ii}^f$ (K^f: volumetric elastic coefficient) into the previous equation leads to the following continuity equation for the liquid phase:

$$-\frac{\partial}{\partial x_i}\left[\frac{k^f}{\gamma_w}\left(\rho^f\ddot{u}_i^s + \frac{\partial p^f}{\partial x_i} - \rho^f b_i\right)\right] + S_r\dot{\varepsilon}_{ii}^s + n\dot{S}_r + nS_r\frac{\dot{p}^f}{K^f} = 0 \tag{2.109}$$

Similarly, we can derive the continuity equation for the air phase by assuming that the spatial gradients of porosity and saturation are sufficiently small:

$$-\frac{\partial}{\partial x_i}\left[\frac{k^a}{\gamma_w}\left(\rho^a\ddot{u}_i^s + \frac{\partial p^a}{\partial x_i} - \rho^a b_i\right)\right] + (1-S_r)\dot{\varepsilon}_{ii}^s - n\dot{S}_r + n(1-S_r)\frac{\dot{\rho}^a}{\rho^a} = 0 \tag{2.110}$$

For the saturation we will use a constitutive equation called water characteristic relation or water retention relation.

Since saturation is a function of suction, that is, the pressure head, the time rate for saturation is given by

$$n\dot{S}_r = n\frac{dS_r}{d\theta}\frac{d\theta}{d\psi}\frac{d\psi}{dp^c}\dot{p}^c = \frac{1}{\gamma_w}\frac{d\theta}{d\psi}\dot{p}^c \tag{2.111}$$

where $\theta = \frac{V_w}{V}$ is the volumetric water content, p^c is the matric suction ($p^c = (p^a - p^f)$), $\psi = p^c/\gamma_w$ is the pressure head for suction, and $C = \left|\frac{d\theta}{d\psi}\right|$ is the specific water content.

And we need the constitutive equation for air phase such as ideal gas.

REFERENCES

Atkin, R.J., and Craine, R.E. 1976. Continuum theories of mixtures: basic theory and historical developments, *Q. J. Mech. Appl. Math.*, 29(2):209–244.

Biot, M.A. 1941. General theory of three-dimensional consolidation, *J. Appl. Phys.*, 12(4):155-164.

Biot, M.A. 1955. Theory of elasticity and consolidation for a porous anisotropic solid, *J. Appl. Phys*, 26(2):182–185.

Biot, M.A. 1956. Theory of propagation of elastic waves in a fluid-saturated porous solid, *J. Acoust. Soc. Amer.*, 28(2):168–191.

Biot, M.A. 1962. Mechanics of deformation and acoustic propagation in porous media, *J. Appl. Phys.*, 33(4):1482–1498.

Bolzon, G., Schrefler, B.A. and Zienkiewicz, O. C. 1996. Elastoplastic soil constitutive laws generalized to partially saturated states. Géotechnique, 46(2):279–289.

Bowen, R.M. 1976. Theory of mixtures, in *Continuum Physics*, Vol. III, A.C. Eringen, ed., Academic Press, 1–127.

Coussy, O. 1995. *Mechanics of Porous Continua*, New York: John Wiley & Sons.

de Boer, R. 2000a. Contemporary progress in porous media theory, *Appl. Mech. Rev.*, 53(12):323–369.

de Boer, R. 2000b. *Theory of Porous Media*, New York: Springer.

Ehlers. W. and Graf, T. 2003. Adaptive computation of localization phenomena in geotechnical applications, *Bifurcation and instabilities in geomechanics*, J. Labuz and A. Drescher eds., Swets & Zeitlinger, Lisse, 247–262.

Ehlers, W., Graf, T., and Ammann, M. 2004. Deformation and localization analysis of partially saturated soil, *Comp. Meth. Appl. Mech. Eng.*, 193:2885–2910.

Eringen, A.C. 2003. Note on Darcy's law, *J. Appl. Phys.*, 94(2):1282.

Fillunger, P. 1913. Der auftrieb in talsperren, *Österreichische Wochenschrift für den öffentrichen baudienst*, 19:532–556, 567–570.

Gallipoli, D., Gens, A., Sharma, R. and Vaunat, J. 2003. An elasto-plastic model for unsaturated soil incoorporating the effects of suction and degree of saturation on mechanical behavour, *Géotechnique*, 53(1):123–135.

Jommi, C. 2000. Remarks on the constitutive modeling of unsaturated soils, experimental evidence and theoretical approaches in unsaturated soil, Proc. Int. workshop on unsaturated soils, Trento, Tarantino, A. and Mancuso, C. eds., Rotterdam, Balkema: 139–153.

Kimoto, S., Oka, F., Fushita, T., and Fujiwaki, M. 2007. A chemo-thermo-mechanically coupled numerical simulation of the subsurface ground deformations due to methane hydrate dissociation, *Computers and Geotechnics*, 34(4):216–228.

Kimoto, S., Oka, F., and Fushita, T. 2010. A chemo-thermo-mechanically coupled analysis of ground deformation induced by gas hydrate dissociation, *Int. J. Mech. Sci.*, 52:365–376.

Lewis, R.W., and Schrefler, B.A. 1998. *The Finite Element Method in the Static and Dynamic Deformation and Consolidation of Porous Media*, 2nd ed., New York: John Wiley & Sons.

LIQCA Research and Development Group (Representative: F. Oka, Kyoto University). 2005. User's Manual for LIQCA2D04 (2005 released print), http://nakisuna2.kuciv.kyoto-u.ac.jp/liqca.htm.

Loret, B., and Khalili, N. 2000. A three-phase model for unsaturated soils, *Int. J. Numer. Anal. Meth. Geomech.*, 24:893–927.

Morland, L.W. 1972. A simple constitutive theory for a fluid-saturated porous solid, *J. Geophy. Res.*, 77:890–900.

Mills, N. 1967. On a theory of multi-component mixture, *Q. J. Mech. Appl. Math.*, 20:449–508.

Oka, F., Yashima, A., Shibata, T., Kato, M., and Uzuoka, R. 1994. FEM-FDM coupled liquefaction analysis of a poroussoil using an elasto-plastic model, *Appl. Sci. Res.*, 52: 209–245.

Oka, F., Kimoto, S., Takada, N., Gotoh, H., and Higo, Y. 2010. A seepage-deformation coupled analysis of an unsaturated river embankment using a multiphase elasto-viscoplastic theory, *Soils and Foundations*, 50(4):483–494.

Sheng, D., Sloan, W., Gens, A., and Smith, D.W. 2003.Finite element formulation and algorithms for unsaturated soils, Part I: Theory. *Int. J. Numer. Anal. Meth. Geomech.*, 27:745–765.

Terzaghi, K. 1943. *Theoretical soil mechanics*, John Wiley & Sons Inc., New York.

Vardoulakis, I. 2002. Dynamic thermo-poro-mechanical analysis of catastrophic landslides, *Géotechnique*, 52(3):157–171.

Zienkiewicz, O.C., Chan, A.H.C., Pastor, M., Schrefler, B.A., and Shiomi, T. 1999. *Computational Geomechanics with Special Reference to Earthquake Engineering*, John Wiley & Sons.

Zienkiewicz, O.C., Chang, C.T., and Bettes, P. 1980. Drained, undrained, consolidating and dynamic behavior assumptions in soils, limits of validity, *Géotechnique*, 30(4):385–395.

Chapter 3

Fundamental constitutive equations

As has been discussed in Chapter 1, Section 1.9, constitutive equations are required to describe material responses to external actions. In this chapter, we will review fundamental constitutive equations such as the elastic model and the elastoplastic model.

3.1 ELASTIC BODY

The reversible deformation of materials is described by elastic constitutive equations. The linear elastic constitutive equation, whose stress–strain is linear, is a generalization of Hooke's law. It is given by

$$\sigma_{ij} = E_{ijkl}\varepsilon_{kl} \tag{3.1}$$

$$E_{ijkl} = \lambda\delta_{ij}\delta_{kl} + \mu(\delta_{ik}\delta_{jl} + \delta_{il}\delta_{jk}) + \xi(\delta_{ik}\delta_{jl} - \delta_{il}\delta_{jk}) \tag{3.2}$$

where E_{ijkl} is a fourth-order isotropic tensor, and λ, μ, ξ are material constants.

Assuming a symmetry for the stress and strain tensors, the elastic coefficient, E_{ijkl}, can be expressed by

$$E_{ijkl} = \lambda\delta_{ij}\delta_{kl} + \mu(\delta_{ik}\delta_{jl} + \delta_{il}\delta_{jk}) \tag{3.3}$$

and Equations (3.1) and (3.3) yield

$$\sigma_{ij} = \lambda\varepsilon_{kk}\delta_{ij} + 2\mu\varepsilon_{ij} \tag{3.4}$$

where λ and μ are called Lamé's constants.

Since the constitutive model for an elastic body depends on the strain, that is, the initial shape, the elastic model is considered to be a model for solids.

The linear elastic constitutive equation can be written in components as

$$
\begin{aligned}
\varepsilon_{xx} &= \frac{1}{E}(\sigma_{xx} - \nu\sigma_{yy} - \nu\sigma_{zz}) & \varepsilon_{xy} &= \frac{1+\nu}{E}\sigma_{xy} \\
\varepsilon_{yy} &= \frac{1}{E}(\sigma_{yy} - \nu\sigma_{zz} - \nu\sigma_{xx}) & \varepsilon_{yz} &= \frac{1+\nu}{E}\sigma_{yz} \\
\varepsilon_{zz} &= \frac{1}{E}(\sigma_{zz} - \nu\sigma_{xx} - \nu\sigma_{yy}) & \varepsilon_{zx} &= \frac{1+\nu}{E}\sigma_{zx}
\end{aligned} \tag{3.5}
$$

where E is Young's modulus and ν is Poisson's ratio.

Under isotropic compressive stress conditions, $\sigma_{xx} = \sigma_{yy} = \sigma_{zz} = c$, $\sigma_{xy} = \sigma_{yz} = \sigma_{zx} = 0$, and strain components are given by

$$\varepsilon_{xx} = \varepsilon_{yy} = \varepsilon_{zz} = \frac{1-2\nu}{E}\sigma_{xx} = \frac{1-2\nu}{E}c$$

Subsequently,

$$\varepsilon_{v} = \varepsilon_{xx} + \varepsilon_{yy} + \varepsilon_{zz} = \frac{3(1-2\nu)}{E}c = \frac{1}{K}c$$

where K is the elastic volumetric modulus, namely,

$$K = \frac{E}{3(1-2\nu)} \tag{3.6}$$

Under shear stress conditions, shear strains develop as

$$\varepsilon_{xy} = \frac{1}{2}\gamma_{xy} = \frac{1+\nu}{E}\sigma_{xy} = \frac{1}{2G}\sigma_{xy}$$

Hence,

$$G = \frac{E}{2(1+\nu)} \tag{3.7}$$

where G is the elastic shear modulus.

Under one-dimensional compressive conditions, for example, $\sigma_{yy} = c$,

$$\varepsilon_{yy} = \frac{1}{E}\sigma_{yy} = \frac{1}{E}c, \quad \varepsilon_{xx} = \varepsilon_{zz} = -\frac{\nu}{E}c = -\nu\varepsilon_{yy} \tag{3.8}$$

where ν is Poisson's ratio.

When elastic stored energy is positive, that is, $W > 0$,

$$W = \frac{1}{2}\sigma_{ij}\varepsilon_{ij} = \frac{1}{2}\lambda\varepsilon_{kk}^2 + \mu\varepsilon_{ij}\varepsilon_{ij} \tag{3.9}$$

where λ and μ are positive. Hence,

$$-1 < \nu < \frac{1}{2} \tag{3.10}$$

3.2 NEWTONIAN VISCOUS FLUID

Fluid is defined as a material whose stress tensor is isotropic at the state of rest. If the material behavior depends on the strain rate, the material exhibits viscosity. The material model in which the shearing stress is linearly proportional to the strain rate is known as a linear Newtonian fluid. The model is expressed as

$$\sigma_{ij} = -p\delta_{ij} + \lambda^* D_{kk}\delta_{ij} + 2\mu^* D_{ij} \tag{3.11}$$

where p is the pressure, and λ^* and μ^* are viscous coefficients.

For incompressible materials,

$$s_{ij} = 2\mu^* D_{ij} \tag{3.12}$$

where D_{ij} is the rate of deformation tensor, s_{ij} is the deviatoric tensor, and μ^* is the viscous coefficient.

Equation (3.12) indicates that the rate of deformation tensor D_{ij} (for small deformation fields, it corresponds to strain rates $\dot{\varepsilon}_{ij}$) is dependent on the present stress, and consequently, the deformation continues even under small deviatoric stresses.

3.3 BINGHAM BODY AND VISCOPLASTIC BODY

The simplest viscoplastic model is a fluid model with a yield stress limit; it is called the Bingham body. For the Bingham model with yield shear stress, k, viscous flow never occurs when the stress is less than the yield stress. The viscoplastic strain rate, $\dot{\varepsilon}$, is proportional to shear stress, τ, when the stress is more than yield shear stress, k.

$$2\eta\dot{\varepsilon} = \langle F \rangle \tau \tag{3.13}$$

$$\langle F \rangle = (F + |F|)/2, \quad F = 1 - k/\tau \tag{3.14}$$

where η is a viscous coefficient.

Hohenemser and Prager (1932) have generalized the Bingham model and have proposed a viscoplastic model, namely,

$$2\eta D_{ij} = \langle F \rangle S_{ij} \tag{3.15}$$

$$\langle F \rangle = 0; \ F < 0, \ \langle F \rangle = F; \quad F \geq 0 \tag{3.16}$$

$$F = 1 - \frac{k}{\sqrt{2J_2}} \tag{3.17}$$

where $J_2 = \frac{1}{2} s_{ij} s_{ij}$ is the second invariant of the deviatoric stress tensor.

From Equation (3.15), we obtain

$$4\eta^2 I_2^2 = 2F^2 J_2 \tag{3.18}$$

where I_2 is defined as

$$I_2 = \sqrt{D_{ij} D_{ij}} \tag{3.19}$$

Substituting Equation (3.17) into Equation (3.14) yields

$$s_{ij} = \left(2\eta + \frac{k}{I_2} \right) D_{ij} = 2\eta D_{ij} + \frac{k}{I_2} D_{ij} \tag{3.20}$$

for F > 0.

Since the second term in Equation (3.20) is the zeroth order homogeneous of degree one with respect to the rate of deformation, the term is rate independent. Therefore, the model has a viscous term and a rate-independent plastic term with yield stress, $k \cdot \left(S_{ij} - k\frac{D_{ij}}{I_2}\right)$ corresponds to the overstress that will be discussed in Section 3.6.

3.4 von MISES PLASTIC BODY

Von Mises (1928) proposed a rigid perfectly plastic model, and the yield condition for the von Mises plastic model and the flow rule are given by

$$\text{Yield condition: } \frac{S_{ij}S_{ij}}{2} = J_2 = k^2/2 \tag{3.21}$$

$$\text{Flow rule: } D_{ij} = \lambda \frac{S_{ij}}{\sqrt{2J_2}} \tag{3.22}$$

Using Equation (3.18), from Equation (3.21) we obtain $\lambda = I_2$
Hence,

$$S_{ij} = \frac{k}{I_2} D_{ij} \tag{3.23}$$

This equation is given by replacing viscous coefficient, η, in the linear Newtonian viscous fluid model with $\frac{k}{I_2}$.

The right-hand term in Equation (3.23) is homogeneous of degree one with respect to D_{ij}; the von Mises model is a rate-independent plastic model.

Hohenemser and Prager's model is similar to the Maxwell type of viscoelastic model that will be explained later, although shear flow never occurs below the yield stress in Hohenemser and Prager's model.

3.5 VISCOELASTIC CONSTITUTIVE MODELS

In general, soils exhibit elasticity, viscosity, and plasticity. Materials with viscosity are affected by the strain rate or the stress rate. In particular, models that can describe viscous and elastic behaviors are called viscoelastic models. The representative viscoelastic models include the Maxwell model and the Kelvin–Voigt model.

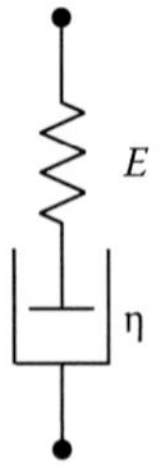

Figure 3.1 Maxwell model.

3.5.1 Maxwell viscoelastic model

The one-dimensional Maxwell type of viscoelastic model can be illustrated by the model with a series of elastic and viscous elements, as shown in Figure 3.1.

Elastic strain, ε^e, and viscous strain rate, $\dot{\varepsilon}^v$, are given by

$$\varepsilon^e = \frac{\sigma}{E} \tag{3.24}$$

$$\dot{\varepsilon}^v = \frac{\sigma}{\eta} \tag{3.25}$$

where σ is the stress, E is the elastic modulus, and η is the viscous coefficient. The total strain rate is given by the sum of the each strain rate, as $\dot{\varepsilon} = \dot{\varepsilon}^e + \dot{\varepsilon}^v$.

Then, the constitutive equation for the Maxwell model is given by

$$\dot{\varepsilon} = \frac{\dot{\sigma}}{E} + \frac{\sigma}{\eta} \tag{3.26}$$

Equation (3.26) can be extended to a multidimensional form as

$$\dot{\varepsilon}_{ij} = C_{ijkl}\dot{\sigma}_{kl} + A_{ijkl}\sigma_{kl} \tag{3.27}$$

Equation (3.27) shows that the strain rate is a function only of the quantities determined by the present states, that is, the stress and the stress rate. This implies that the model is for liquid or fluid. For this reason, the model is called the Maxwell fluid model.

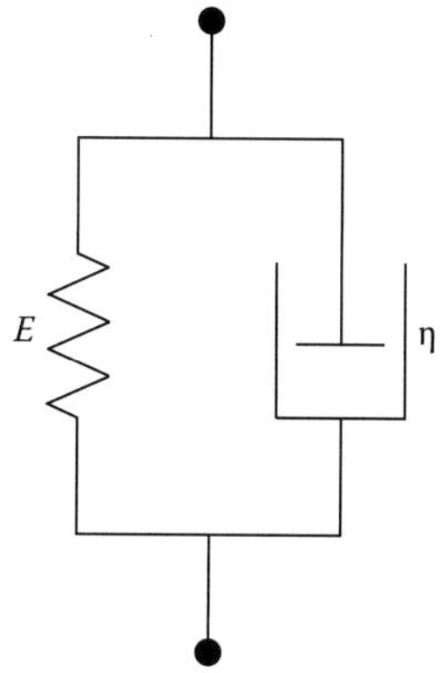

Figure 3.2 Voigt model.

3.5.2 Kelvin–Voigt model

The Kelvin–Voigt model consists of an elastic spring element and a viscous dashpot element, which are connected in a parallel manner, as shown in Figure 3.2.

Since strain is for two elements and the total stress is broken into two parts, as $\sigma = \sigma^e + \sigma^v$, the constitutive equation is given by

$$\varepsilon = \frac{\sigma^e}{E} \tag{3.28}$$

$$\dot{\varepsilon} = \frac{\sigma^v}{\eta} \tag{3.29}$$

$$\sigma = E\varepsilon + \eta\dot{\varepsilon} \tag{3.30}$$

Equation (3.30) can be extended to a multidimensional form as

$$\sigma_{ij} = E_{ijkl}\varepsilon_{kl} + \eta_{ijkl}\dot{\varepsilon}_{kl} \tag{3.31}$$

For the Kelvin–Voigt model, the stress depends not only on the strain rate, which is determined by the current states but also on the strain, which is affected by the value at the reference state. This indicates why the Kelvin–Voigt model is called the Voigt solid model.

3.5.3 Characteristic time

When strain, ε, is applied to the Maxwell type of viscoelastic body at $t = 0$ instantaneously, we have

$$\frac{1}{E}\dot{\sigma} + \frac{\sigma}{\eta} = 0 \tag{3.32}$$

Solving Equation (3.29) with the initial stress of σ_0, we obtain the stress–time profile:

$$\sigma = \sigma_0 exp(-t/\tau') \tag{3.33}$$

where $\sigma = \sigma_0/e$ at $t = \tau'$.

Then, $\tau' = \eta/E$ denotes the time when the stress becomes 1/*e* of the initial value and is called the relaxation time, where *e* is the natural logarithm (Figure 3.3).

Now let us consider the behavior of the Kelvin–Voigt model under constant stress conditions, that is, the creep process. If the applied stress is constant as σ_0, Equation (3.30) yields

$$\sigma_0 = E\varepsilon + \dot{\varepsilon}\eta \tag{3.34}$$

The solution for the preceding equation is given by

$$\varepsilon = \frac{\sigma_0}{E}(1 - exp(-t/\tau)) \tag{3.35}$$

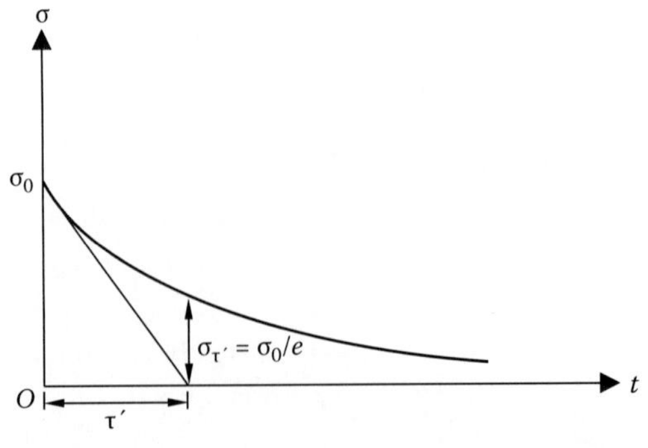

Figure 3.3 Stress relaxation.

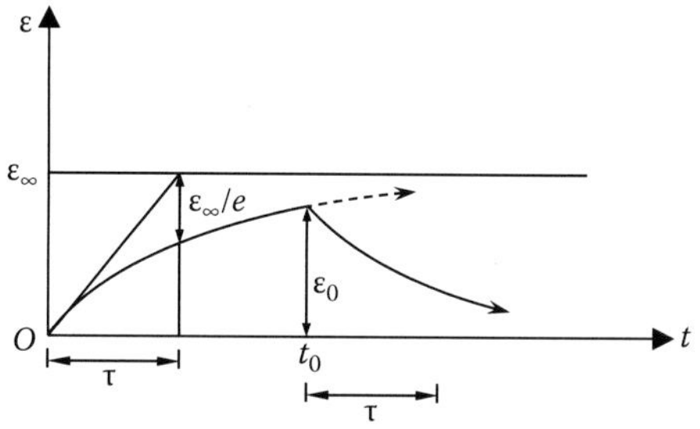

Figure 3.4 Strain during the creep.

in which the initial strain is zero and $\tau = \eta/E$, and just after the loading at time $t = 0$, the strain approaches σ_0/E as $t \to \infty$. τ is a parameter that indicates the rate at which the strain exponentially attains equilibrium and is called retardation time (see Figure 3.4).

The two aforementioned viscoelastic models can be formulated in an integral form, namely, a convolution integral, which uses the relaxation and creep functions as

$$\sigma_{ij}(t) = \int_{-\infty}^{t} f(t - t')\frac{d\varepsilon}{dt}dt' \tag{3.36}$$

$$\varepsilon_{ij}(t) = \int_{-\infty}^{t} g(t - t')\sigma(t')dt' \tag{3.37}$$

where $f(t)$ and $g(t)$ are relaxation and creep functions.

Exercise

Derive the constitutive equation for the three-element model consisting of the elastic element and the Voigt element.

3.6 ELASTOPLASTIC MODEL

As has been described, the features of materials such as mechanical, thermal, and electromagnetic characters are expressed by constitutive equations. When we apply loads to the materials, the materials deform. If the strain induced by the loading is small, the materials behave elastically. However, part of the deformation cannot be recovered if the strain level is

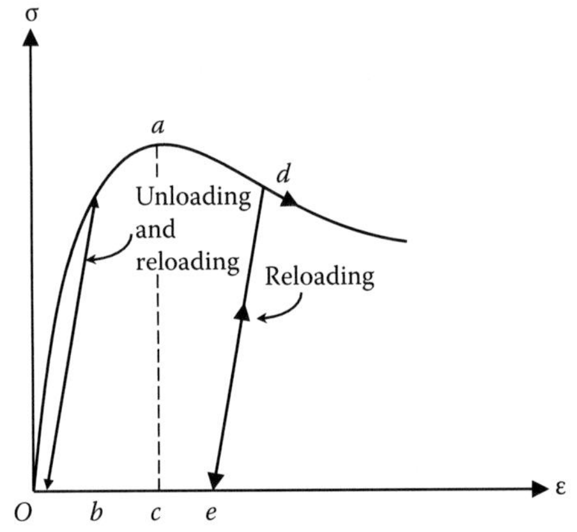

Figure 3.5 Stress–strain relation for elastoplastic material.

high (Figure 3.5). This type of behavior is elastoplastic and it is modeled by an elastoplastic constitutive equation. In the elastoplastic model, time-dependent behavior such as creep is not considered. Classical elastoplastic models are generalizations of the von Mises plasticity model and have been developed by Hill (1948), Drucker (1951), Prager (1949), and Mandel (1964). In this section, elastoplastic models will be explained within the framework of a small strain field with the yield function, loading criteria, the plastic potential, and hardening rules.

3.6.1 Yield conditions

The yield conditions prescribe the elastic limit and can be classified as initial yield conditions and subsequent yield conditions. The yield conditions prescribe the elastic limit under which materials behave elastically. The initial yield conditions are for the initial yield, and the subsequent yield conditions are for the yield conditions after the initial yielding (Figure 3.5).

For a perfectly elastoplastic body, deformation increases without the need for any change in stress after the initial yielding, while for the hardening (i.e., work-hardening) materials, it is necessary for the stress to be increased for the deformation to continuously increase after the initial yielding. The yield conditions after the material has been initially yielded are called the subsequent yield conditions (Figure 3.6). For one-dimensional problems, the yield conditions correspond to the simple yield limit stress. For multi-dimensional problems, on the other hand, the yield conditions have to be described within the multidimensional stress space.

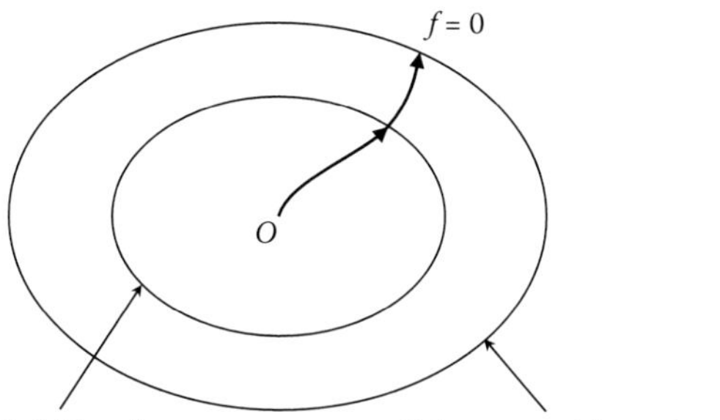

Figure 3.6 Yield surfaces.

The yield function in the multidimensional stress space is written with hardening, that is, an increase in the yield limit as

$$f(\sigma_{ij}, k_{ij}) = 0 \tag{3.38}$$

where σ_{ij} is the stress tensor and k_{ij} is a hardening-softening parameter.

When we take account of the strain hardening, it is usual to assume that k_{ij} is a function of the plastic strain or the plastic work.

In the case of plastic strain hardening,

$$f(\sigma_{ij}, k_{ij}(\varepsilon_{kl}^p)) = 0 \tag{3.39}$$

in which $\varepsilon_{ij}^p = \int d\varepsilon_{ij}^p$ is the plastic strain tensor.

For isotropic materials, the yield function is expressed by the scalar function of the stress and the plastic strain as

$$f(I_1, I_2, I_3, k(\varepsilon_{ij}^p)) = 0 \tag{3.40}$$

where $k(\varepsilon_{ij}^p)$ is the scalar function, namely, I_1, I_2, and I_3 are the first, the second, and the third invariant of the stress tensor, respectively.

3.6.2 Additivity of the strain

The total strain increment is assumed to be the sum of plastic strain increment, $d\varepsilon_{ij}^p$, and elastic strain increment tensor, $d\varepsilon_{ij}^e$:

$$d\varepsilon_{ij} = d\varepsilon_{ij}^e + d\varepsilon_{ij}^p \tag{3.41}$$

3.6.3 Loading conditions

The loading indicates the changes in stress that induce the plastic strain. For the strain-hardening case, the yield surface expands, that is, the stress point moves to a point outside the current yield surface. Then,

$$f(\sigma_{ij} + d\sigma_{ij}, k(\varepsilon_{ij}^p)) > 0 \tag{3.42}$$

Since the current stress point is on the yield surface,

$$f(\sigma_{ij}, k(\varepsilon_{ij}^p)) = 0 \tag{3.43}$$

Hence,

$$\frac{\partial f}{\partial \sigma_{ij}} d\sigma_{ij}\bigg|_{k(\varepsilon_{ij}^p)=const.} > 0 \tag{3.44}$$

or

$$df_{k(\varepsilon_{ij}^p)=const.} > 0 \tag{3.45}$$

Next, we consider the stress change, $d\sigma_{ij}^*$, parallel to the tangent of the yield surface. Then,

$$\frac{\partial f}{\partial \sigma_{ij}} d\sigma_{ij}^*\bigg|_{k(\varepsilon_{ij}^p)=const.} = 0 \tag{3.46}$$

Hence,

$$df_{\varepsilon_{ij}^p=const.} = 0 \tag{3.47}$$

In this case, the yield surface does not expand and the plastic strain does not yield, and the stress change is called neutral loading. When the direction of the stress change is toward the inside of the yield surface, plastic strain does not occur:

$$\frac{\partial f}{\partial \sigma_{ij}} d\sigma_{ij}\bigg|_{k(\varepsilon_{ij}^p)=const.} < 0 \tag{3.48}$$

Then, the plastic strain does not occur and the process is elastic. This process is called unloading:

$$df_{k(\varepsilon_{ij}^{p})=const.} < 0 \tag{3.49}$$

3.6.4 Stability of elastoplastic material

Drucker (1951) defined stable material by the following remark called stability postulates: "The plastic work done by an external agency during the application of additional stress is positive."

$$d\sigma_{ij} d\varepsilon_{ij} > 0 \tag{3.50}$$

This condition is called the local stability or the stability in the small. This condition is satisfied in the strain-hardening range, but $d\sigma_{ij} d\varepsilon_{ij} < 0$ in the strain-softening range (Figure 3.7). Equation (3.50) is called the positiveness of the second-order work. In fact, it has been pointed out that Drucker's stability postulate is a sufficient condition for the stability in the sense of Lyapunov, which was derived by Hill (1958). This means that if the second order work by Equation (3.50) is violated, the material is potentially unstable. The violation of the Lyapunov stability causes the materials to experience unstable behavior such as strain localization, that is, shear band formation and diffuse failure. Strain localization indicates the instability with preferred orientation. Diffuse failure denotes the instability with no preferred orientation and manifests the barreled type of failure mode. Darve et al. (2001) pointed out that the diffuse failure

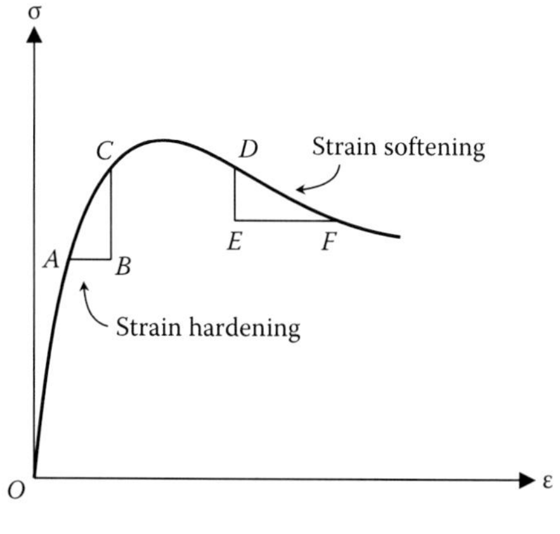

Figure 3.7 Strain hardening and strain softening.

model for loose sand is due to the violation of Hill's stability criteria, that is, second-order work criteria. Nova (1994) showed that Hill's condition is equivalent to the controllability condition.

Hill (1957, 1958) proposed a stability condition for elastoplastic bodies at the equilibrium state in the sense of Lyapunov. Hill's stability condition is a sufficient condition for stability and has been explained by Nakamura and Uetani (1981) as "the stability of the body system in the sense of Lyapunov is that the displacement is small when the disturbance is small."

The stability of the solution of differential equation in the sense of Lyapunov is defined as follows: A solution $f(x,t)$ at point x is stable in the sense of Lyapunov if for $\forall \varepsilon > 0$. There exists $\delta > 0$ such that if $|x - y| < \delta$, then $\| f(x,t) - f(y,t) \| < \varepsilon$ for all t, where t is a time and $\| \cdot \cdot \|$ is a norm for the solution.

Hill's theorem is described as follows: Let state S_{II} be near equilibrium state S_I. If the following inequality is satisfied, S_I is stable in the sense of Lyapunov:

$$\Delta U - \Delta W_{ex} > 0 \tag{3.51}$$

where ΔU is the sum of the stored energy in the body and the energy consumed for the plastic deformation during the deformation from S_I to S_{II}, and ΔW_{ex} is the work to the body by external energy (Figure 3.8).

During the process from S_I to S_{II}, the following energy balance holds:

$$\Delta K + \Delta U = \Delta I + \Delta W_{ex} - \Delta D \tag{3.52}$$

where ΔK is the kinetic energy at S_{II}, ΔI is the input energy in the body system, and ΔD is the consumed energy caused by viscous damping.

From Equation (3.51), there exists a positive number δ that satisfies $(\Delta U - \Delta W_{ex})_{\min} > \delta$. For that δ we take ΔI as $\Delta I < \delta$. Since ΔD is nonnegative, the following inequality holds:

$$\Delta K = -(\Delta U - \Delta W_{ex}) + \Delta I - \Delta D \leq -\delta + \Delta I < 0 \tag{3.53}$$

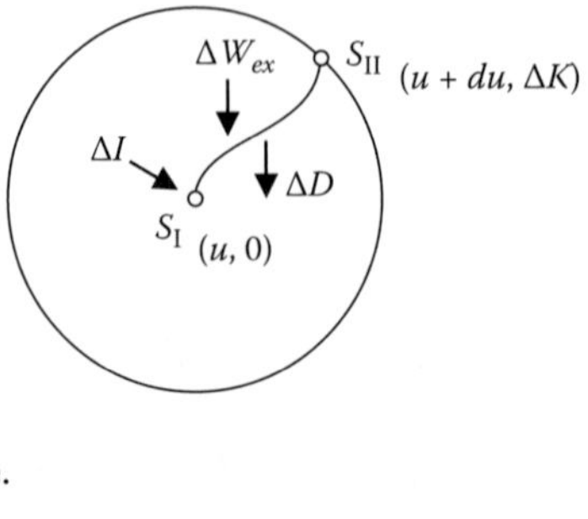

Figure 3.8 Energy balance.

Since kinetic energy, ΔK, is nonnegative, Equation (3.53) does not hold. It means the state S_{II} cannot be reached due to small disturbance δ, which is smaller than input energy, ΔI. The previous discussions indicate that the elastoplastic body system is always stable in the sense of Lyapunov if the body system satisfies Equation (3.51). Hence, if the body does not satisfy Equation (3.51), the body system is potentially unstable. In other words, a small disturbance may grow. Let us explain Equation (3.51) by the principle of virtual work within the framework of a small displacement field:

$$\Delta U = \int_v (\sigma_{ik} + \delta\sigma_{ik})\delta u^*_{i,k} dv \tag{3.54}$$

where $\delta\sigma_{ik}$ is the change in stress due to the small disturbance.

$$\Delta W_{ex} = \int_s t_i \delta u^*_i ds + \int_v \rho g_i \delta u^*_i dv \tag{3.55}$$

Since

$$\int_v \sigma_{ik}\delta u^*_{i,k} = \int_s t_i \delta u^*_i ds + \int_v \rho g_i \delta u^*_i dv \tag{3.56}$$

and Equation (3.51) holds, we obtain

$$\Delta^2 W = \int_v \delta\sigma_{ij}\delta\varepsilon_{ij} dv > 0 \tag{3.57}$$

It is worth noting that this condition is a sufficient condition for stability and is called the stability in the sense of second-order work.

3.6.5 Maximum work theorem

The maximum work theorem is given by the following proposition by Drucker (1951): "The net work done by the external agency during the cycle of adding and removing stress is non-negative" (Figure 3.9).

Let us consider a stress cycle where the stress is at point *a* at the initial state, then the stress changes to point *b* on the yield surface, then the stress changes further to point *c* outside the yield surface, then the stress changes to point *d* on the yield surface, and finally the stress returns to the initial point by the unloading process (Figure 3.9).

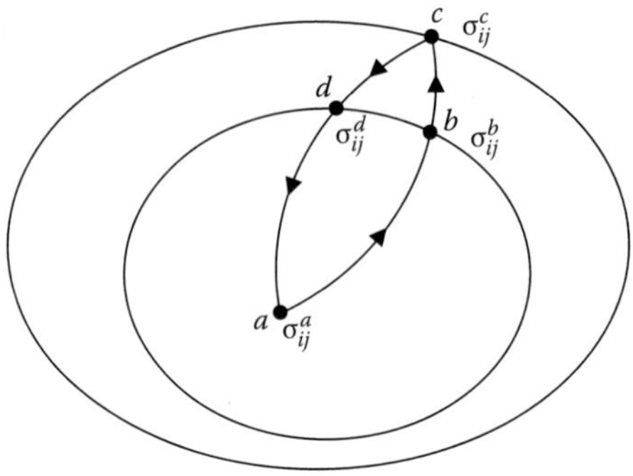

Figure 3.9 Stress cycle and work.

Since the process from point a to point b and from point d to point a is elastic, the total work during the cycle is given by

$$w = \int_{cycle} \sigma_{ij}\, d\varepsilon_{ij} \tag{3.58}$$

$$\begin{aligned} w &= \int_a^b \sigma_{ij} d\varepsilon_{ij}^e + \int_b^c \sigma_{ij}(d\varepsilon_{ij}^e + d\varepsilon_{ij}^p) + \int_c^a \sigma d\varepsilon_{ij}^e \\ &= \int_{cycle} \sigma\, d\varepsilon_{ij}^e + \int_b^c \sigma_{ij} d\varepsilon_{ij}^p \end{aligned} \tag{3.59}$$

Since the elastic deformation is reversible, the first term on the right-hand side is zero. Hence, the total work done by the stress is

$$w = \int_b^c \sigma_{ij} d\varepsilon_{ij}^p = D(d\varepsilon_{ij}^p) \tag{3.60}$$

In Equation (3.60), $D(d\varepsilon_{ij}^p)$ is called the internal dissipation.

From the above, the work done by the stress that causes the change in the system becomes

$$\int_b^c (\sigma_{ij} - \sigma_{ij}^a) d\varepsilon_{ij}^p \tag{3.61}$$

In addition, for the preceding equation to be held to the loading from b to c during the arbitrary short time,

$$(\sigma_{ij} - \sigma_{ij}^a) d\varepsilon_{ij}^p \geq 0 \tag{3.62}$$

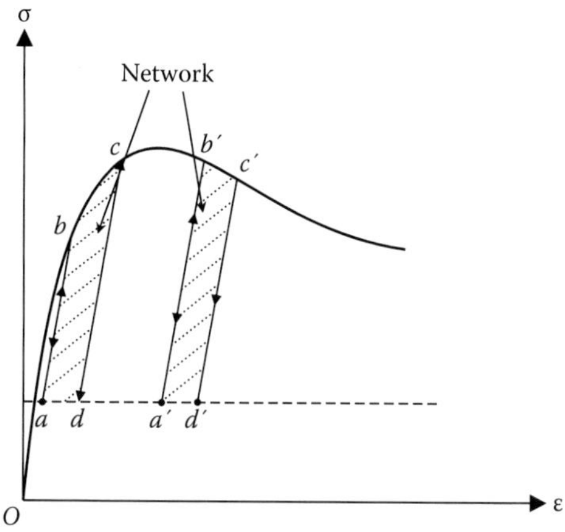

Figure 3.10 Network during the stress cycles.

where we have equality if and only if the process is in the elastic range. Equation (3.62) is known as the maximum plastic work theorem. Equation (3.62) can be written as

$$\sigma_{ij} d\varepsilon_{ij}^{p} \geq \sigma_{ij}^{a} d\varepsilon_{ij}^{p} \tag{3.63}$$

Accordingly, the maximum work theorem indicates that the work during the plastic deformation is the maximum compared with the work done by the stress that does not violate the yield conditions. In other words, internal dissipation, $D(d\varepsilon_{ij}^{p})$, is the maximum during the plastic process. It is worth noting that stress, σ_{ij}, causes plastic deformation and the maximum work theorem holds in the strain-softening process (Figure 3.10).

3.6.6 Flow rule and normality (evolutional equation of plastic strain)

When the yield surface is convex with no cusp and it is continuous, and the direction of the plastic stress increment depends on current stress, σ_{ij}, the flow rule is given by

$$d\varepsilon_{ij}^{p} = h \frac{\partial f}{\partial \sigma_{ij}} \tag{3.64}$$

$$h > 0 \tag{3.65}$$

where f is the plastic potential and h is the scalar parameter. f is called the plastic potential simply because of its similarity to the potential in classical mechanics and $f = const.$ surface is the equipotential surface. Now let us consider the yield surface and the flow rule that satisfies the maximum plastic work theorem.

The sufficient conditions for the yield surface and the flow rule to satisfy the maximum work theorem have been derived by Drucker (1951) as follows:

1. Convexity of the yield surface
2. Plastic potential is given by the yield function

In this case, the flow rule is called the associated flow rule. This condition is also called the normality of the strain rate tensor onto the yield surface.

It can be shown that the two aforementioned conditions are sufficient conditions for the maximum work theorem, as follows: As shown in Figure 3.11, if the yield surface is not convex in the six-dimensional stress space, the angle between $(\sigma_{ij}^p - \sigma_{ij}^o)$ and $d\varepsilon_{ij}^p$ is obtuse when we consider the strain space and the stress space to be the same.

Then, the inner products of the vectors show that $(\sigma_{ij}^p - \sigma_{ij}^o)d\varepsilon_{ij}^p < 0$; this violates the maximum work theorem. Moreover, even if the yield surface is 凸 convex, the angle between $\sigma_{ij}^p - \sigma_{ij}^o$ and $d\varepsilon_{ij}^p$ is obtuse if the plastic strain increment vector is not perpendicular to the yield surface. When the two aforementioned conditions are satisfied, the maximum work theorem always holds. Hence, the two conditions are sufficient conditions.

The normality rule can be directly obtained from the maximum work theorem (Maugin 1992). In addition, Collins and Houlsby (1997) presented that the normal flow rule is derived based on the thermodynamical consideration by Ziegler (1983), which will be shown in Equation (3.130).

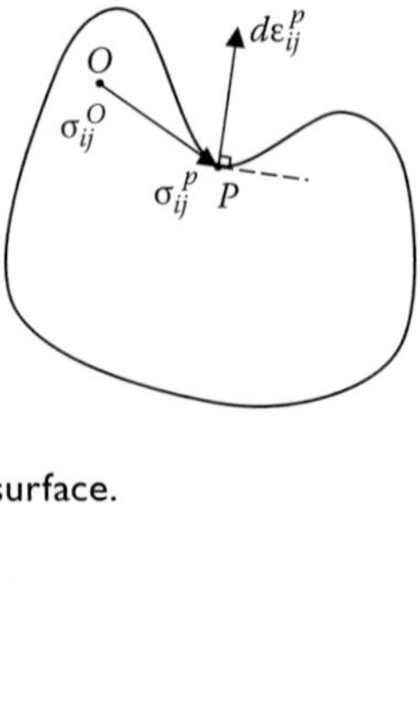

Figure 3.11 Convexity of yield surface.

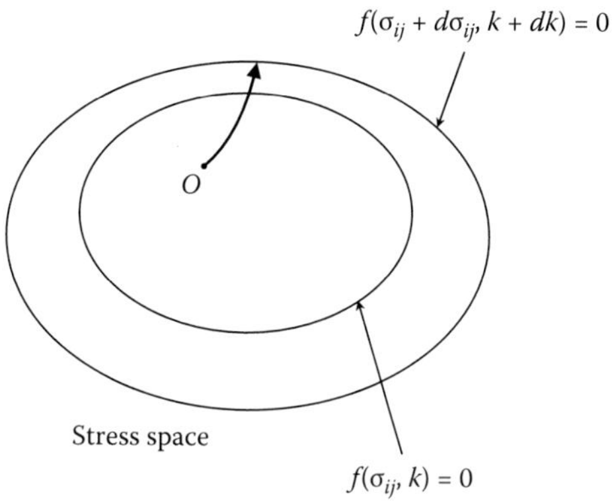

Figure 3.12 Subsequent yield surface.

3.6.7 Consistency conditions

When stress, σ_{ij}, on the yield surface moves onto the new stress, $\sigma_{ij} + d\sigma_{ij}$, the following equation must hold for the case of strain hardening (see Figure 3.12) as:

$$f(\sigma_{ij} + d\sigma_{ij}, k(\varepsilon_{ij}^p) + dk(\varepsilon_{ij}^p)) = 0 \tag{3.66}$$

$$f(\sigma_{ij}, k(\varepsilon_{ij}^p)) = 0 \tag{3.67}$$

Then,

$$f(\sigma_{ij} + d\sigma_{ij}, k(\varepsilon_{ij}^p) + dk(\varepsilon_{ij}^p)) - f(\sigma_{ij}, k(\varepsilon_{ij}^p)) = 0 \tag{3.68}$$

Taking a first-degree Taylor series expansion of the yield function, with respect to the plastic strain, the stress to the first-order term yields

$$f(\sigma_{ij} + d\sigma_{ij}, k(\varepsilon_{ij}^p) + dk(\varepsilon_{ij}^p)) = f(\sigma_{ij}, \varepsilon_{ij}^p) + \frac{\partial f}{\partial \sigma_{ij}} d\sigma_{ij} + \frac{\partial f}{\partial k} \frac{\partial k}{\partial \varepsilon_{ij}^p} d\varepsilon_{ij}^p = 0 \tag{3.69}$$

Hence,

$$\frac{\partial f}{\partial \sigma_{ij}} d\sigma_{ij} + \frac{\partial f}{\partial k}\frac{\partial k}{\partial \varepsilon_{ij}^p} d\varepsilon_{ij}^p = df = 0 \quad or \ \dot{f} = 0 \tag{3.70}$$

This equation is called Prager's consistency condition, and it indicates that the stress points during the loading process must always be on the yield surface.

From this Prager's condition, we can determine parameter h in the flow rule as

$$h = -\frac{\frac{\partial f}{\partial \sigma_{ij}} d\sigma_{ij}}{\frac{\partial f}{\partial k}\frac{\partial k}{\partial \varepsilon_{ij}^p}\frac{\partial f}{\partial \sigma_{ij}}} \tag{3.71}$$

These loading and unloading conditions can be alternatively expressed by the following Kuhn–Tucker complementary condition:

$$h \geq 0, \quad f(\sigma_{kl}, k) \leq 0, \quad hf(\sigma_{kl}, k) = 0 \tag{3.72}$$

$h \geq 0$ is determined by the following consistency condition:

$$\text{Consistency condition: } h\dot{f}(\sigma_{kl}, k) = 0 \tag{3.73}$$

Strain hardening is given by

$$k = k(\varepsilon_{ij}^p) \ \text{or} \ \dot{k} = \frac{\partial k}{\partial \varepsilon_{ij}^p}\dot{\varepsilon}_{ij}^p \tag{3.74}$$

In other expressions, the loading and unloading conditions are expressed by

$$f < 0 \ \rightarrow \ h = 0 \ \text{(elastic)}$$

$$f = 0, \quad \dot{f} = 0 \quad \text{and} \quad h > 0 \rightarrow \quad \dot{\varepsilon}_{kl}^p \neq 0, \dot{\kappa} \neq 0 \quad \text{(loading)}$$

$$f = 0, \quad \dot{f} = 0 \ \text{ and } \ h = 0 \ \rightarrow \quad \dot{\varepsilon}_{kl}^p = 0, \dot{\kappa} = 0 \quad \text{(neutral loading)} \tag{3.75}$$

$$f = 0, \quad \dot{f} < 0 \ \rightarrow \ h = 0 \quad \dot{\varepsilon}_{kl}^p = 0, \dot{\kappa} = 0 \ \text{(unloading)}$$

It is worth noting that $\dot{f} > 0$ is excluded since $f \leq 0$ for a subsequent short increment.

$$
\begin{aligned}
\dot{f} &= \frac{\partial f}{\partial \sigma_{ij}} \dot{\sigma}_{ij} + \frac{\partial f}{\partial \kappa} \dot{\kappa} = \frac{\partial f}{\partial \sigma_{ij}} C_{ijkl} (\dot{\varepsilon}_{kl} - \dot{\varepsilon}^{p}_{kl}) + \frac{\partial f}{\partial \kappa} \dot{\kappa} \\
&= \frac{\partial f}{\partial \sigma_{ij}} C_{ijkl} \dot{\varepsilon}_{kl} - \frac{\partial f}{\partial \sigma_{ij}} C_{ijkl} h \frac{\partial f}{\partial \sigma_{kl}} + \frac{\partial f}{\partial \kappa} \frac{\partial \kappa}{\partial \varepsilon^{p}_{ij}} \dot{\varepsilon}^{p}_{ij} \\
&= \frac{\partial f}{\partial \sigma_{ij}} C_{ijkl} \dot{\varepsilon}_{kl} - \frac{\partial f}{\partial \sigma_{ij}} C_{ijkl} h \frac{\partial f}{\partial \sigma_{kl}} + \frac{\partial f}{\partial \kappa} \frac{\partial \kappa}{\partial \varepsilon^{p}_{ij}} h \frac{\partial f}{\partial \sigma_{ij}} \\
&= \frac{\partial f}{\partial \sigma_{ij}} C_{ijkl} \dot{\varepsilon}_{kl} - h \left(\frac{\partial f}{\partial \sigma_{ij}} C_{ijkl} \frac{\partial f}{\partial \sigma_{kl}} + \frac{\partial f}{\partial \kappa} \frac{\partial \kappa}{\partial \varepsilon^{p}_{ij}} \frac{\partial f}{\partial \sigma_{ij}} \right) \leq 0
\end{aligned}
\tag{3.76}
$$

Let us assume that the following inequality holds:

$$
\left(\frac{\partial f}{\partial \sigma_{ij}} C_{ijkl} \frac{\partial f}{\partial \sigma_{kl}} + \frac{\partial f}{\partial \kappa} \frac{\partial \kappa}{\partial \varepsilon^{p}_{ij}} \frac{\partial f}{\partial \sigma_{ij}} \right) > 0 \tag{3.77}
$$

From $\dot{f} = 0$,

$$
h = \frac{\dfrac{\partial f}{\partial \sigma_{ij}} C_{ijkl} \dot{\varepsilon}_{kl}}{\left(\dfrac{\partial f}{\partial \sigma_{ij}} C_{ijkl} \dfrac{\partial f}{\partial \sigma_{kl}} + \dfrac{\partial f}{\partial \kappa} \dfrac{\partial \kappa}{\partial \varepsilon^{p}_{ij}} \dfrac{\partial f}{\partial \sigma_{ij}} \right)} \tag{3.78}
$$

Tensors of tangent elastoplastic moduli are obtained by

$$
\begin{aligned}
\dot{\sigma}_{ij} &= C_{ijkl} (\dot{\varepsilon}_{kl} - \dot{\varepsilon}^{p}_{kl}) \\
&= C^{ep}_{ijkl} \dot{\varepsilon}_{kl}
\end{aligned}
\tag{3.79}
$$

$$
C^{ep}_{ijkl} = C^{e}_{ijkl} \qquad \text{if} \quad h = 0 \tag{3.80}
$$

If plastic potential function g is not equal to yield function f, that is,

$$
d\varepsilon^{p}_{ij} = h \frac{\partial g}{\partial \sigma_{ij}} \tag{3.81}
$$

we obtain non-associated flow rule, and

$$C^{ep}_{ijkl} = C^{e}_{ijkl} - \frac{C^{e}_{ijmn}\dfrac{\partial g}{\partial \sigma_{mn}} C^{e}_{klpq}\dfrac{\partial f}{\partial \sigma_{pq}}}{\left(\dfrac{\partial f}{\partial \sigma_{ij}} C_{ijkl}\dfrac{\partial g}{\partial \sigma_{kl}} + \dfrac{\partial f}{\partial \kappa}\dfrac{\partial \kappa}{\partial \varepsilon^{p}_{ij}}\dfrac{\partial g}{\partial \sigma_{ij}}\right)} \quad if \quad h > 0 \tag{3.82}$$

For the case in which $g = f$, the theory is called the associated flow rule.

3.7 OVERSTRESS TYPE OF ELASTOVISCOPLASTICITY

3.7.1 Perzyna's model

Perzyna (1963) has formulated a viscoplastic model called the overstress model in order to generalize the linear viscoplastic model by Hohenemser and Prager (1932). Perzyna adopted a nonlinear functional of F and proposed the following flow rule. He has tried to take the nonlinear form of function F in Equation (3.84) and has proposed the following viscoplastic flow rule:

$$\dot{\varepsilon}^{vp}_{ij} = \gamma\langle\Phi(F)\rangle\frac{\partial f}{\partial \sigma_{ij}} \tag{3.83}$$

$$F = \frac{f - \kappa}{\kappa} \tag{3.84}$$

where $\dot{\varepsilon}^{vp}_{ij}$ is the viscoplastic strain rate tensor, f is a dynamic loading function or dynamic viscoplastic potential, γ is a viscosity coefficient, Φ is a functional of F that denotes a material viscoplasticity, and κ is a hardening parameter.

$$\langle\Phi(F)\rangle = \Phi(F); \quad F > 0, \quad \langle\Phi(F)\rangle = 0; \quad F \le 0 \tag{3.85}$$

For the work-hardening case,

$$\dot{\kappa} = \sigma_{ij}\dot{\varepsilon}^{vp}_{ij}, \qquad \kappa = \int_0^{\varepsilon^{vp}} \sigma_{ij}\, d\varepsilon^{vp}_{ij} \tag{3.86}$$

and for the strain-hardening case,

$$\dot{\kappa} = A_{kl}\dot{\varepsilon}^{vp}_{kl}, \qquad \kappa = \int_0^{\varepsilon^{vp}} A_{kl}\, d\varepsilon^{vp}_{kl} \tag{3.87}$$

where A_{kl} is a material tensor.

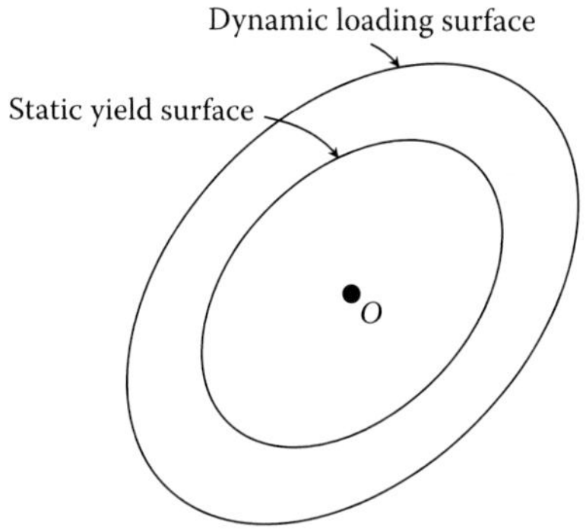

Figure 3.13 Dynamic loading surface.

Since $F = 0$ denotes a static yield function, dynamic loading function f takes the same form as the static yield function (Figure 3.13).

If we solve Equation (3.83) with respect to f, we have

$$f(\sigma_{ij}, \varepsilon_{ij}^{vp}) = \kappa\left[1 + \Phi^{-1}\left(\frac{I_2^{vp}}{\gamma}\left\{\frac{\partial f}{\partial \sigma_{ij}}\frac{\partial f}{\partial \sigma_{ij}}\right\}^{1/2}\right)\right] \tag{3.88}$$

Equation (3.88) indicates that the yield function depends on the viscoplastic strain rate; $I_2^{vp} = \left(\dot{\varepsilon}_{ij}^{vp}\,\dot{\varepsilon}_{ij}^{vp}\right)^{1/2}$.

In addition, the dynamic yield function coincides with the static yield function as $\gamma \rightarrow \infty$, that is, the coefficient of viscosity becomes infinite, ∞.

Total strain rate tensor, $\dot{\varepsilon}_{ij}$, is composed of viscoplastic strain rate tensor, $\dot{\varepsilon}_{ij}^{vp}$, and elastic strain rate tensor, $\dot{\varepsilon}_{ij}^{e}$, as

$$\dot{\varepsilon}_{ij} = \dot{\varepsilon}_{ij}^{e} + \dot{\varepsilon}_{ij}^{vp} \tag{3.89}$$

The graphical presentation of the Perzyna's model leads to Figure 3.14.

3.7.2 Duvaut and Lions' model

Duvaut and Lions (1972) proposed a viscoplastic model as follows:

$$\dot{\varepsilon}_{ij} = \dot{\varepsilon}_{ij}^{e} + \dot{\varepsilon}_{ij}^{vp} \tag{3.90}$$

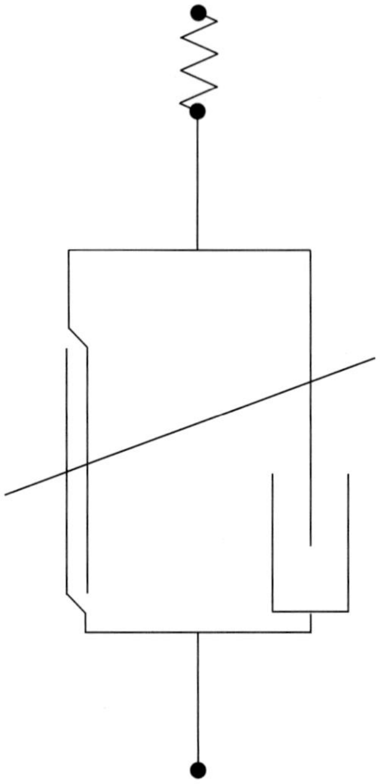

Figure 3.14 Perzyna-type viscoplastic model.

When $F < 0$,

$$\dot{\varepsilon}_{ij}^{vp} = 0 \tag{3.91}$$

When $F \geq 0$,

$$\dot{\varepsilon}_{ij}^{vp} = \frac{1}{2\mu}[\sigma_{ij} - (P_K\sigma)_{ij}] \tag{3.92}$$

where μ is a viscosity coefficient and $(P_K\sigma)_{ij}$ is the stress obtained by the orthogonal projection of the current stress on the closed convex 凸 set K.

$$K = \{\sigma_{ij}, F(\sigma_{ij}) \leq 0\} \tag{3.93}$$

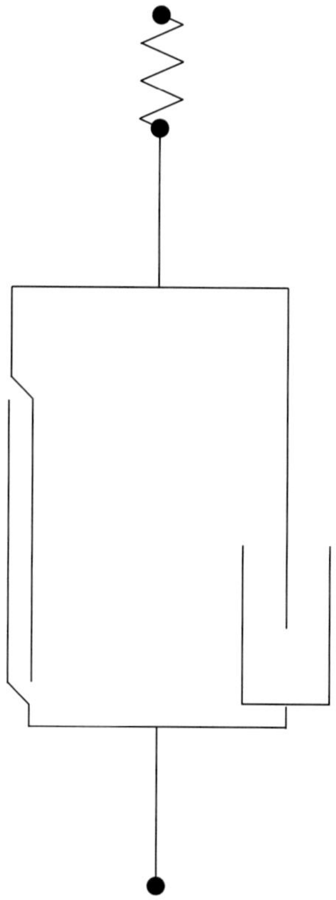

Figure 3.15 Duvaut and Lion-type viscoplastic model.

where F is a continuous a convex function and corresponds to the static yield surface.

Rewriting Equation (3.92) yields

$$\sigma_{ij} = (P_K\sigma)_{ij} + 2\mu\dot{\varepsilon}_{ij}^{vp} \tag{3.94}$$

In the model, the stress is broken into viscous and plastic elements, as shown in Figure 3.15. The feature of the model is that the viscoplastic model can be constructed only by introducing a viscous coefficient, not by introducing a nonlinear functional. The limit of the model leads to the linear viscoplastic model.

3.7.3 Phillips and Wu's model

Phillips and Wu (1973) proposed a viscoplastic model that is similar to Duvaut and Lions' model and at almost the same time. In Philips and Wu's theory, an excess stress is defined by h, which is obtained as the perpendicular distance from the current stress point to the quasi-static yield stress in the deviatoric stress space. k_{ij} is given by the stress at the intersection of the perpendicular line to the quasi-static yield surface from current stress, s_{ij}, and the quasi-static yield surface.

$$h_{ij} = hn_{ij} = s_{ij} - k_{ij} \tag{3.95}$$

The direction of the viscoplastic strain rate is parallel to the direction of h_{ij}.
Subsequently, the deviatoric flow rule is given by

$$\dot{e}_{ij}^{vp} = \gamma(k_{rs}, e_{rs}^{vp}, I_2^{vp})\Phi[h]n_{ij} : \ h > 0 \tag{3.96}$$

$$\dot{e}_{ij}^{vp} = 0 : \ h = 0 \tag{3.97}$$

where $\gamma(k_{rs}, e_{rs}^{vp}, I_2^{vp})$ is a viscosity function that depends on position k_{rs}, the viscoplastic deviatoric strain e_{rs}^{vp}, and its second invariant of viscoplastic deviatoric strain rate, I_2^{vp}. Function $\Phi[h]$ is a scalar function of h.

If we obtain stress point k_{ij} by Duvaut and Lions' theory as $(P_K\sigma)_{ij}$, it is not always possible to take a stress point on the quasi-static yield surface because the line from the current stress point is perpendicular to the dynamic yield function.

3.8 ELASTOVISCOPLASTIC MODEL BASED ON STRESS HISTORY TENSOR

Oka (1985) has proposed a model that can describe the continuous transition from the viscoplastic theory to the plastic theory. He constructed an elastoviscoplastic theory based on the concept of memory and internal variables by Perzyna (1971) using stress history tensor, σ_{ij}^*. Using this theory, it is possible to express strain softening and a continuous transition between plastic and viscoplastic models.

It is assumed that total strain rate, $\dot{\varepsilon}_{ij}$, can be composed of two parts: elastic component, $\dot{\varepsilon}_{ij}^{e}$; and viscoplastic or plastic component, $\dot{\varepsilon}_{ij}^{vp}$.

$$\dot{\varepsilon}_{ij} = \dot{\varepsilon}_{ij}^{e} + \dot{\varepsilon}_{ij}^{vp} \tag{3.98}$$

3.8.1 Stress history tensor and kernel function

Total stress history, σ_{ij}^{z}, is a union of the two sets of the present stress and reduced stress history, σ_{rij}^{z}, which is defined by the history of the stress without the present stress. Following Eringen's manner (1967), the total history tensor and the reduced stress history tensor are given by

$$\sigma_{ij}^{z} = [\sigma_{ij}(z), \sigma_{rij}^{z}, 0 < z' \leq z] \tag{3.99}$$

$$\sigma_{rij}^{z} = [\sigma_{ij}(z - z'), 0 < z' \leq z] \tag{3.100}$$

where time measure z is defined by

$$dz = g(\dot{\varepsilon}_{ij})dt \tag{3.101}$$

Taking account of the integral, in a wider sense, the total stress history tensor can be expressed by the functional of the reduced stress history as

$$\sigma_{ij}^{*} = \hat{\sigma}_{ij}^{*}(\sigma_{rij}^{z}) \tag{3.102}$$

The total stress history tensor can be expressed by the convolution integral, with respect to the generalized strain measure, as

$$\sigma_{ij}^{*} = \int_{0}^{z} K(z - z')\sigma_{ij}(z')dz' \tag{3.103}$$

where K is a continuous kernel function.

It is assumed that $\partial K/\partial z < 0$. This assumption provides that the stress history tensor satisfies the principle of fading memory with respect to the stress history. Namely, the effect of the older stress on the stress history tensor becomes small (Figure 3.16).

Based on the theorem of the integral, in a wider sense, the stress history is defined in the closed interval as $0 \leq z' \leq z$.

3.8.2 Flow rule and yield function

$$d\varepsilon_{ij}^{p} = H \frac{\partial f_p}{\partial \sigma_{ij}} df_y \tag{3.104}$$

where f_p is the plastic potential function, f_y is the yield function, and H is the hardening–softening function.

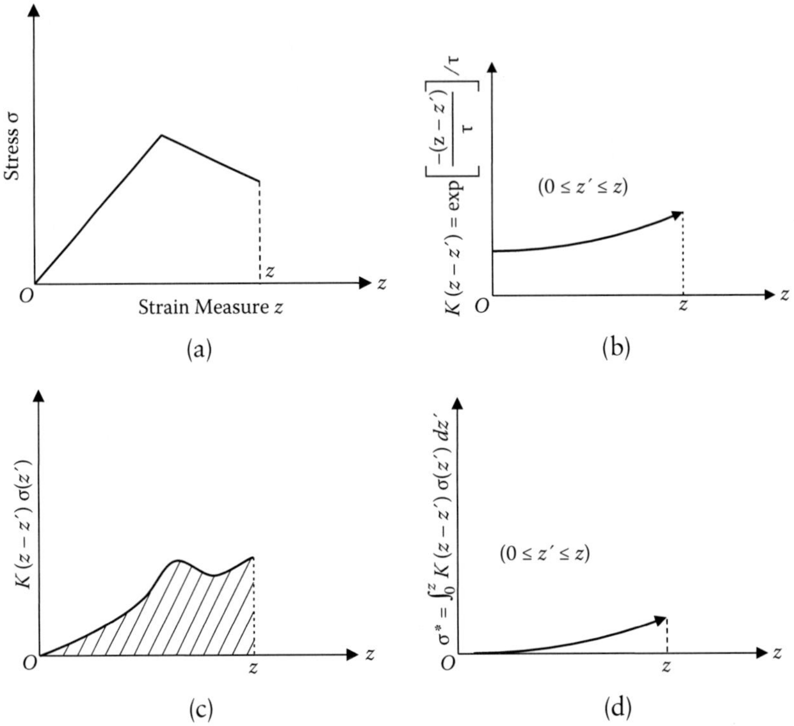

Figure 3.16 (a–d) Stress and stress history tensor.

The yield function is a function of the stress history tensor and hardening and softening parameter, κ.

$$f_y(\sigma^*_{ij}, \kappa) = 0 \tag{3.105}$$

Loading conditions are given by

$$\text{When } f_y = 0, \quad df_y[=(\tfrac{\partial f_y}{\partial \sigma^*_{ij}})d\sigma^*_{ij}] > 0, \text{ then } d\varepsilon^p_{ij} \neq 0 \tag{3.106}$$

$$\text{When } f_y = 0, \quad df_y = 0, \text{ then } d\varepsilon^p_{ij} = 0 \tag{3.107}$$

$$\text{When } f_y = 0, \quad df_y < 0, \text{ then } d\varepsilon^p_{ij} = 0 \tag{3.108}$$

The plastic potential function is a function of the current stress and the other history parameter, L.

$$f_p(\sigma_{ij}, L) = 0 \tag{3.109}$$

3.9 OTHER VISCOPLASTIC AND VISCOELASTIC–PLASTIC THEORIES

Naghdi and Murch (1963) proposed a viscoelastic–plastic model considering the yielding of viscoelastic materials. Tateishi and Miyamoto (1974) also constructed a viscoelastic yield theory. Kremple (1986) proposed an overstress theory using total strain rates, not using the concept of viscoplastic strain for metallic materials. On the other hand, Chaboche and Rousellier (1983a,b) derived a more general kinematic viscoplasticity theory based on the nonlinear hardening theory advocated by Armstrong and Frederick (1966) and Perzyna's type of viscoplasticity theory. Chaboche's type of viscoplasticity theory will be explained in Chapter 6. Valanis (1971) has initiated an integral type of viscoplasticity theory without the yield surface in which he defined an inherent time measure. In his theory, Valanis has used a convolution integral of strain with respect to an internal time; hence, his theory is called the endochronic theory. Watanabe and Alturi (1986) generalized it as an internal time theory.

3.10 CYCLIC PLASTICITY AND VISCOPLASTICITY

Cyclic constitutive models for materials are of great importance in the numerical simulation of behavior during cyclic loading or dynamic loading, such as for vibrational problems.

Prager (1949) is the first to have proposed a cyclic plasticity model with a linear kinematical hardening rule. Prager's linear kinematical hardening model can reproduce the Bauschinger effect, cyclic hardening, and the cyclic hysteresis loop for the stress–strain curves.

The Bauschinger effect indicates the lowering of the yield stress upon reverse loading behavior, such as compression after loading in tension, as shown in Figure 3.17. The Bauschinger effect is named after the German engineer Johann Bauschinger (1860–1934).

Prager's yield function and kinematical hardening rule are given by

$$f(\sigma_{ij} - \chi_{ij}) - R = 0 \tag{3.110}$$

$$d\chi_{ij} = c\,d\varepsilon_{ij}^{p} \tag{3.111}$$

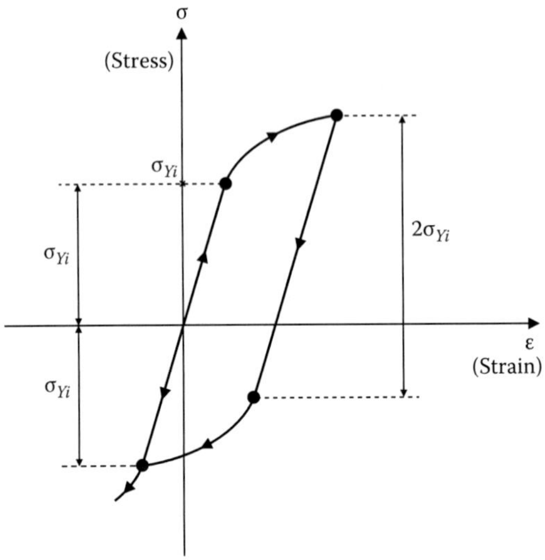

Figure 3.17 Bauschinger effect.

where $f(\sigma_{ij} - \chi_{ij}) - R = 0$ is the yield function and χ_{ij} is the kinematical hardening parameter, indicating the size of an initial yield surface, R is the plastic strain increment, and c is a material parameter. Schematic figure for the kinematical hardening is shown in Figure 3.18.

Prager's model has a shortcoming regarding the reduction in dimension of the stress space, such as in plane stress problems. To improve this point, Ziegler (1959) proposed another type of kinematical hardening

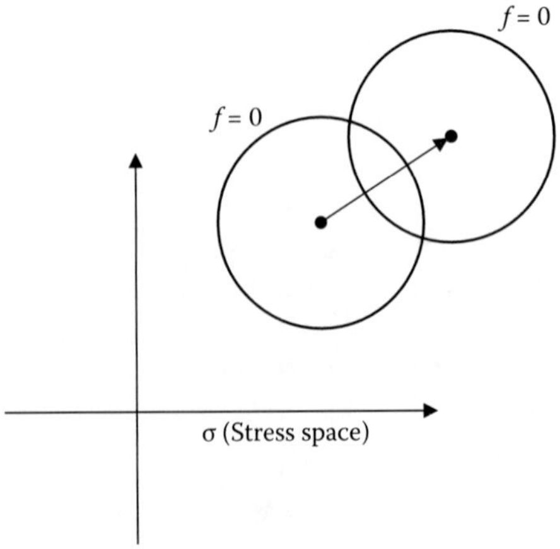

Figure 3.18 Kinematic hardening.

for the cyclic plasticity model. Ziegler's kinematical hardening rule is given by

$$d\chi_{ij} = cd\sigma_{ij} \tag{3.112}$$

where c is a constant hardening modulus.

Both Prager's and Ziegler's models are restricted to linear hardening, although experiments denote nonlinear hardening. To solve this problem, many other models have been proposed. Mroz (1981) proposed a cyclic plasticity model based on the "field of work-hardening moduli" to describe the nonlinear hardening shown in the experiments using a large number of yield surfaces.

Dafalias and Popov (1976) proposed a nonlinear cyclic plasticity model introducing a continuously varying plastic modulus in which the hardening modulus depends on the distance between the present stress state and the stress state on the bounding surface, which is called the bounding surface model or the two-surface model.

Chaboche (1977) newly proposed a nonlinear hardening plasticity model based on the theory by Armstrong and Frederick's (1966) nonlinear kinematical hardening rule (Figure 3.19). In the nonlinear hardening model, the hardening parameter follows the differential equations as

$$dX_{ij} = \frac{2}{3}cd\varepsilon_{ij}^{p} - \gamma X_{ij}dp \tag{3.113}$$

$$dp = \left(\frac{2}{3}d\varepsilon_{ij}^{p}d\varepsilon_{ij}^{p}\right)^{1/2} \tag{3.114}$$

where X_{ij} is a kinematical hardening parameter.

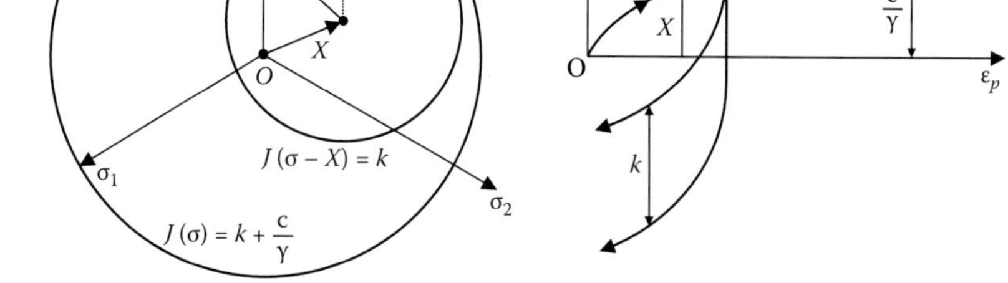

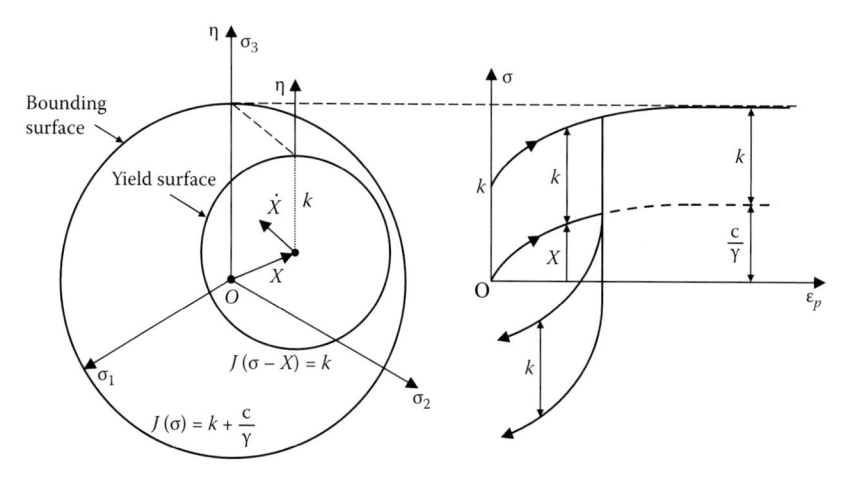

Figure 3.19 Nonlinear kinematic hardening rule (Chaboche and Rousselier 1983).

The plastic flow rule is given by

$$d\varepsilon_{ij}^{p} = d\lambda \frac{\partial f}{\partial \sigma_{ij}} \tag{3.115}$$

$$dp = d\lambda = \frac{1}{h}\left\langle \frac{\partial f}{\partial \sigma_{ij}} d\sigma_{ij} \right\rangle = \frac{1}{h}\left\langle \frac{3}{2}\frac{\sigma_{ij} - X_{ij}}{R+k} d\sigma_{ij} \right\rangle \tag{3.116}$$

$$h = \frac{2}{3} c \frac{\partial f}{\partial \sigma_{ij}} \frac{\partial f}{\partial \sigma_{ij}} - \gamma X_{ij} \frac{\partial f}{\partial \sigma_{ij}} \left(\frac{2}{3} \frac{\partial f}{\partial \sigma_{kl}} \frac{\partial f}{\partial \sigma_{kl}} \right)^{1/2} \tag{3.117}$$

Chaboche and Rousselier developed a viscoplastic cyclic model based on the nonlinear kinematical hardening rule (1983).

For the case where the von Mises criterion is used, the viscoplastic flow rule is given by

$$d\varepsilon_{ij}^{vp} = \left\langle \frac{J(\sigma_{ij} - X_{ij}) - R - k}{K} \right\rangle^{n} \frac{\partial f}{\partial \sigma_{ij}} \tag{3.118}$$

$$dX_{ij} = C\left(\frac{2}{3} a d\varepsilon_{ij}^{vp} - X_{ij} dp \right) \tag{3.119}$$

where <> is Macauley's bracket and n, K, R, k, C, and a are material constants.

During cyclic plastic loading, progressive deformation is, in general, observed. However, if there is no progressive plastic deformation, we can describe the cyclic hysteresis loop by the following type of one-dimensional stress–strain relation as

$$\frac{\sigma - \sigma_r}{\delta} = f\left(\frac{\varepsilon - \varepsilon_r}{\delta} \right) \tag{3.120}$$

where δ is a scaling factor and σ_r and ε_r are the values at the stress reversal point.

When δ is 2 in Equation (3.110), it is called the Masing rule (1927).

3.11 DISSIPATION AND THE YIELD FUNCTIONS

The energy dissipation has been dealt with with the Clausius–Duhem inequality as shown in Chapter 1. For the dissipation, the following important results have been obtained. For the existence of the yield functions,

Collins and Houlsby (1997) adopted Ziegler's thermomechanical approach (Ziegler 1983) and presented a thermodynamical background. They are based on the several thermodynamical potentials such as Helmholtz free energy and a dissipation function, Gibbs free energy function, and entropy function. Collins and Houlsby (1997) obtained a new potential function through Legendre transformations and then they derived a yield function through a singular transformation for the case where the dissipation function is homogeneous of degree one in the rate of internal variables. Houlsby and Puzrin (2000) presented constitutive models for geomaterials based on a thermomechanical approach by Collins and Houlsby (1997).

From the first law of thermodynamics Equations (1.138) and (1.150) and the definition of the free energy, the energy dissipation, D, in an isothermal process is given by

$$\sigma_{ij}\dot{\varepsilon}_{ij} = \rho\dot{\Psi} + D \tag{3.121}$$

where $\dot{\Psi}$ is a rate of free energy.

Following the Collins and Houlsby (1997), the internal dissipation function, D, is a rate at which energy dissipated.

Based on the work by Ziegler (1983), Collins and Houlsby (1997) assumed that the dissipation function, D, arises from the internal variables and corresponding stresslike variables χ_{ij} as:

$$D = \chi_{ij}\dot{\alpha}_{ij} \geq 0 \tag{3.122}$$

The dissipation function, D, corresponds to the mechanical dissipation by Equation (1.149).

In the following, we will derive the yield function. Ziegler (1983) assumed that the dissipation, D, is a function of the internal variable, α_{ij}, and its time rate, $\dot{\alpha}_{ij}$.

When D is homogeneous function of degree one in $\dot{\alpha}_{ij}$, the dissipation function can be described by the orthogonality condition; the difference between χ_{ij} and $\frac{\partial D}{\partial \dot{\alpha}_{ij}}$ is orthogonal to $\dot{\alpha}_{ij}$, or if it is zero, we have

$$D = \chi_{ij}\dot{\alpha}_{ij} = \frac{\partial D}{\partial \dot{\alpha}_{ij}}\dot{\alpha}_{ij} \tag{3.123}$$

$$\chi_{ij} = \frac{\partial D(\alpha_{ij}, \dot{\alpha}_{ij})}{\partial \dot{\alpha}_{ij}} \tag{3.124}$$

where χ_{ij} is a stress-like variable.

Through Legendre transformation, we obtain the dual Legendre function, Ω, as

$$D(\alpha_{ij}, \dot{\alpha}_{ij}) + \Omega(\alpha_{ij}, \chi_{ij}) = \chi_{ij}\dot{\alpha}_{ij} \tag{3.125}$$

It is worth noting that in general, Legendre transformation is defined as $F(p) = \max_x(px - f(x))$; p is a gradient of $f(x)$and $\max_x$ indicates that maximization in x, and $f(x)$ is convex downward.

When D is a homogeneous function of degree one in $\dot{\alpha}_{ij}$,

$$\Omega(\alpha_{ij}, \chi_{ij}) = 0 \tag{3.126}$$

In addition from Equation (3.123), Equation (3.125), and Equation (3.126), we have,

$$d\Omega = \frac{\partial \Omega}{\partial \chi_{ij}} d\chi_{ij} + \frac{\partial \Omega}{\partial \alpha_{ij}} d\alpha_{ij} = 0 \tag{3.127}$$

and

$$dD + d\Omega = \frac{\partial D}{\partial \alpha_{ij}} d\alpha_{ij} + \frac{\partial D}{\partial \dot{\alpha}_{ij}} d\dot{\alpha}_{ij} = d\chi_{ij}\dot{\alpha}_{ij} + \chi_{ij} d\dot{\alpha}_{ij} \tag{3.128}$$

From Equation (3.124),

$$\frac{\partial D}{\partial \alpha_{ij}} d\alpha_{ij} = d\chi_{ij}\dot{\alpha}_{ij} \tag{3.129}$$

From Equations (3.127) and (3.129), the following relations are derived:

$$\dot{\alpha}_{ij} = \dot{\lambda} \frac{\partial \Omega}{\partial \chi_{ij}} \tag{3.130}$$

and

$$\frac{\partial D}{\partial \alpha_{ij}} = -\dot{\lambda} \frac{\partial \Omega}{\partial \alpha_{ij}} \tag{3.131}$$

where $\dot{\lambda}$ is a multiplier. This function Ω corresponds to the yield function for plasticity theory and Equation (3.130) is called a flow rule in plasticity theory.

REFERENCES

Armstrong, P.J., and Frederick, C.O. 1966. A mathematical representation of the multiaxial Baushinger effect, C.E.G.B., Report RD/B/N 731.

Bingham, E.C. 1922. *Fluidity and Plasticity*, New York: McGraw-Hill.

Chaboche, J. I. 1977. Viscoplastic constitutive equations for the description of cyclic and anisotropic behavior of material, bulletin de l'Academie Polonaise des Sciences, *Série Sc. Et tech.*, 251(1):33–42.

Chaboche, J.L. 1986. Time-independent constitutive theories for cyclic plasticity, *Int. J. Plasticity*, 2(2):149–188.

Chaboche, J.L., and Rousselier, G. 1983a. On the plastic and viscoplastic constitutive equations, Part I: Rules developed with internal variable concept, *J. Pressure Vessel Tech.*, ASME, 105:153–158.

Chaboche, J.L., and Rousselier, G. 1983b. On the plastic and viscoplastic constitutive equations, Part II: Application of internal variable concepts to the 316 stainless steel, *J. Pressure Vessel Tech.*, ASME, 105:159–164.

Christensen, R.M. 1982. *Theory of Viscoelasticity: An Introduction*, 2nd ed., New York: Academic Press.

Collins, I.F., and Houlsby, G.T. 1997. Application of thermomechanical principles to the modeling of geotechnical materials, Proc. Roy. Soc. London A, 453:1975–2001.

Dafalias, Y.F., and Popov, E.P. 1976. Plastic internal variables formalism of cyclic plasticity, *J. Appl. Mech.*, 98:645–651.

Darve, F., Laouafa, F., and Servant, G. 2001. Mode of rupture in geomaterials, Proceedings of International Workshop on Deformation of Earth Materials, TC34 of ISSMGE, May 23, 2001, Sendai, Japan, 39–50.

Drucker, D.C. 1951. A more fundamental approach to plastic stress-strain relations, Proc. First U.S. Natl. Congr. Appl. Mech., 487–491.

Duvaut, G., and Lions, J.L. 1976. *Inequalities in Mechanics and Physics*, Springer-Verlag, Berlin (Les inequations en mechanique et en physique, Paris: Dunod, 1972).

Eringen, A.C. 1967. *Mechanics of Continua*, New York: John Wiley & Sons.

Hill, R. 1948. A variational principle of maximum plastic work in classical plasticity, *Q. J. Mech. Appl. Math.*, 1:18–28.

Hill, R. 1957. On uniqueness and stability in the theory of finite elastic strain, *J. Mech. Phys. Solids*, 5:229–241.

Hill, R. 1958. A general theory of uniqueness and stability in elastic-plastic solids, *J. Mech. Phys. Solids*, 6:236–249.

Hohenemser, K., and Prager, W. 1932. Über die ansätze der mechanik isotroper kontinua, ZAMM, 12:216–226.

Houlsby, G.T., and Puzrin, A.M. 2000. A mechanical framework for constitutive models for rate independent dissipative materials, *Int. J. Plasticity*, 16:1017–1047.

Ionescu, I.R., and Sofonea, M. 1993. *Functional and Numerical Methods in Viscoplasticity*, Oxford University Press.

Lubliner, J. 1990. *Plasticity Theory*, New York: Macmillan.

Kremple, E. 1986. Viscoplasticity based on overstress with a differential growth law for the equilibrium stress, *Mech. Mater.*, 5:35–48.

Lyapunov, A.M. 1907. Problème général de la stabilité des mouvements, *Ann. Fac. Sci. Toulouse*, 9:203–274.

Malvern, L.E. 1969. *Introduction to the Mechanics of a Continuous Media*, New York: Prentice-Hall.

Mandel, J. 1964. Conditions de Stabilite' et Postulat de Drucker, In Proceedings of IUTAM Symposium on Rheology and Soil Mechanics, J. Kravtchenko, ed., Springer-Verlag, 58–68.

Martin, J.B. 1927. *Plasticity: Fundamentals and General Results*, Cambridge, MA: The MIT Press, 1975.

Masing, G., Veroff Siemens-Konzen, III Band.

Maugin, G.A. 1992. *The Thermodynamics of Plasticity and Fracture*, Cambridge University Press.

Mroz, Z. 1981. On generalized kinematic hardening rule with memory of maximal prestress, *Journal de Mécanique Appliquée*, 5:241–259.

Naghdi, P.M., and Murch, S.A. 1963. On the mechanical behavior of viscoelastic/ plastic solids, *J. Appl. Mech.*, 321–328.

Nakamura, T., and Uetani, K. 1981.Critical behaviours of elasto-plastic structures, Review paper, Materials, *Japanese Soc. Mater. Sci.*, 30(333):535–548 (In Japanese).

Nova, R. 1994. Controllability of the incremental response of soil specimens subjected to arbitrary loading programs. *J. Mech. Behav. Mater.*, 5(2):193–201.

Oka, F. 1985. Elasto/viscoplastic constitutive equations with memory and internal variables, *Comput. Geotech.*, 1:59–69.

Perzyna, P. 1963. The constitutive equations for work-hardening and rate sensitive plastic materials, Proc. Vibrational Probl., Warsaw, 4(3):281–290.

Perzyna, P. 1971. Memory effects and internal changes of a material, *Int. J. Nonlinear Mech.*, 6:707–746.

Phillips, A., and Wu, H.-C. 1973. A theory of viscoplasticity, *Int. J. Solids Struct.*, 9:15–30.

Prager, W. 1949. Recent developments in the mathematical theory of plasticity, *J. Appl. Phys.*, 20:235–241.

Prager, W. 1961. *Introduction to Mechanics of Continua*, Boston: Ginn and Company.

Simo, J.C., and Hughes, T.J.R. 1998. *Computational Inelasticity* (Interdisciplinary Applied Mathematics), New York: Springer. (Corrected 2nd printing, 2006)

Tateishi, T., and Miyamoto, H. 1974. On the yield conditions of viscoelastic body, *J. JSME*, 40(339):3005–3016 (In Japanese).

Valanis, K.C. 1971. A theory of viscoplasticity without a yield surface, *Arch. Mech. Stos.*, 23(4):517–533.

von Mises, R. 1928. Mechanik der plastischen formanderung von kristallen, Zeitschrift für Angewandte Mathematik und Mechanik, 8:161–185.

Watanabe, O., and Alturi, S.N. 1986. Internal time, general internal variable, and multi-yield-surface theories of plasticity and creep: a unification of concepts, *Int. J. Plasticity*, 2:107–134.

Ziegler, H. 1959. A modification of Prager's hardening rule, *Q. Appl. Math.*, 17:55–65.

Ziegler, H. 1983. *An Introduction to Thermomechanics*, 2nd ed., North-Holland.

Chapter 4

Failure conditions and the Cam-clay model

4.1 INTRODUCTION

Various constitutive models for soil have been developed over the last four decades. In particular, many elastoplastic models have been proposed since Cam-clay models were established by Roscoe et al. (1963; Roscoe and Burland 1968). The constitutive model for soil should be able to describe all types of soil behavior. The behavior of soil is complex, however, due to its nature, that is, its granularity, its multiphase structure, and its inhomogeneity. The typical characteristics of soil can be listed as follows:

1. Multiphase mixture of soil particles, pore water and pore air, saturated and unsaturated soil, and effective stress
2. Elasticity and hypoelasticity
3. Plasticity, hypoplasticity, and dilatancy characteristics
4. Rate sensitivity, viscoelasticity and viscoplasticity
5. Density dependency and confining pressure dependency
6. Strain-hardening and strain-softening characteristics
7. Cyclic deformation characteristics
8. Structural and induced anisotropy
9. Noncoaxiality
10. Deformation localization, bifurcation, and instability
11. Discontinuity
12. Degradation and the growth of microstructures
13. Inhomogeneity and nonlocality
14. Temperature dependency
15. Electromagnetic characteristics, the dieletric constant, and conductivity

Some of the characteristics have been included in the formulation of the constitutive models. In particular, elastoplasticity and dilatancy are now included in almost all of the models. However, some of them are not well incorporated into the models.

In this chapter, we begin with failure criterion and the Cam-clay model. In Chapters 5 and 9 the characteristics of soil, such as rate dependency, structural degradation, and cyclic plasticity will be discussed with regard to constitutive modeling and its use in numerical analyses.

4.2 FAILURE CRITERIA FOR SOILS

When we apply shear stress to soil, the shear stress increases, exceeds the elastic limit (initial yield stress), and then reaches the peak stress, as shown in Figure 4.1. The state at the peak stress is called the "failure" state. Thereafter, when strain has been successively applied to the soil, the stress decreases and reaches the stress state at a large strain, which may be called the residual strength. The behavior of soil is strongly affected by the confining pressure and the stress components. Based on the many experimental works and measurements in failure tests, several failure hypotheses have been proposed. The following hypotheses are representative of them.

4.2.1 Failure criterion by Coulomb

Leonardo da Vinci found that the friction between the contact areas of two bodies does not depend on the contact areas themselves but on the normal stress applied to the contact areas. Thereafter, Amonton (1699) re-found it.

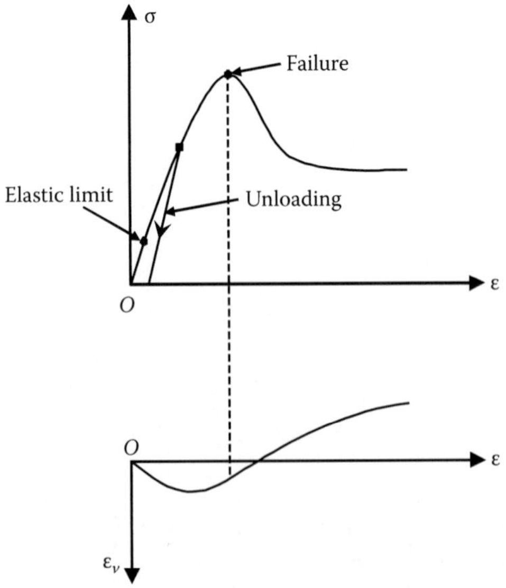

Figure 4.1 Schematic stress–strain relations for inelastic geomaterials.

Following these studies, Charles-Augustin de Coulomb (1773) proposed a failure criterion, namely,

$$\tau = \sigma_N \tan\phi + c \tag{4.1}$$

where τ is the shear stress, σ_N is the normal stress, c is the cohesive strength that is independent of the normal stress, and ϕ is the frictional angle.

4.2.2 Failure criterion by Tresca

Tresca's criterion (1884) is called the maximum shear stress hypothesis

$$\tau = \frac{\sigma_1 - \sigma_3}{2} = c \tag{4.2}$$

where c is the maximum shear strength.

When we consider the effect of the confining pressure, Equation (4.2) becomes the extended Tresca's criterion, namely,

$$\frac{\sigma_1 - \sigma_3}{2\sigma_m} = c \tag{4.3}$$

where c is a constant and σ_m is the mean stress. For saturated soil, σ_m is replaced by the mean effective stress.

4.2.3 Failure criterion by von Mises

The von Mises (1913) yield criterion is given by

$$(\sigma_1 - \sigma_3)^2 + (\sigma_2 - \sigma_3)^2 + (\sigma_1 - \sigma_2)^2 = c \tag{4.4}$$

where c is a material constant.

Elastic shear energy, W, is given by

$$W = \frac{1}{12G}\left[(\sigma_1 - \sigma_3)^2 + (\sigma_2 - \sigma_3)^2 + (\sigma_1 - \sigma_2)^2\right] \tag{4.5}$$

where G is the elastic shear modulus.

Equation (4.5) indicates that failure occurs when the elastic energy reaches the specific value of the material. In addition, the features of the von Mises criterion include the intermediate principal stress.

The extended von Mises criterion considers the effect of the mean effective stress and is given by

$$\left[(\sigma_1 - \sigma_2)^2 + (\sigma_2 - \sigma_3)^2 + (\sigma_3 - \sigma_1)^2\right] / \sigma_m = c \tag{4.6}$$

where c is a material constant.

4.2.4 Failure criterion by Mohr

Otto Mohr (1900) assumed that the shear strength, τ, of the mobilized slip plane is a function of the normal stress perpendicular to the slip plane, namely,

$$\tau = F(\sigma_N) \tag{4.7}$$

Then, the shear strength criterion is given by the envelope of Mohr's circles at failure since the stress point on the slip plane is on Mohr's circle at failure.

4.2.5 Mohr–Coulomb failure criterion

When the envelope in Mohr failure criterion is expressed by a straight line, the criterion is called Mohr–Coulomb failure criterion. The criterion is given by

$$\frac{\sigma_1 - \sigma_3}{2} = \frac{\sigma_1 + \sigma_3}{2} \sin\phi + c\cos\phi \tag{4.8}$$

where σ_1, σ_3 are the maximum and the minimum principal stresses, and c is the cohesive strength.

In cases where the cohesive strength can be disregarded, Equation (4.8) becomes

$$\frac{\sigma_1}{\sigma_3} = \frac{1 + \sin\phi}{1 - \sin\phi} \tag{4.9}$$

where ϕ is the angle of the internal friction.

In Mohr–Coulomb failure criterion, the intermediate principal stress is not taken into account but has been widely applied to soil.

4.2.6 Matsuoka–Nakai failure criterion

Matsuoka and Nakai (1974) generalized Mohr–Coulomb criterion considering the intermediate stress as

$$\frac{\sigma_3}{\sigma_1} + \frac{\sigma_1}{\sigma_3} + \frac{\sigma_2}{\sigma_3} + \frac{\sigma_3}{\sigma_2} + \frac{\sigma_2}{\sigma_1} + \frac{\sigma_1}{\sigma_2} = c \tag{4.10}$$

where c is a material constant.

Using the stress invariants, Equation (4.10) can be written as

$$\frac{I_1 I_2}{I_3} - 3 = c \tag{4.11}$$

where I_1, I_2, I_3 are the invariants of the stress tensor (Equations 1.67, 1.68, and 1.69) as

$$I_1 = \sigma_1 + \sigma_2 + \sigma_3, I_2 = \sigma_1\sigma_2 + \sigma_2\sigma_3 + \sigma_3\sigma_1, I_3 = \sigma_1\sigma_2\sigma_3.$$

When we can ignore the cohesive strength, Equation (4.11) can be rewritten as

$$\frac{I_1 I_2}{I_3} - (9 + 8\tan^2\phi) = 0 \tag{4.12}$$

where ϕ is the angle of the internal friction.

4.2.7 Lade failure criterion

Lade (1977) proposed a failure criterion for geomaterials using the first and the third stress invariants as

$$\frac{I_1^3}{I_3} = const. \quad (1977) \tag{4.13}$$

Then, Lade and Kim (1988) generalized it as

$$\left(\frac{I_1^3}{I_3} - 27\right)\left(\frac{I_1}{p_a}\right)^m = const. \quad (2003) \tag{4.14}$$

where p_a is the atmospheric pressure.

4.2.8 Failure criterion on π plane

Let us illustrate the shape of the failure criterion on the π plane. Considering the coordinates on the π plane shown in Figure 4.2 and the fact that the z axis is perpendicular to the π plane, the following relation holds between two coordinates.

The angle between the hydrostatic axis (the space diagonal) and any of the coordinates of the principal stress space ($\sigma_1, \sigma_2, \sigma_3$) is $\theta_0 = \cos^{-1}(1/\sqrt{3})$, as shown in Figure 4.2, and the plane perpendicular to the hydrostatic axis is called the π plane (or the octahedral plane).

$$\begin{pmatrix} \sigma_1 \\ \sigma_2 \\ \sigma_3 \end{pmatrix} = \begin{pmatrix} 0 & 2/\sqrt{6} & 1/\sqrt{3} \\ -1/\sqrt{2} & -1/\sqrt{6} & 1/\sqrt{3} \\ 1/\sqrt{2} & -1/\sqrt{6} & 1/\sqrt{3} \end{pmatrix} \begin{pmatrix} x \\ y \\ z \end{pmatrix},$$

$$\begin{pmatrix} x \\ y \\ z \end{pmatrix} = \begin{pmatrix} 0 & -1/\sqrt{2} & 1/\sqrt{2} \\ 2/\sqrt{6} & -1/\sqrt{6} & -1/\sqrt{6} \\ 1/\sqrt{3} & 1/\sqrt{3} & 1/\sqrt{3} \end{pmatrix} \begin{pmatrix} \sigma_1 \\ \sigma_2 \\ \sigma_3 \end{pmatrix} \tag{4.15}$$

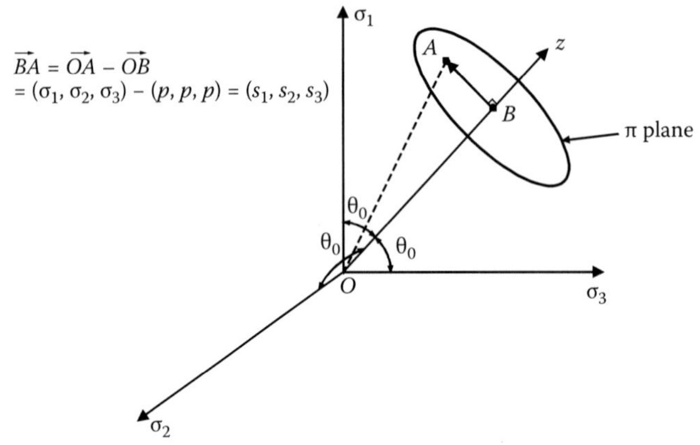

Figure 4.2 π plane.

From Equation (4.15), the distance between the point (x,y) and the origin (0,0) is given by $\sqrt{x^2+y^2}$ as

$$R = \sqrt{x^2+y^2} = \frac{1}{\sqrt{3}}\sqrt{(\sigma_1-\sigma_2)^2+(\sigma_2-\sigma_3)^2+(\sigma_3-\sigma_1)^2} \tag{4.16}$$

On the other hand, the component of the stress vector by the projection of the stress vector to the hydrostatic axis (z axis), σ_{oct} (octahedral normal stress), and the components parallel to the π plane, τ_{oct} (octahedral shear stress), are given as

$$\sigma_{oct} = \frac{\sigma_1+\sigma_2+\sigma_3}{3} = \sigma_m \tag{4.17}$$

$$\tau_{oct} = \frac{1}{3}\sqrt{(\sigma_1-\sigma_2)^2+(\sigma_2-\sigma_3)^2+(\sigma_3-\sigma_1)^2} \tag{4.18}$$

Let us determine the shape of Mohr–Coulomb's failure criterion on the π plane in the case of $c = 0$.

If we set σ_1 as the maximal stress and σ_3 as the minimal stress,

$$\sigma_1-\sigma_3 = (\sigma_1+\sigma_3)\sin\phi \tag{4.19}$$

Using Equation (4.13), we have

$$\sigma_1 - \sigma_3 = 3y/\sqrt{6} - x/\sqrt{2} \tag{4.20}$$

$$\sigma_1 + \sigma_3 = x/\sqrt{2} + y/\sqrt{6} + 2z/\sqrt{3} \tag{4.21}$$

Hence, from Equation (4.19),

$$y = ax + \frac{2\sqrt{2}\sin\phi}{3 - \sin\phi} z \tag{4.22}$$

$$a = \frac{\sqrt{3}(1 + \sin\phi)}{3 - \sin\phi} \tag{4.23}$$

The magnitude of a is

$$\frac{1}{\sqrt{3}} \leq a \leq \sqrt{3} \tag{4.24}$$

As shown in Figure 4.3, the slope of the failure line through point P is larger than that of line PQ, which is $\frac{1}{\sqrt{3}}$ parallel to the σ_2 axis on the z plane. Consequently, the shape of the failure conditions is deformative hexagonal.

In Figure 4.4, four failure criteria are shown on the π plane. The relationship between Tresca and von Mises criteria corresponds to the relationship between Mohr–Coulomb and Matsuoka–Nakai criteria. Figure 4.5 illustrates the failure criteria in the principal stress space.

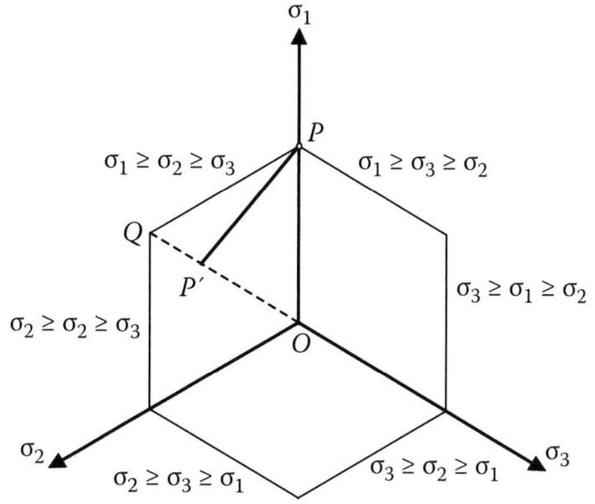

Figure 4.3 Mohr–Coulomb failure criterion on π plane.

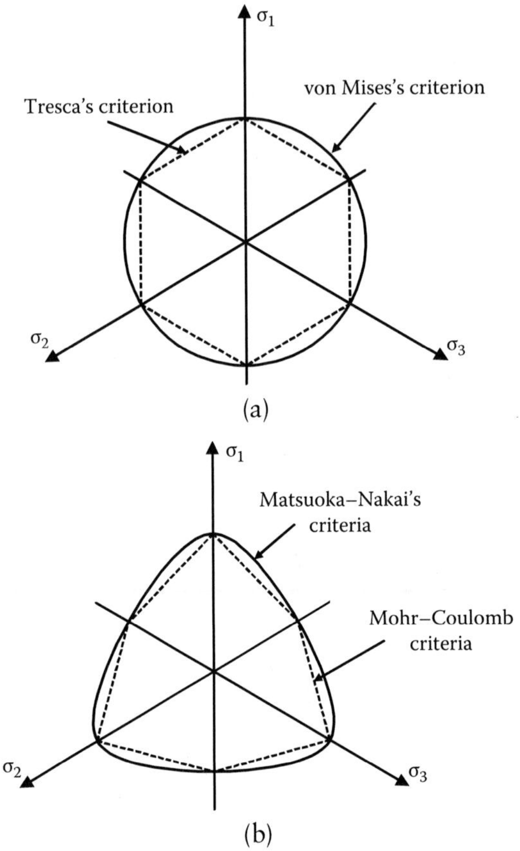

Figure 4.4 (a), (b) Failure criteria on π plane.

4.2.9 Lode angle and Mohr–Coulomb failure condition

In the following, the invariants of the deviatoric stress tensor are used as

$$J_1 = s_{ii} = s_1 + s_2 + s_3 = 0 \tag{4.25}$$

$$J_2 = \frac{1}{2} s_{ij} s_{ij} = \frac{1}{2}\left(s_1^2 + s_2^2 + s_3^2\right) = -(s_1 s_2 + s_2 s_3 + s_3 s_1) \tag{4.26}$$

$$J_3 = \frac{1}{3} s_{ij} s_{jk} s_{ki} = \det[s_{ij}] = s_1 s_2 s_3 \tag{4.27}$$

where s_i is the principal value of deviatoric stress tensor s_{ij}.

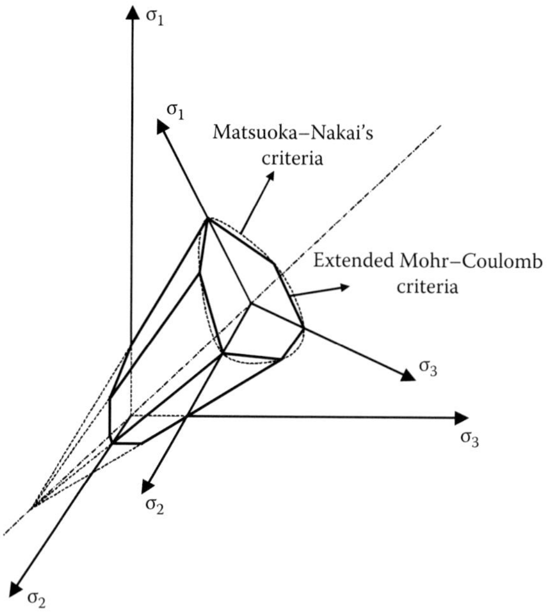

Figure 4.5 Extended Mohr–Coulomb and Matsuoka–Nakai failure criteria in the principal stress space.

Expressing the mean stress by p, the coordinates of point B in Figure 4.2 are (p, p, p).

Let an arbitrary point in the principal stress space be $(\sigma_1, \sigma_2, \sigma_3)$. The vector BA is given by

$$(\sigma_1 - p, \sigma_2 - p, \sigma_3 - p) = (s_1, s_2, s_3) \tag{4.28}$$

where $p = \sigma_m$.

The magnitude of vector OB is given by the inner product as

$$(\sigma_1, \sigma_2, \sigma_3) * \left(\frac{1}{\sqrt{3}}, \frac{1}{\sqrt{3}}, \frac{1}{\sqrt{3}} \right) = \sqrt{3}p \tag{4.29}$$

Hence, the magnitude of vector BA is given by R in Equation (4.14) as

$$\frac{1}{\sqrt{3}} \sqrt{(\sigma_1 - \sigma_2)^2 + (\sigma_2 - \sigma_3)^2 + (\sigma_3 - \sigma_1)^2} = \sqrt{2J_2} = R \tag{4.30}$$

Since the unit vector on the y axis is expressed by

$$n_y = \frac{1}{\sqrt{6}}(2, -1, -1) \tag{4.31}$$

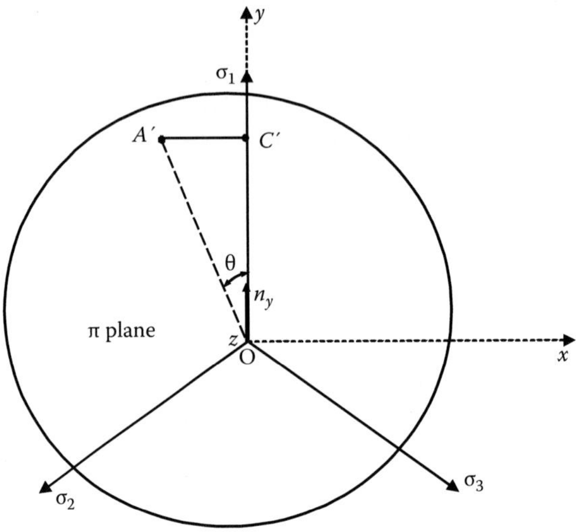

Figure 4.6 π plane.

OC′ in Figure 4.6 is given by

$$OC' = (s_1, s_2, s_3) * \frac{1}{\sqrt{6}}(2,-1,-1) = R\cos\theta \tag{4.32}$$

Consequently,

$$R\cos\theta = \frac{1}{\sqrt{6}}(2s_1 - s_2 - s_3) \tag{4.33}$$

In view of $s_1 + s_2 + s_3 = 0$, $R\cos\theta = \sqrt{3/2}s_1$. Since $R = \sqrt{2J_2}$, $\cos\theta = \sqrt{\frac{3}{2}}\frac{s_1}{\sqrt{2J_2}}$.

Using the formula for trigonometrical functions, $\cos 3\theta = 4\cos^3\theta - 3\cos\theta$, we obtain

$$\cos 3\theta = \frac{3\sqrt{3}}{2}\frac{1}{J_2^{3/2}}\left(s_1^3 - s_1 J_2\right) \tag{4.34}$$

In view of $J_2 = -(s_1 s_2 + s_2 s_3 + s_3 s_1)$, Equation (4.34) becomes

$$\cos 3\theta = \frac{3\sqrt{3}}{2}\frac{J_3}{J_2^{3/2}} \tag{4.35}$$

where θ is called Lode's angle that indicates the angle on the π plane.

For example, $0 \le \theta \le \frac{\pi}{3}$ for the region $\sigma_1 \ge \sigma_2 \ge \sigma_3$. Lode's angle can be expressed by the principal stresses, $\sigma_1, \sigma_2, \sigma_3$, $\sigma_1 \ge \sigma_2 \ge \sigma_3$, as

$$\tan\theta = \frac{\sqrt{3}(\sigma_2 - \sigma_3)}{2\sigma_1 - \sigma_2 - \sigma_3} \tag{4.36}$$

or

$$\cos 3\theta = \frac{(2-b)(1-2b)(1+b)}{2(b^2 - b + 1)^{3/2}} \tag{4.37}$$

where $b = \frac{\sigma_2 - \sigma_3}{\sigma_1 - \sigma_3}$, $\sigma_1 \ge \sigma_2 \ge \sigma_3$. $b = 0$ corresponds to the triaxial compression conditions and $b = 1$ corresponds to the triaxial extension conditions.

Using Lode's angle, Mohr–Coulomb's failure criterion is expressed in the following. Using the xyz coordinate system stress, the point on the π plane is expressed by

$$\begin{aligned} x &= -R\sin\theta \\ y &= R\cos\theta \\ z &= \sqrt{3}p \end{aligned} \tag{4.38}$$

where $R = \sqrt{2J_2}$.

Using the expression in Equation (4.15), we can write down the principal stresses for the case $\sigma_1 \ge \sigma_2 \ge \sigma_3$ as

$$\begin{aligned} \sigma_1 &= \frac{2}{\sqrt{6}} R\cos\theta + p \\ \sigma_2 &= \frac{1}{\sqrt{2}} R\sin\theta - \frac{1}{\sqrt{6}} R\cos\theta + p \\ \sigma_3 &= -\frac{1}{\sqrt{2}} R\sin\theta - \frac{1}{\sqrt{6}} R\cos\theta + p \end{aligned} \tag{4.39}$$

Substituting Equation (4.39) into Equation (4.19), we have a Mohr–Coulomb failure criterion expressed by the stress invariants and Lode's angle as

$$p\sin\phi - \frac{R}{2\sqrt{6}}\left((3 - \sin\phi)\cos\theta + \sqrt{3}(1 + \sin\phi)\sin\theta\right) + c\cos\phi = 0 \tag{4.40}$$

where c is the cohesion and ϕ is the internal friction angle.

When $c = 0$,

$$\frac{\sqrt{2J_2}}{p} = \frac{6\sqrt{2}\sin\phi}{3(1+\sin\phi)\sin\theta + \sqrt{3}(3-\sin\phi)\cos\theta} \tag{4.41}$$

4.3 CAM-CLAY MODEL

As has been previously described, the features of materials such as mechanical, thermal, and electromagnetic characters are expressed by constitutive equations. When we apply loads to these materials, the materials deform. If the strain induced by the loading is small, the materials behave elastically, that is, the deformation can be recovered. However, part of the deformation cannot be recovered if the strain level is high as shown in Figure 4.1. This type of behavior is elastoplastic, and the model that can describe such elastoplastic behavior is called an elastoplastic constitutive equation.

In addition, if the model can account for time-dependent behavior, such as creep, it is called an elastoviscoplastic model. Herein, we will disregard the time-dependent behavior.

4.3.1 Original Cam-clay model

Roscoe and the research group at Cambridge University (1963) developed an elastoplastic constitutive model for clay, which is called the Cam-clay model after the name of the river flowing through the Cambridge University campus. In the derivation of the Cam-clay model, the following assumptions have been used with the concept of critical state:

1. Existence of yield function
2. Drucker type of elastoplastic theory; flow rule and normality rule for strain rate
3. Internal dissipation (or stress–dilatancy relation)
4. Hardening rule; plastic void ratio e^p is used as a hardening parameter

According to the work by Roscoe et al. (1963), we will consider the axisymmetric stress conditions and use the following quantities and symbols:

σ_i: effective stress tensor, $i = 1,2,3, \sigma_2 = \sigma_3,\ \sigma_i' = \sigma_i - u_w$
u_w: pore water pressure
σ_i: principal stress
q: deviator stress, $q = \sigma_1 - \sigma_3, \sigma_1 > \sigma_3$
p: mean total stress, $p = (\sigma_1 + 2\sigma_3)/3$

p': mean effective stress, $p' = (\sigma_1' + 2\sigma_3')/3$
M: slope of the critical state line on p'–q plane or value of q/p' at critical state
Under triaxial compression conditions, $\sigma_1' > \sigma_2' = \sigma_3'$

The stress ratio at the critical state, that is, the failure state, is

$$q/p' = \frac{\sigma_1' - \sigma_3'}{(\sigma_1' + 2\sigma_3')/3}$$

$$= \frac{3(\sigma_1'/\sigma_3' - 1)}{(\sigma_1'/\sigma_3' + 2)} \tag{4.42}$$

If we set the internal friction angle $=\phi'$, $\sigma_1/\sigma_3 = (1+\sin\phi')/(1-\sin\phi')$. Then,

$$M = (q/p') = \frac{6\sin\phi'}{3-\sin\phi'} \tag{4.43}$$

Under triaxial extension conditions, $\sigma_1' = \sigma_2' > \sigma_3'$, then

$$q/p' = \frac{\sigma_1' - \sigma_3'}{(2\sigma_1' + \sigma_3')/3} \tag{4.44}$$

and

$$M = \frac{6\sin\phi'}{3+\sin\phi'} \tag{4.45}$$

where ϕ' is the internal friction angle.

$$d\varepsilon_v^p \text{ : plastic volumetric strain increment, } d\varepsilon_v^p = d\varepsilon_{11}^p + 2d\varepsilon_{33}^p \tag{4.46}$$

$$d\varepsilon_d^p \text{ : plastic shear strain increment, } d\varepsilon_d^p = d\varepsilon_{11}^p - \frac{1}{3}d\varepsilon_v^p = \frac{2}{3}\left(d\varepsilon_{11}^p - d\varepsilon_{33}^p\right) \tag{4.47}$$

The yield function of the Cam-clay model is derived in the following. From the plastic flow rule in Equation (3.64),

$$d\varepsilon_v^p = h\frac{\partial f}{\partial p'} \tag{4.48}$$

$$d\varepsilon_d^p = h\frac{\partial f}{\partial q} \tag{4.49}$$

$$\frac{d\varepsilon_v^p}{d\varepsilon_d^p} = \frac{\frac{\partial f}{\partial p'}}{\frac{\partial f}{\partial q}} \tag{4.50}$$

From the normality of the strain increment vector to the yield surface given by assumption 2, as Equations (4.48) and (4.49),

$$\left(d\varepsilon_v^p, d\varepsilon_d^p\right) * (dp', dq) = 0 \tag{4.51}$$

where * denotes the inner product.

Then,

$$\frac{d\varepsilon_v^p}{d\varepsilon_d^p} = -\frac{dq}{dp'} \tag{4.52}$$

From Equations (4.50) and (4.52), we have

$$\frac{\frac{\partial f}{\partial p'}}{\frac{\partial f}{\partial q}} = -\frac{dq}{dp'} \tag{4.53}$$

Integrating Equation (4.53) provides the yield surface.

The internal dissipation energy increment, dW^p, has been assumed to be given by

$$dW^p = \sigma_{ij} d\varepsilon_{ij}^p = \sigma'_{11} d\varepsilon_{11}^p + 2\sigma'_{33} d\varepsilon_{33}^p = p' d\varepsilon_v^p + q d\varepsilon_d^p = Mp' d\varepsilon_d^p \tag{4.54}$$

This dissipation function is related to the dissipation function shown in Chapter 3.

By rewriting Equation (4.54), we obtain the following stress–dilatancy relation:

$$\frac{d\varepsilon_v^p}{d\varepsilon_d^p} = M - \frac{q}{p'} \tag{4.55}$$

Integrating Equation (4.50) with Equations (4.53) and (4.55) leads to a yield function as

$$f = \frac{q}{Mp'} + ln(p'/p'_0) - ln(p_y/p'_0) = 0 \tag{4.56}$$

where p_y is the hardening parameter and p'_0 is the initial mean effective stress.

It is worth noting that deviator stress, q, corresponds to plastic deviatoric strain, ε_d^p, under axisymmetric stress conditions, as shown in the definition for plastic work increment, dW^p. This plastic work increment corresponds to the dissipation function shown in Chapter 3.

The yield function is illustrated in the (p', q) space shown in Figure 4.7.

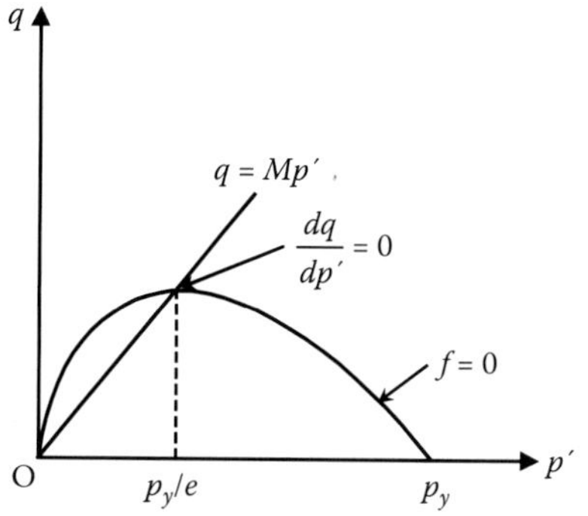

Figure 4.7 Yield surface for Cam-clay model.

When we assume that the hardening parameter is given by plastic void ratio, e_p, and its evolution equation is given by Equation (4.57), parameter h in the flow rule is obtained.

$$de^p = -(\lambda - \kappa)\frac{dp_y}{p_y} \tag{4.57}$$

where λ is the compression index, the slope of the compression curve (e–lnp'); κ is the swelling index, the slope of the swelling curve (e–lnp'); and ln indicates the natural logarithm.

The relationship between the plastic volumetric strain and the plastic specific volume (v^p) is expressed by

$$d\varepsilon_v^p = -\frac{dv^p}{v} = -\frac{de^p}{1+e} \tag{4.58}$$

where v is the specific volume (= 1 + e).

Upon substitution of Equation (4.57) into Equation (4.58) we obtain

$$\frac{dp_y}{p_y} = \frac{1+e}{\lambda - \kappa} d\varepsilon_v^p \tag{4.59}$$

If we assume that the initial plastic volumetric strain is zero at $p_y = p_0'$, the yield function is rewritten as

$$f = \frac{q}{Mp'} + ln(p'/p_0) - \frac{1+e}{\lambda - \kappa}\varepsilon_v^p = 0 \tag{4.60}$$

Then,

$$\frac{\partial f}{\partial p'} = -\frac{q}{Mp^2} + \frac{1}{p'} \tag{4.61}$$

$$\frac{\partial f}{\partial q} = \frac{1}{Mp'} \tag{4.62}$$

$$\frac{\partial f}{\partial \varepsilon_v^p} = -\frac{1+e}{\lambda - \kappa} \tag{4.63}$$

Subsequently, using Prager's consistency condition, we have h as

$$h = \frac{-\frac{\partial f}{\partial p'}dp' - \frac{\partial f}{\partial q}dq}{\frac{\partial f}{\partial \varepsilon_v^p}\frac{\partial f}{\partial p'}} = \frac{dq + (M - \frac{q}{p'})dp'}{\frac{1+e}{\lambda-\kappa}(M - \frac{q}{p'})} \tag{4.64}$$

For the Cam-clay model, the strain increments under triaxial stress conditions are given by

$$d\varepsilon_v^p = \frac{\lambda - \kappa}{(1+e)Mp'}[dq + (M-\eta)dp'] \tag{4.65}$$

$$d\varepsilon_q^p = \frac{\lambda - \kappa}{(1+e)Mp'}\left[\frac{dq + (M-\eta)dp'}{M-\eta}\right] \tag{4.66}$$

where $\eta = q/p'$, $d\varepsilon_v^p$ is the plastic volumetric strain, and $d\varepsilon_q^p$ is the plastic shear strain.

Let us obtain the state boundary surface using the Cam-clay model discussed earlier. Since the total volumetric strain increment is zero under undrained conditions,

$$\frac{dq}{dp'} = \frac{q}{p'} - \frac{\lambda M}{\lambda - \kappa} \tag{4.67}$$

The elastic void ratio increment, de^e, is given by

$$de^e = -\kappa\frac{dp'}{p'} \tag{4.68}$$

Then, the plastic volumetric increment, $d\varepsilon_v^p$, is obtained by

$$d\varepsilon_v^p = -\frac{de - de^e}{1+e} = -\frac{1}{1+e}\left(de + \kappa\frac{dp'}{p'}\right) \tag{4.69}$$

By substituting Equation (4.69) into the total differentiation of Equation (4.60) and integrating it, we obtain

$$\frac{q}{Mp'} + ln(p'/p_0') = \frac{1}{\lambda - \kappa}[e_0 - e - \kappa ln(p'/p_0') + \lambda - \kappa] \tag{4.70}$$

where we set $e = e_0$ and $p' = p_0'$ when $q/p' = M$.

If we set the specific volume at $v \equiv 1 + e = \Gamma$, when $p' = 1(kN/m^2)$, and $q/p' = M$ at the critical state, we have the following equation from Equation (4.70):

$$\Gamma = v_0 + \lambda ln p_0' \tag{4.71}$$

Then, we get the state boundary surface from Equation (4.70) as

$$q = \frac{Mp'}{\lambda - \kappa}[\Gamma + \lambda - \kappa - v - \lambda ln p'] \tag{4.72}$$

The state boundary surface is shown in Figures 4.8 and 4.9.

From the state boundary surface, we obtain the undrained stress path under the undrained condition of $e = const.$ and the yield surface by the condition of ε_v^p = constant.

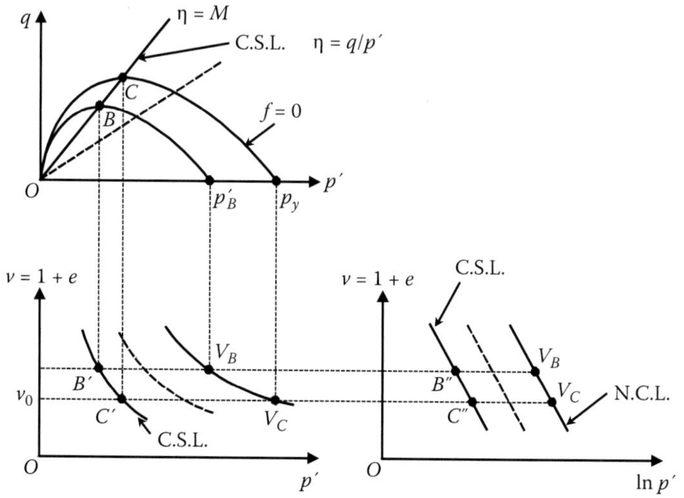

Figure 4.8 Yield surface and critical state line.

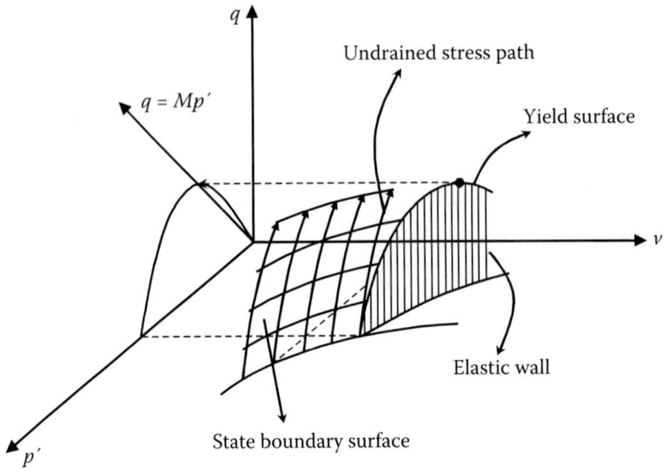

Figure 4.9 State surface.

From the void ratio constant condition, that is, the undrained condition, we have the undrained stress path. And from the plastic volumetric strain constant condition, ε_v^p = constant, we obtain the yield surface in the e–p'–q space.

It is worth noting that there is a unique relationship between the stress states and the void ratio for the critical state from $q/p' = M$ and Equation (4.72). In addition, if this unique relation is observed when we perform triaxial tests on geomaterials, we can verify that the Cam-clay model concept is applicable to geomaterials. For sand, Been and Jeffries (1985) found that the steady state concept, which is equivalent to the critical state concept, holds and proposed state parameter that controls dilatancy characteristics.

The projection of the state boundary surface on the p'-v plane is

$$v = \Gamma - \lambda lnp' \tag{4.73}$$

and it is illustrated in Figure 4.10.

From Equations (4.65) and (4.66), $d\varepsilon_v^p = 0$ at the critical state, $q = Mp'$, and $d\varepsilon_d^p \to \infty$.

The normally consolidated line is given by $v = N - \lambda lnp'$; it is parallel to the critical state line in Equation (4.73), as shown in Figures 4.10 and 4.11.

4.3.2 Ohta's theory

Ohta and Hata (1971) derived a change in void under general stress conditions by the stress–dilatancy relation (Shibata, 1963) and the e–λlnp relation during the compression process.

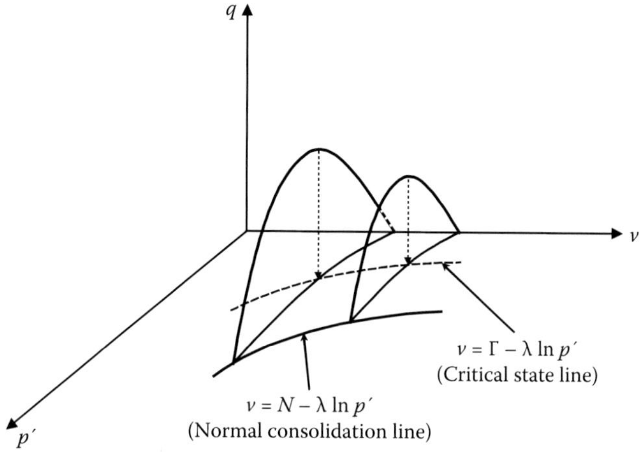

Figure 4.10 Normally consolidated line and critical state line.

The stress–dilatancy relation is given by

$$-\frac{de}{1+e_0} = \mu d\left(\frac{q}{p'}\right) \tag{4.74}$$

where μ is a material constant and e_0 is the initial void ratio.

From isotropic compression tests, we have

$$de = -\lambda\frac{dp}{p'} \tag{4.75}$$

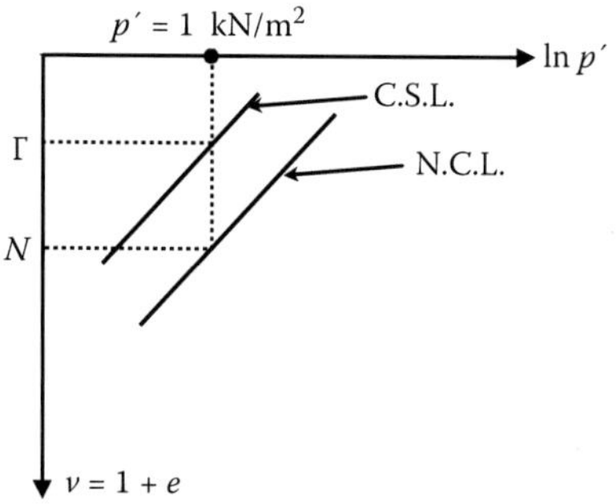

Figure 4.11 Critical state line and normally consolidated line.

Adding two equations, we obtain

$$de = -\lambda \frac{dp}{p'} - \mu(1+e_0)d\left(\frac{q}{p'}\right) \tag{4.76}$$

Integrating the preceding equations on the normally consolidated curve, under the conditions of $p' = p_0', e = e_0, q/p' = 0$, produces

$$e_0 - e = \lambda ln(p'/p_0') + (1+e_0)\mu \frac{q}{p'} \tag{4.77}$$

If we set

$$(\lambda - \kappa)/M = (1+e_0)\mu, \tag{4.78}$$

we obtain

$$e_0 - e = \lambda ln(p'/p_0') + (\lambda - \kappa)/M \frac{q}{p'} \tag{4.79}$$

This equation is equal to the equation obtained by substituting Equation (4.70) into Equation (4.60) and integration under the conditions of $p' = p_0', e = e_0, q/p' = 0$ on the normally consolidated curve. Namely, the preceding equation coincides with Equation (4.79).

Ohta and Hata (1971) obtained the yield surface by the projection of the intersection of Equation (4.79) and the elastic wall $(e_0 - e = \kappa ln(p'/p_0'))$ onto the p'–q plane.

Consequently, Ohta and Hata (1971) derived an elastoplastic theory equivalent to the Cam-clay model.

Equation (4.56) can be generalized under the general three-dimensional stress conditions as

$$f = \frac{\sqrt{2J_2}}{M^* \sigma_m'} + \ln\left(\frac{\sigma_m'}{\sigma_{my}'}\right) = 0 \tag{4.80}$$

where J_2 is the second invariant of the deviatoric stress tensor, σ_m' is the mean effective stress, σ_{my}' is the hardening parameter, and M^* is a value of $\sqrt{2J_2}/\sigma_m'$ at critical state.

4.3.3 Modified Cam-clay model

Using the Cam-clay model yields shear strain during isotropic compression due to the shape of the yield surface. In order to modify this point, Roscoe and Burland (1968) proposed an elliptic-shaped type of yield function by generalizing the energy equation.

In the original Cam-clay theory, the increment of work dissipated per unit bulk volume of an isotropic continuum during deformation was expressed as

$$\delta W^p = p'\delta\varepsilon_{vol}^p + q\delta\varepsilon_q^p \tag{4.81}$$

where p' and q are the mean effective stress and the shear stress, respectively, and $\delta\varepsilon_{vol}$ and $\delta\varepsilon_q$ are the corresponding strain increments associated with p' and q.

In the derivation of the modified Cam-clay theory, Roscoe and Burland (1968) suggested that it was reasonable to assume that under an isotropic stress of $q = 0$, there is no distortion ($\delta\varepsilon_q = 0$) in normally consolidated clay. Since normally consolidated clay can deform irrecoverably under isotropic stress, Equation (4.81) becomes

$$(\delta W_{dissipated})_{q=0} = p'\delta\varepsilon_{vol}^p \tag{4.82}$$

At the critical state condition, $q/p' = M$ and $\delta\varepsilon_{vol,p} = 0$, so that Equation (4.81) becomes

$$(\delta W_{dissipated})_{q=Mp} = Mp'\delta\varepsilon_q^p \tag{4.83}$$

A generalization of the conditions in Equations (4.82) and (4.83) is

$$\delta W_{dissipated} = p'\sqrt{(\delta\varepsilon_{vol}^p)^2 + (M\delta\varepsilon_q^p)^2} \tag{4.84}$$

This is the expression for work dissipation used in the modified Cam-clay theory.

Employing Equation (4.84) and the normality condition (Roscoe and Burland, 1968), the equation for the yield locus of the modified Cam-clay can be obtained as

$$f_y = \frac{p'}{p'_y} - \frac{M^2}{\eta^2 + M^2} = 0 \tag{4.85}$$

where η is the value of the stress ratio (q/p) and M is the value when the stress ratio reaches the critical state. $p' = p'_y$ at $\eta = 0$ (isotropic normal compression).

Equation (4.85) represents an ellipse in the q, p space with its center at $p_y/2$ shown in Figure 4.12.

From the flow rule, if we set $\eta = q/p'$, the stress–dilatancy equation is obtained as

$$\frac{d\varepsilon_p^p}{d\varepsilon_d^p} = \frac{M^2 - \eta^2}{2\eta} \tag{4.86}$$

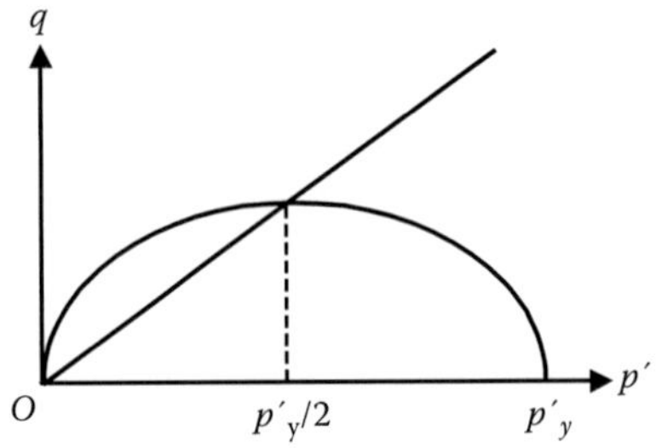

Figure 4.12 Yield surface for modified Cam-clay model.

This equation indicates that the shear strain does not occur when the stress ratio is zero, that is, $\eta = 0$.

As is similar to Equation (4.80), Equation (4.85) can be generalized under the general three-dimensional stress conditions as

$$f = \frac{M^{*2}}{2J_2 + M^{*2}} - \frac{\sigma'_m}{\sigma'_{my}} = 0 \tag{4.87}$$

The Cam-clay model has been used as a basic model for geomaterials. Then, other types of inelastic models for describing plastic behaviors of geomaterials have been developed over the last 20 years such as nonlinear incremental model (Chambon 1986; Darve 1990). The hypoplastic model is also a nonlinear incremental model (Kolymbas 1991; Gudehus 1996). As for the cyclic plasticity theory, subloading surface models have been developed by Hashiguchi and Ueno (1977), and Hashiguchi (1989) considering the expansion and contraction of loading surface. Oka et al. (1999) developed a cyclic model based on the nonlinear kinematical hardening theory. A rate-dependent model such as an elastoviscoplastic model will be presented in Chapter 5.

4.3.4 Stress–dilatancy relations

The stress–dilatancy equations such as Equations (4.52) and (4.86) can be brought back to Row's stress–dilatancy theory. Row (1962) derived a stress–dilatancy relation by minimizing the ratio of the rate of work done on an assembly of particles by major principal stress to the rate of work done by the minor principal stress as

$$\frac{\sigma_1}{\sigma_3} = KD; \quad K = \tan^2\left(45° + \frac{\varphi_\mu}{2}\right), \; D = 1 - \frac{d\varepsilon_v^p}{d\varepsilon_1^P} \tag{4.88}$$

where φ_μ is the interparticle friction angle, $d\varepsilon_v^p$ is the plastic volumetric strain increment, $d\varepsilon_1^p$ is the plastic axial strain increment, and σ_1, σ_3 is the major and minor principal stress respectively.

Equations (4.52) and (4.58) are the variants of Row's relation and it has been pointed out that the dilatancy is a function of fabric, density of packing and stress level, and the void ratio (Wan and Guo 1999).

REFERENCES

Amontons, G. 1699. Method of substituting the force of fire for horse and man power to move machines, Histoire et Mémoires de l'Académie Royale des Sciences.

Atkinson, J.H., and Bransby, P.L. 1981. The Mechanics of Soils, In *An Introduction to Critical State Soil Mechanics*, London: McGraw-Hill.

Been, K., and Jeffries, M. 1985. A state parameter for sands, *Géotechnique*, 35(2):99–112.

Chambon, R. 1986. Bifurcation en bande de cisaillement incrementalement nonlinearire, *J. Mec. Theor. Appl.*, 5(2):277–298.

Chen, W.F., and Mizuno, E. 1990. *Nonlinear analysis in soil mechanics*, Elsevier Science Publishers B.V.

Coulomb, C.A. 1773. Essai sur application des regles des maiximas et minimis a quelques problems de statique relatifs a l'architecture, Mem. Acad. Roy. Pres. Divers Savants, 7:342–382.

Darve, F. 1990. The expression of rheological laws in incremental form and the main classes of constitutive equations, In Geomaterials Constitutive Equations and Modelling, F. Darve, ed., *Elsevier Appl. Science*, London, 123–148.

Darve, F., and Larouafa, F. 2000. Instabilities in granular materials and application to land slides, *Mech. Cohesive-Frictional Mater.*, 5(8):627–652.

Drucker, D.C. 1951. A more fundamental approach to plastic stress-strain relations, Proc. 1st U.S. Natl. Congr. Appl. Mech., 487.

Gudehus, G. 1996. A comprehensive constitutive equation for granular materials, *Soils and Foundations*, 36(1):1–12.

Hashiguchi, K. 1989. Subloading surface model in unconventional plasticity, *Int. J. Solids Struct.*, 25:917–945.

Hashiguchi, K., and Ueno, M. 1977. Elasto-plastic constitutive laws of granular materials, In Constitutive Equations of Soils, S. Murayama and A.N. Schofield, eds., Proc. 9th Int. Conf. SMFE Spec. Session 9, Tokyo, JSSMFE, 73–82.

Hill, R. 1957. On uniqueness and stability in the theory of finite elastic strain, *J. Mech. Phys. Solids*, 5:229–241.

Hill, R.1958. A general theory of uniqueness and stability in elastic-plastic solids, *J. Mech. Phys. Solids*, 6:239–249.

Kolymbas, D. 1991. An outline of hypoplasticity, *Arch. Appl. Mech.*, 61:143–154.

Lade, P. 1977. Elasto-plastic stress-strain theory for cohesionless soil with curved yield surfaces, *Int. J. Solids Struct.*, 13:1019–1035.

Lade, P. V. and Kim, M. K. 1988. Single hardening constitutive model for frictional mateials II. Yield criterion and plastic work contours, *Computers and Geotechnics*, 6:13–29.

Lode, W. 1925. Versuche uber den Einfluß der mittleren Hauptspannung auf die Fließgrenze, ZAMM, 5:142-144.

Lubliner, J. 1990. *Plasticity Theory*, New York: Macmillan.

Malvern, L. 1969. *Introduction to the Mechanics of a Continuous Medium*, Prentice-Hall.

Martin, J.B. 1975. *Plasticity: Fundamentals and General Results*, Cambridge, MA: The MIT Press.

Matsuoka, H., and Nakai, T. 1974. Stress-deformation and strength characteristics of soil under three different principal stresses, *Proc. JSCE*, 232:59–70.

Mohr, O. 1900. Welche Umstande bedingen die Elastizitatsgrenze und den Bruch eines Materials?, *Zeitschrift des vereines deutscher Ingenieure*, 44:1524–1530.

Muir Wood, D. 1991. *Soil Behaviour and Critical State Soil Mechanics*, Cambridge University Press.

Nova, R. 1994. Controllability of the incremental response of soil specimens subjected to arbitrary loading programmes, *J. Mech. Behav. Mater.*, 5(2):193–201.

Ohta, H., and Hata, S. 1971. A theoretical study of the stress-strain relation for clays, *Soils and Foundations*, 11(3):45–70.

Oka, F., Yashima, A., Tateishi, A., Taguchi, Y.. and Yamashita, S. 1999. A cyclic elasto-plastic constitutive model for sand considering a plastic-strain dependence of the shear modulus, *Géotechnique*, 49(5):661–680.

Prager, W. 1961. *Introduction to Mechanics of Continua*, Boston: Ginn and Company.

Roscoe, K.H., and Burland, J.B. 1968. On the generalized stress-strain behaviour of "wet" clay, in *Engineering Plasticity*, J. Heyman and F.A. Leckie, eds., Cambridge University Press, 535–609.

Roscoe, K.H., Schofield, A.N., and Wroth, C.P. 1963. Yielding of clays in states wetter than critical, *Géotechnique*, 13(3):211–240.

Row, P.W. 1962. The stress-dilatancy relation for static equilibrium of an assembly of particle in contact, Proc. Royal Soc. London, 269(A):500–527.

Schofield, A.N., and Wroth, C.P. 1968. *Critical State Soil Mechanics*, London: McGraw-Hill.

Shibata, T. 1963. On the dilatancy, *Bulletin of the DPRI of Kyoto University*, 6: 128–134 (In Japanese).

Tresca, H. 1864. Memoir sur l'ecoulement des corps solides soumis a de fortes pression, Comptes Rendus Acad. Sci. Paris, 59:754–758.

von Mises, R. 1913. Mechanik der festen Korper im plastisch-deformablen Zustand, nachrichten von der koniglichten Gesellschaft der Wissenschaft zu Gottingen, 582–592.

Wan, R., and Guo, P.J. 1999. A pressure and density dependent dilatancy model for granular materials, *Soils and Foundations*, 39(6):1–12.

Chapter 5

Elastoviscoplastic modeling of soil

It is well known that the behavior of soil, in particular cohesive soil, is affected by the loading time and the strain rate (e.g., Adachi et al. 1996). Hence, it is necessary to include the viscous effect in the derivation of the constitutive equations. Studies on the effect of the viscosity of soil have been performed with viscoelastic or viscoplastic models. Over the last 30 years, viscoplastic models have been proposed following the development of elastoplastic models. In this chapter, we will discuss both the viscous nature and the plastic nature of soil.

5.1 RATE-DEPENDENT AND TIME-DEPENDENT BEHAVIOR OF SOIL

5.1.1 Strain rate-dependent behavior of clayey soil

Richardson and Whitman (1963) studied the effect of the strain rate on remolded and reconstituted Mississippi River Valley clay by conducting undrained triaxial compression tests. Figure 5.1 shows the results of triaxial compression tests that were done under a strain rate of about 1%/min for high strain rate tests and under a strain rate of $2\times10^{-3}\,\%/\mathrm{min}$ for low strain rate tests. It can be seen that the stress ratio is higher in the case of higher strain rates in the early stage of loading and that the decrease in mean effective stress is larger in the case of lower strain rates. The slope of the stress ratio-axial strain curve is larger in the case of higher strain rates. Yong and Japp (1969) performed triaxial tests on water-saturated kaolin under the high strain rate of 100%–2000%/min, as shown in Figure 5.2, and they derived the following phenomenological relation:

$$\sigma(\varepsilon,\dot{\varepsilon}) = \sigma_0(\varepsilon,\dot{\varepsilon}_0) + \phi\, log(\dot{\varepsilon}/\dot{\varepsilon}_0) \tag{5.1}$$

where $\sigma = \sigma_{11} - \sigma_{33}$ is the deviator stress, ε is the axial strain, and $\dot{\varepsilon}$ is the axial strain rate. Subscript 0 indicates the values at the referential state,

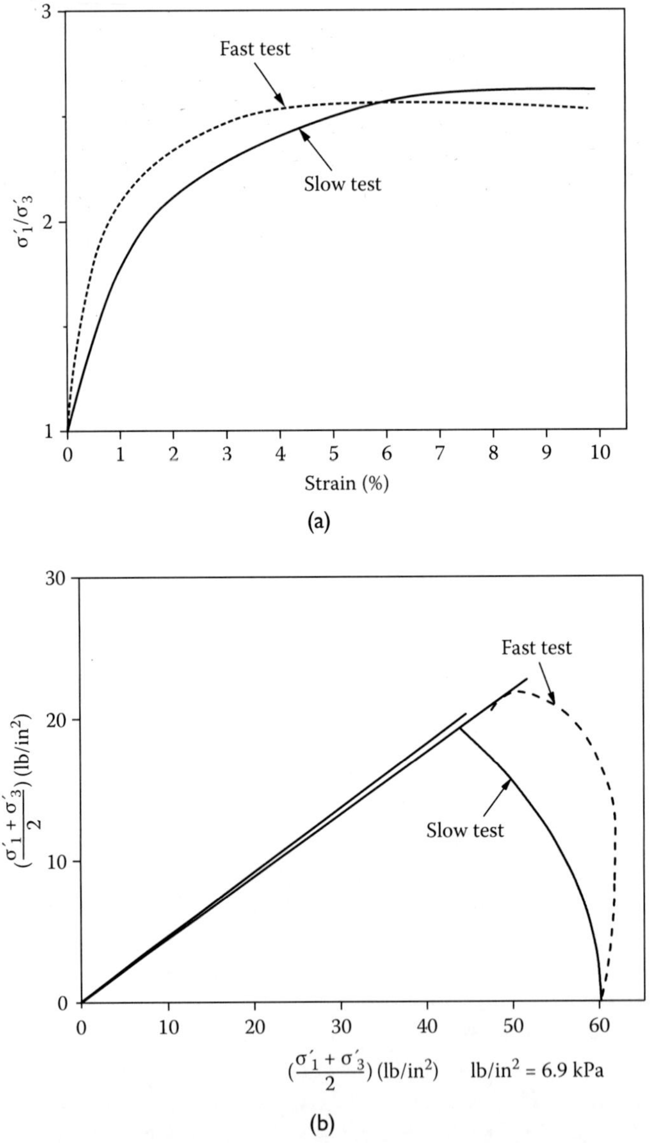

Figure 5.1 (a) Stress–strain relations with different strain rates. (b) Stress paths with different strain rates. (After Richardson, A.M. Jr., and Whitman, R., 1963, *Géotechnique*, 13(4):310–324.)

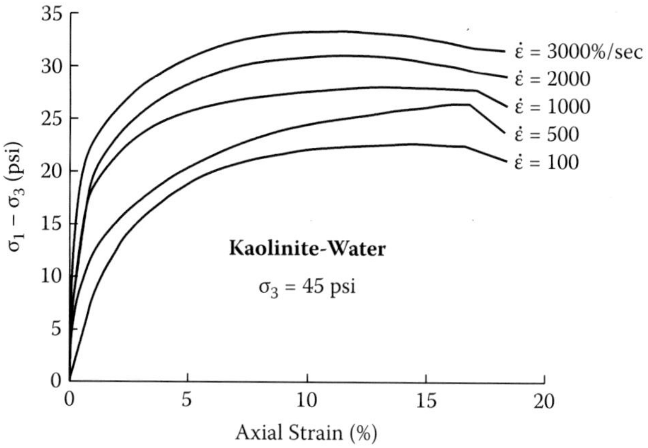

Figure 5.2 Stress–strain relations for different strain rates. (After Yong, R.N., and Japp, R.D., 1969, ASTM, STP, 450:233–262.)

that is, the referential strain rate, σ_0 is the stress at a strain by the reference strain rate, and ϕ is a parameter.

5.1.2 Creep deformation and failure

Creep deformation is a typical time-dependent behavior of soil. The successive deformation of soil, called creep, can be observed during constant stress conditions. Finn and Shead (1973) carried out undrained creep tests for normally consolidated and overconsolidated clay in which the deviator stress was kept constant, but the mean effective stress varied. Figure 5.3 shows the undrained triaxial creep test results for an isotropically consolidated specimen. Figure 5.4 illustrates the deformation–time relationship during creep. In the early stage of creep, the strain rate decreases; this is called the preliminary creep process. Then, the strain rate becomes a constant minimum strain rate; this is called the secondary creep process. Finally, the strain rate increases with time; this is called acceleration or tertiary creep, and the soil reaches the failure state. Finn and Shead (1973) found the following relationship between the minimum strain rate and the failure time, as shown in Figure 5.5:

$$Time\,to\,failure \times \dot{\varepsilon}_{\min} = \text{constant} \tag{5.2}$$

Saito and Uezawa (1969) found the following empirical relationship based on many observed results:

$$\dot{\varepsilon} \times (t_f - t) = \text{constant} \tag{5.3}$$

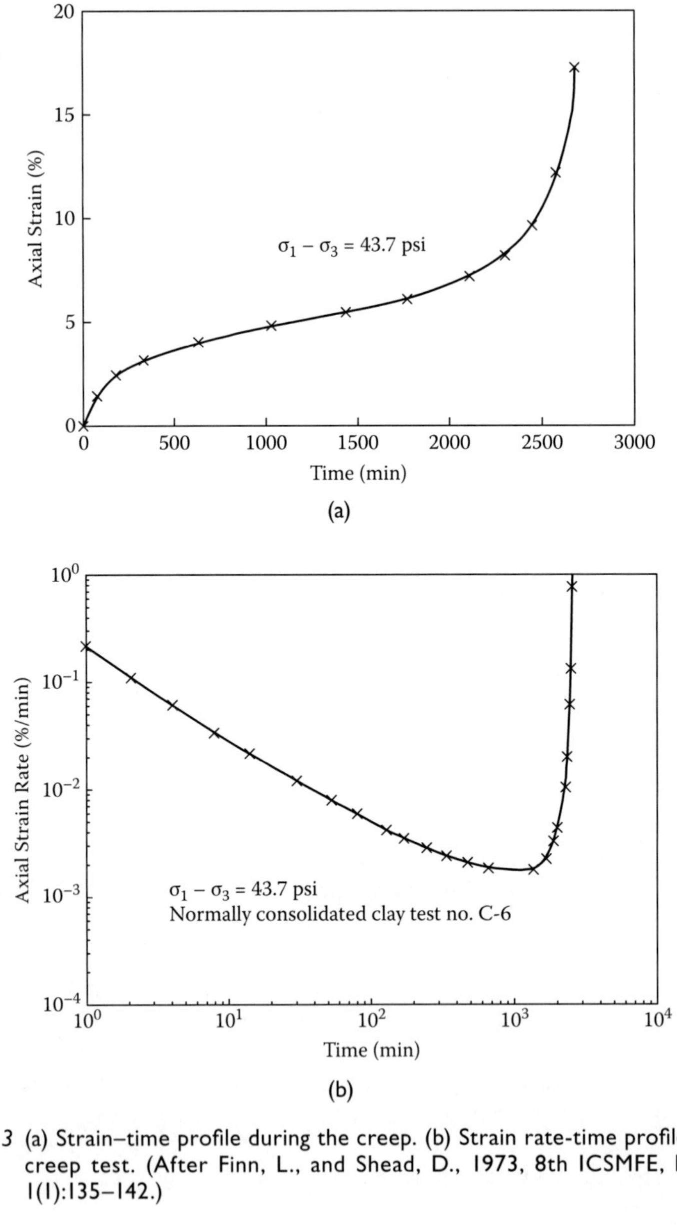

Figure 5.3 (a) Strain–time profile during the creep. (b) Strain rate-time profile for the creep test. (After Finn, L., and Shead, D., 1973, 8th ICSMFE, Moscow, 1(1):135–142.)

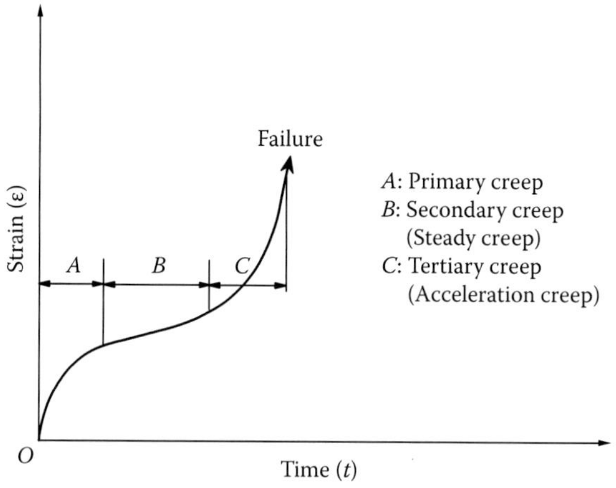

Figure 5.4 Schematic stress–strain relationship during the creep.

where t_f is the failure time and $\dot{\varepsilon}$ is the strain rate at time t during the acceleration creep process. Figure 5.6 is a measured example; it is seen that Equation (5.3) is clearly effective.

Since it is difficult to determine the time of the onset of acceleration creep in practical cases, Equation (5.3) is proposed. It is interesting to note that Equation (5.3) is similar to Equation (5.2).

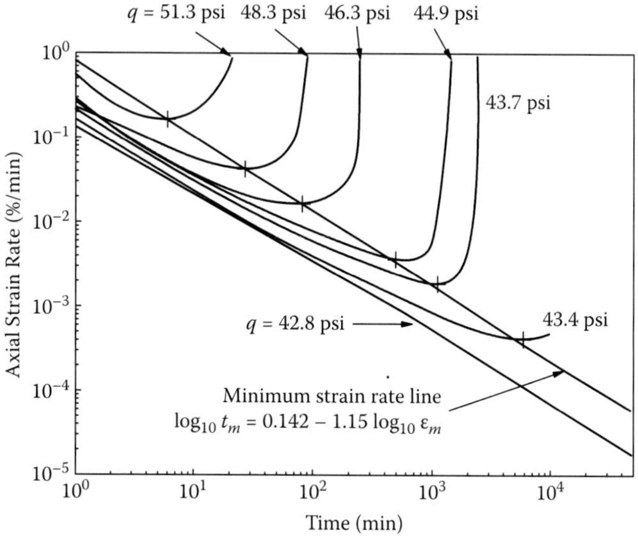

Figure 5.5 Strain rates–time profile during the creep. (After Finn, L., and Shead, D., 1973, 8th ICSMFE, Moscow, 1(1):135–142.)

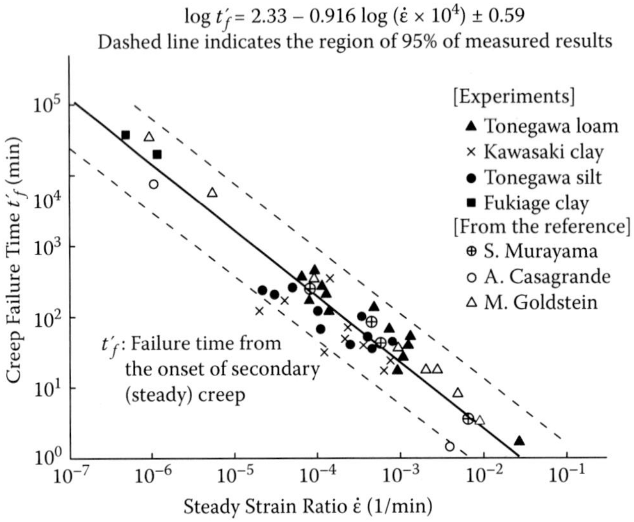

Figure 5.6 Creep failure time versus strain rate at steady state. (After Saito, M., 1992, Jissho Doshitsu Kogaku, Gihodo Syuppan Co. Ltd., 156.)

5.1.3 Stress relaxation behavior

It is observed that the stress decreases when the strain is fixed. This behavior is called the stress relaxation phenomenon. Figure 5.7 indicates stress relaxation during undrained triaxial compression tests (Akai et al. 1974). In this figure, it is seen that the stress decreases, but the pore water pressure remains almost constant.

5.1.4 Strain rate-dependent compression

Figure 5.8 shows the strain–time relationship during consolidation tests on samples with different thicknesses (Aboshi 1973). The strain does not converge with time and it increases in proportion to $\log t$. However, the strain does converge during consolidation for elastic material. This type of behavior is called secondary consolidation. It is a typical type of time-dependent behavior for soil. Figure 5.9 illustrates the results of one-dimensional consolidation tests with different strain rates. CRS tests, that is, constant rate of strain tests, are shown in Figure 5.9 (Sällfors 1975), in which the strain rates (%/min) are as follows: C7-1(0.003), C7-2(0.006), C7-3A(0.012), C7-4(0.020), C7-5(0.045), and C7-6(0.100). In the case of higher strain

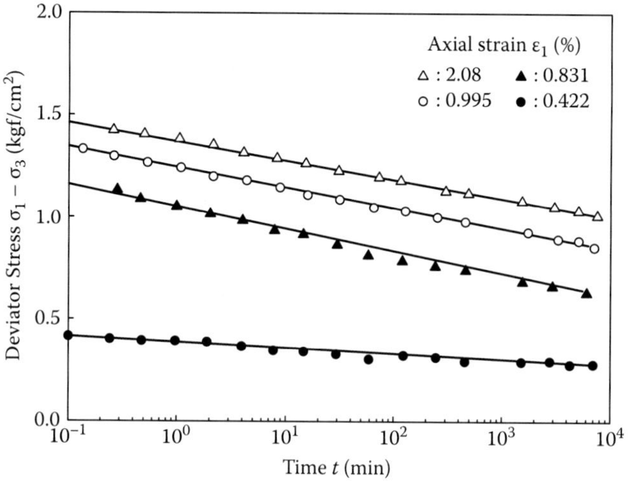

(a)

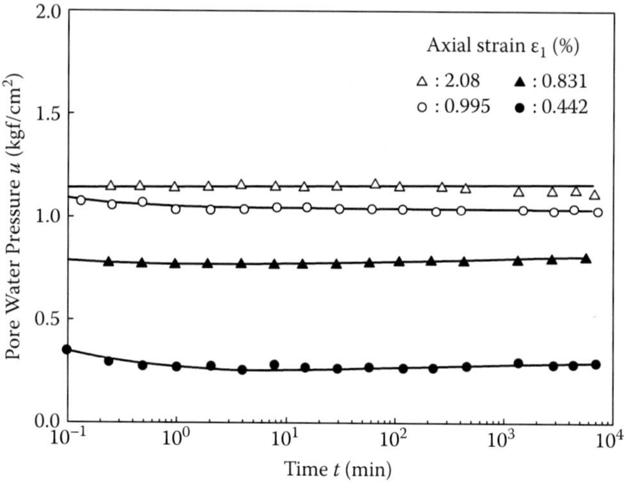

(b)

Figure 5.7 (a) Deviator stress–time profile during the stress relaxation. (b) Pore water pressure–time profile during the stress relaxation. (After Akai, K., Adachi, T., and Ando, N., 1974, Proc. JSCE, 225:53–61.) (× 98 kPa).

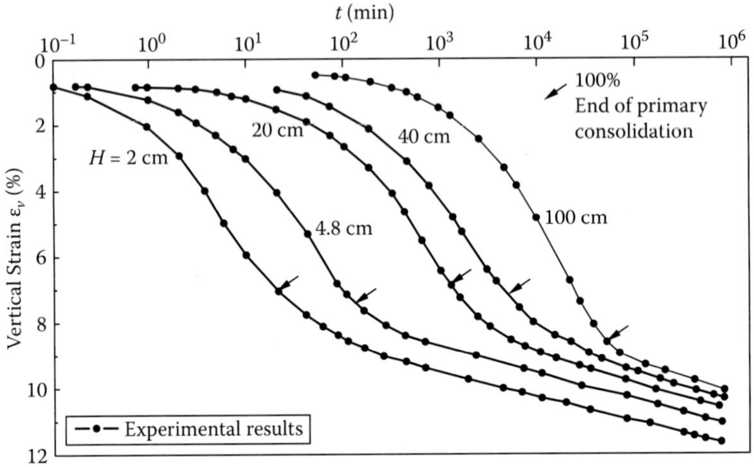

Figure 5.8 Settlement–time curves during the one-dimensional consolidation with different thickness. (After Aboshi, H., 1973, 8th ICSMFE, Moscow, 4(3):88.)

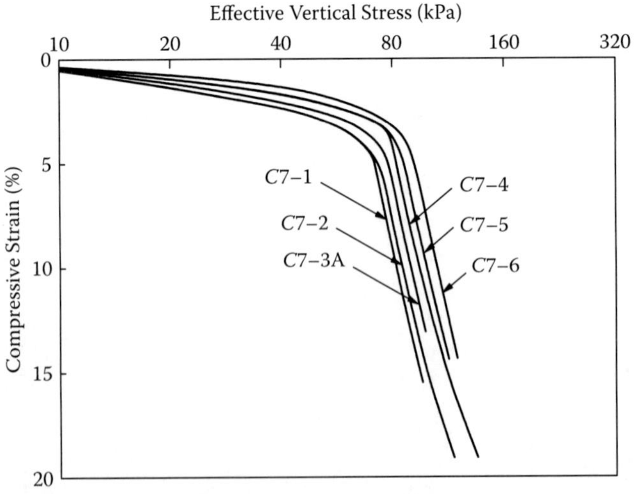

Figure 5.9 Vertical effective stress–compression strain relations for different strain rates. (After Sällfors, G., 1975, Preconsolidation pressure of soft high-plastic clays, PhD thesis, Chalmers University of Technology, Gothenburg, Sweden.)

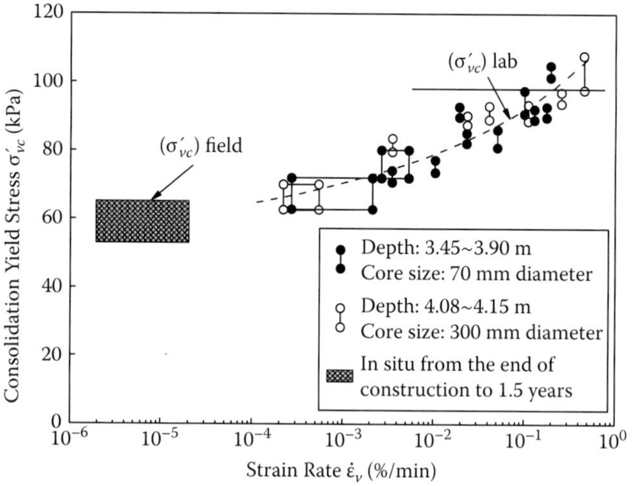

Figure 5.10 Preconsolidation pressure–strain rate relationship in laboratory and in situ. (After Leroueil, S., Samson, L., and Bozozuk, M., 1983, *Can. Geotech. J.*, 20(3):477–490.)

rates, the strain is smaller at the same level of stress and the inflection point of the stress–strain curve, that is, the consolidation yield stress or the compression yield stress is larger. From this figure, we can observe the time-dependency of the compression characteristics. Figure 5.10 illustrates the relationship between the preconsolidation pressure, that is, the consolidation yield stress, and the strain rate by Leroueil et al. (1983). This figure clearly shows the strain–rate dependency of the consolidation yield stress. From earlier, it is concluded that clayey soil exhibits time-dependent behavior such as rate sensitivity, creep, and relaxation.

5.1.5 Isotaches

The stress–strain characteristics, in which the stress–strain curves are the same for the equi-void ratio and the equi-strain rates, are called isotaches (Suklje 1957, 1969). *Iso* means "equivalent" and *tache* indicates the rate. Isotaches include a unique relation among the stress, the strain, and the strain rate, that is, the existence of stress–strain–strain rate relations. Experimental results by Graham et al. (1983) and one-dimensional consolidation test results with different strain rates shown by Leroueil et al. (1985) are shown in Figures 5.11 and 5.12, respectively. These results support the concept of isotaches.

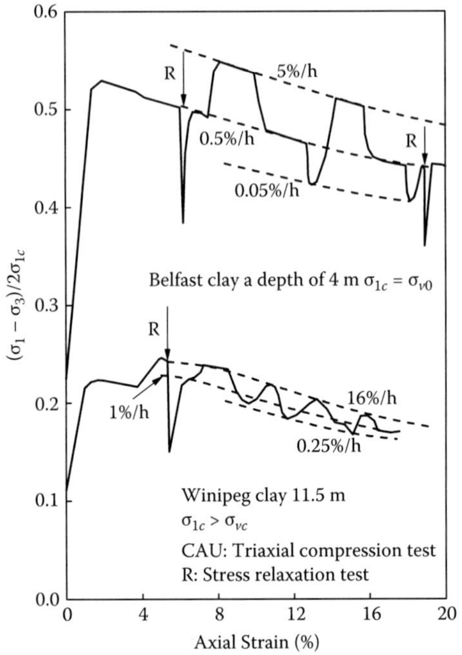

Figure 5.11 Stress–strain curves for triaxial compression tests with step changed strain rates and relaxation procedure. (After Graham, J., Crooks, J.H.A., and Bell, A.L., 1983, *Géotechnique*, 33(3):327–340.)

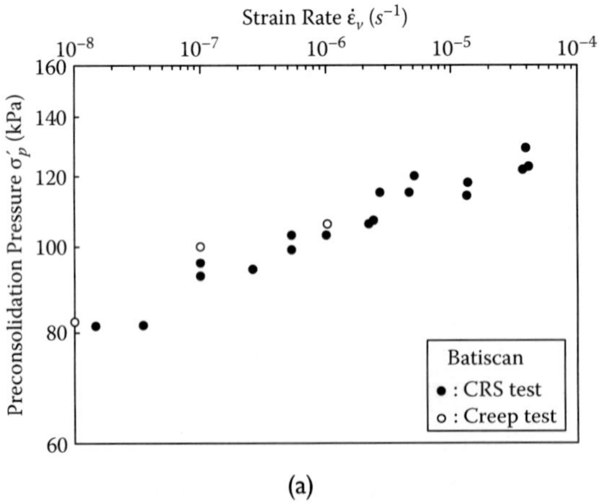

Figure 5.12 (a) Variation in preconsolidation pressure with strain rate for Batiscan clay. (b) Normalized effective stress–strain relationship deduced from CRS odometer tests on Batiscan clay. (After Leroueil, S., Kabbaj, M., Tavenas, F., and Bouchard, R., 1985, *Géotechnique*, 36(2):288–290.)

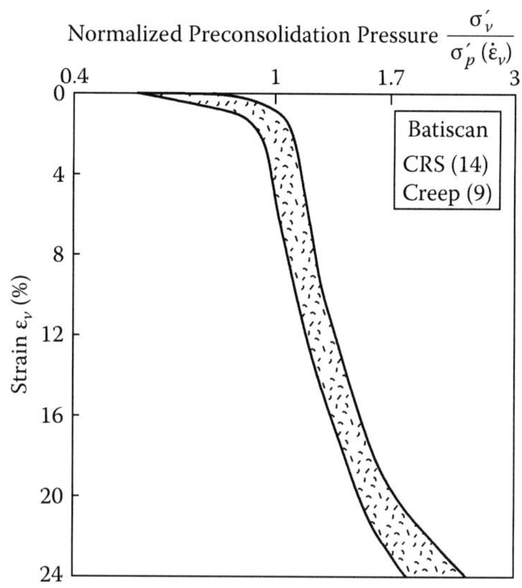

(b)

Figure 5.12 (Continued)

5.2 VISCOELASTIC CONSTITUTIVE MODELS

The modeling of many kinds of materials has been carried out within the framework of viscoelasticity for polymers, metals, concrete, soil, and rock. The well-known linear models are the Maxwell model, the Kelvin–Voigt model, and the spring–Voigt three-parameter model. It has been reported that the linear spring–Voigt model can describe the dynamic nature of soil (Kondner and Ho 1965; Hori 1974). By introducing the concept of the distribution of relaxation time into the linear model, it is possible to model the wide range of time-dependent behavior of soil. Murayama and Shibata (1964) have proven the time-dependency of clay in high-frequency regions by considering the distribution of relaxation time. Murayama (1983) proposed nonlinear viscoelastic and viscoplastic models based on the original model (Murayama and Shibata 1964).

Di Benedetto et al. (1997) proposed a simple asymptotic body (SAB) for the simplification of the viscoelastic model for soil that can be classified into three-parameter models. In the range of small strain, the linear viscoelastic approach is valid. In the range of large strain, however, the features include both viscoelasticity and visoplasticity. Oka, Kodaka, and Kim (2004) have succeeded in describing the behavior by a viscoelastic–viscoplastic model for clay, which can explain the dynamic behavior of clay for a wide range of strain levels.

5.3 ELASTOVISCOPLASTIC CONSTITUTIVE MODELS

5.3.1 Overstress models

To describe both the viscous nature and the plastic nature of soil, viscoplastic modeling is necessary. Perzyna (1963) proposed a viscoplastic theory that generalizes the linear theory for the viscoplasticity theory by Hohenemser and Prager (1932). Prager and Hohenemser's model, a linearly extended viscoplastic model, is based on Bingham's fluid and plasticity model.

Yong and Japp (1969) indicated the possibility of applying the viscoplasticity theory to the dynamic behavior of clay. Then, Adachi and Okano (1974) first proposed an elastoviscoplastic theory for clay based on Perzyna's theory and the original Cam-clay model (1963). They assumed that the hardening parameter was axial strain. They showed that viscoplasticity was an applicable theory to the rate-dependent behavior of water-saturated clay. However, no quantitative description had been successively given for the model. Oka's (1981), Adachi and Oka's (1982) newly proposed elastoviscoplastic model was based on both the Perzyna model and the Cam-clay model. It incorporates the assumption that in the stress state, after the completion of consolidation, the soil has not yet reached the equilibrium state but is still in a nonequilibrium state with the strain-hardening parameter for the inelastic volumetric strain, although the inelastic void ratio has been taken as a hardening parameter in the Cam-clay model. The model is capable of describing the rate sensitivity, the creep, and the relaxation of cohesive soil, and in particular, the volumetric relaxation behavior reported by Arulanandan et al. (1971).

The model is a rigorous combination of two theories, namely, the Cam-clay model and Perzyna model. However, the model has a shortcoming; namely, it cannot describe conventional accelerated creep behavior, that is, creep failure. Professor S. Sture of the University of Colorado pointed out this shortcoming at the International Workshop in Grenoble (Oka 1982). Aubry et al. (1985) experimentally showed that the critical state line is not rate sensitive. It can be understood that rate dependency fades out at the critical state. Giving consideration to the rate independency at the critical state leads to the fact that the viscosity asymptotically becomes zero when approaching the critical state. Following the above point, Adachi, Oka, and Mimura (1987) constructed an improved viscoplastic model by considering the variations in viscosity. The derived model is very capable of describing creep failure, that is, accelerating creep behavior. The prediction obtained through this model indicates that the drop in stress is rather small in comparison to the experimental evidence for sensitive clay and natural soil during strain softening. During the strain-softening behavior of natural clay, it is observed that strain softening follows a rather large decrease in mean effective stress. This indicates that the soil exhibits both shear and volumetric softening. To incorporate these features, a new model has been developed considering the degradation of soil structures and rate dependency by Kimoto (2002) and Kimoto and Oka (2005). This new viscoplastic model will be introduced in the following section.

Many other models have been proposed to describe the time-dependent behavior of soil. For the overstress models, Dafalias (1982), Katona (1984), Baladi and Rohani (1984), and Zienkiewicz et al. (1975) have proposed elastoviscoplastic models within the framework of an overstress type of theory. Another type of overstress model has been proposed by Duvaut and Lions (1976). Although their model is a linear overstress type of model, the Duvaut–Lions model is advantageous in that the plasticity model can easily be transferred into a viscoplastic one using the projection rule. Phillips and Wu's (1973) model is a nonlinear viscoplastic model using a similar projection technique to obtain the overstress. Sawada et al. (2001) proposed a Cosserat viscoplasticity model for clay. Yin and Karstunen (2008) proposed an anisotropic elastoviscoplastic model based on the modified Cam-clay model and Perzyna's type of viscoplasticity for the analysis of a clay foundation beneath an embankment.

5.3.2 Time-dependent model

Sekiguchi (1977) proposed an elastoviscoplastic model that clearly includes real time. Sekiguchi's model was originally proposed as a creep model that included failure. Nova (1982), Dragon and Mroz (1979), and Matsui and Abe (1985) derived time-dependent models that are called nonstationary models. It should be pointed out that these models include time explicitly and violate the principle of objectivity. Yin and Graham (1999) proposed an elastoviscoplastic model based on the modified Cam-clay model and the flow surface. Kutter and Sathialingam (1992) proposed an elastoviscoplastic model for clay with a reference surface and discussed various rate-dependent properties and overconsolidation ratios, for which they used a time-dependent hardening law. The aforementioned models are consistent with the delayed compression concept by Bjerrum (1973) in which equi-time lines are incorporated since the equi-time line includes the solution for secondary compression.

5.3.3 Viscoplastic models based on the stress history tensor

Oka (1985) proposed a viscoplastic model with the stress history tensor, which is based on the assumption that the state of materials depends on the stress and the stress history. He assumed that the yield function depends on the stress history tensor and not on just the current stress or the internal variables. The stress history tensor is given by the convolution integral of the stress tensor with respect to the generalized time measure, which is inherent to the materials. Oka and Adachi (1985) developed an elastoviscoplastic model using a stress history tensor for the analysis of the strain-softening behavior of soft rock and frozen sand (Adachi et al. 1990; Oka et al. 1994), and generalized it as the viscoplastic model (Adachi and Oka 1995; Adachi et al. 2003, 2005). This type of model can be applied to the rate-independent behavior by adopting a special time measure for defining the stress history tensor.

5.4 MICRORHEOLOGY MODELS FOR CLAY

The viscous behavior of clay has been analyzed in the field of microrheology. Murayama and Shibata (1964) applied the rate process theory by Eyring (1936) to clay and derived a rheological model. Then, Singh and Mitchell (1968, 1969) and Mitchell et al. (1968) successfully described the creep behavior of clay based on the rate process theory. Using the rate process theory, an exponential type of nonlinear flow law, between the shear force acting on each flow unit and the strain rate when the shear force is found to be larger than the thermal energy, was created.

5.5 ADACHI AND OKA'S VISCOPLASTIC MODEL

Oka (1981) and Adachi and Oka (1982) developed an elastoviscoplastic constitutive model for clay based on the Cam-clay model and an overstress type of viscoplastic theory (Perzyna 1963). The important assumption taken in the derivation of the model is that "at the end of consolidation, the state of the clay does not reach the static equilibrium state, but is in a non-equilibrium state." In the following, Terzaghi's effective stress is used as

$$\sigma'_{ij} = \sigma_{ij} - u_w \delta_{ij} \tag{5.4}$$

where σ'_{ij} is the effective stress and u_w is the pore water pressure.

It is assumed that the strain rate tensor consists of elastic strain rate tensor, $\dot{\varepsilon}^e_{ij}$, and viscoplastic strain rate tensor, $\dot{\varepsilon}^{vp}_{ij}$, such that

$$\dot{\varepsilon}_{ij} = \dot{\varepsilon}^e_{ij} + \dot{\varepsilon}^{vp}_{ij} \tag{5.5}$$

The elastic strain rate is given by a generalized Hooke type of law as

$$\dot{\varepsilon}^e_{ij} = \frac{1}{2G}\dot{s}_{ij} + \frac{\kappa}{3(1+e)\sigma'_m}\dot{\sigma}'_m \delta_{ij} \tag{5.6}$$

where s_{ij} is the deviatoric stress tensor, σ'_m is the mean effective stress, G is the elastic shear modulus, and e is the void ratio; in the infinitesimal theory, initial void ratio, e_0, is used, κ is the swelling index, and the superimposed dot denotes the time differentiation.

In Equation (5.6), the elastic shear modulus, G, is assumed to be proportional to the square root of the mean effective stress and the initial void ratio can be replaced by the current void ratio for the finite deformation analysis.

As has been presented in Chapter 3, the viscoplastic flow rule by Perzyna is given by

$$\dot{\varepsilon}_{ij}^{vp} = \gamma\langle\Phi(F)\rangle\frac{\partial f}{\partial\sigma'_{ij}} \tag{5.7}$$

$$\langle\Phi(F)\rangle = \begin{cases} \Phi(F) & : F > 0 \\ 0 & : F \le 0 \end{cases} \tag{5.8}$$

$$F = \frac{f - \kappa_s}{\kappa_s} \tag{5.9}$$

where $\dot{\varepsilon}_{ij}^{vp}$ is the viscoplastic strain rate tensor, γ is the viscosity parameter, Φ is a rate-sensitivity function, < > are Macaulay's brackets, f is the loading function, $F = 0$ is a static yield function, and κ_s is the strain-hardening parameter.

To construct a viscoplastic constitutive model for soil, we have to present the yield function. Herein, we assume that the mechanical behavior of soil at the static equilibrium state can be described by the original critical state energy theory developed by Roscoe et al. (1963). According to the Cam-clay model, the following yield function is assumed to be valid for isotropically consolidated clay as

$$f_s = \frac{\eta^*}{M_m^*} + \ln\frac{\sigma'_m}{\sigma'^{(s)}_{my}} = 0 \tag{5.10}$$

in which $\sigma'^{(s)}_{my}$ is the hardening parameter and M_m^* is the value of

$$\eta^* = \sqrt{\eta^*_{ij}\eta^*_{ij}} = \sqrt{2J_2}/\sigma'_m (J_2 = s_{ij}s_{ij}/2) \tag{5.11}$$

$$\eta^*_{ij} = s_{ij}/\sigma'_m \tag{5.12}$$

at the critical state.

f in Equation (5.7) is a function that depends on the viscoplastic strain and the stress; it can be called a dynamic yield function and it takes the same form as the static yield function.

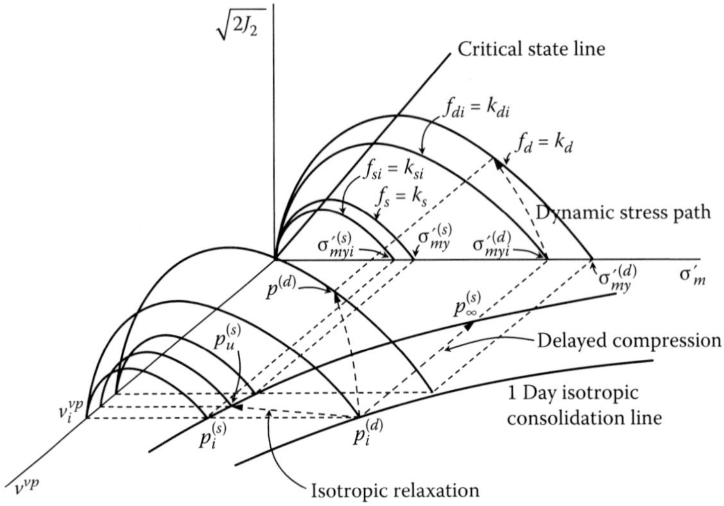

Figure 5.13 Schematic diagram of the yield surface and the loading path.

Figure 5.13 is a schematic diagram of the static and the dynamic loading (or yield) surface. In the figure, $P_i^{(d)}$ is a dynamic state at the end of one day of isotropic consolidation under prescribed pressure, $\sigma'^{(d)}_{myi}$. On the other hand, $P_i^{(s)}$ is the corresponding static state to $P_i^{(d)}$ with the same strain hardening, namely, in the same inelastic volumetric strain state, and it lies on the static isotropic consolidation line that is attained by the infinite time duration of isotropic consolidation. The state path, $P_i^{(d)} \to P^{(d)}$, represents a shear deformation process with an inelastic strain rate. State path $P_i^{(d)} \to P_u^{(s)}$ shows the increase in pore water pressure when returning to the undrained condition after the end of one day of consolidation, and path $P_i^{(d)} \to P_\infty^{(s)}$ represents secondary consolidation (delayed compression).

Since the static yield function is a nondimensional function, we can use $\Phi(F) = \Phi(f_s)$ in Equation (5.7). Based on the experimental results, Oka (1981) and Adachi and Oka (1982) assumed the functional form as

$$\gamma\Phi(f_y) = C_0\sigma'_m M^*_m \exp\left\{m'\left(\frac{\sqrt{2J_2}}{M^*_m\sigma'_m} + \ln\frac{\sigma'_m}{\sigma'^{(s)}_{my}}\right)\right\} \tag{5.13}$$

In the following section, it will be explained why Equation (5.13) is effective.

The derivative of Equation (5.10), with respect to the stress tensor, is given by

$$\frac{\partial f}{\partial \sigma'_{ij}} = \frac{1}{M^*_m\sigma'_m}\left\{\frac{s_{ij}}{\sqrt{2J_2}} + \left(M^*_m - \frac{\sqrt{2J_2}}{\sigma'_m}\right)\frac{1}{3}\delta_{ij}\right\} \tag{5.14}$$

The introduction of Equations (5.13) and (5.14) into Equation (5.7) yields

$$\dot{\varepsilon}_{ij}^{vp} = C_0 \exp\left\{ m'\left(\frac{\sqrt{2J_2}}{M_m^* \sigma'_m} + \ln \frac{\sigma'_m}{\sigma'_{my}} \right) \right\} \left\{ \frac{s_{ij}}{\sqrt{2J_2}} + \left(M_m^* - \frac{\sqrt{2J_2}}{\sigma'_m} \right) \frac{1}{3} \delta_{ij} \right\} \tag{5.15}$$

In Equation (5.15), we omit superscript s for simplicity.

Although the plastic void ratio is adopted as a hardening parameter in the Cam-clay model, the strain-hardening parameter, σ'_{my}, herein follows the following hardening rule as

$$dv^p = \frac{\lambda - \kappa}{1+e} \frac{d\sigma'_{my}}{\sigma'_{my}} \tag{5.16}$$

where dv^p is the viscoplastic volumetric strain increment.

Integrating Equation (5.16) under the initial conditions of $v^p = v_i^p$ and $\sigma'_{my} = \sigma'_{myi}$, we have

$$v^p - v_i^p = \frac{\lambda - \kappa}{1+e} \ln \frac{\sigma'_{my}}{\sigma'_{myi}} \tag{5.17}$$

Line I in Figure 5.14 shows this relation and how the strain-hardening parameter, $\sigma'^{(s)}_{my}$, increases with an increase in inelastic volumetric strain, v^p. It is found that the strain-hardening parameter, $\sigma'^{(s)}_{my}$, namely, yield function

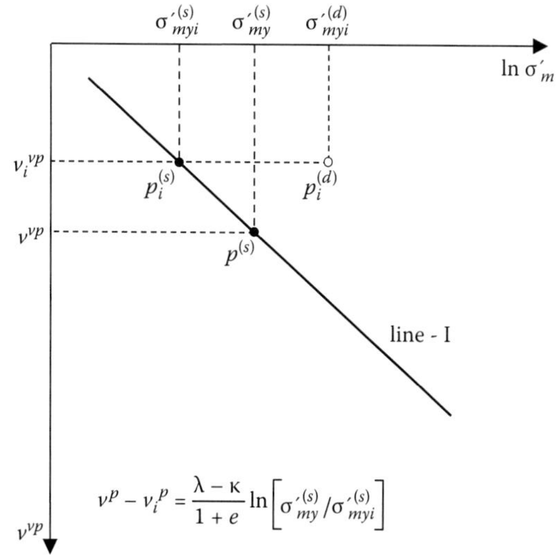

Figure 5.14 Volumetric hardening rule.

F, can be determined only if the values of v^p and $\sigma'^{(s)}_{myi}$ are given. In principle, it is impossible to obtain the value of $\sigma'^{(s)}_{my}$, since line I, which is the infinite time isotropic consolidation line, cannot be obtained. Fortunately, however, it is not necessary to know the exact value of $\sigma'^{(s)}_{myi}$.

Using the consolidation stress at the end of consolidation, σ'_{me}, we have

$$\ln\frac{\sigma'_{my}}{\sigma'_{me}} = \ln\frac{\sigma'_{myi}}{\sigma'_{me}} + \frac{1+e}{\lambda-\kappa}(v^p - v_i^p) \tag{5.18}$$

Then, if $v_i^p = 0$, Equation (5.15) becomes

$$\dot{\varepsilon}_{ij}^{vp} = C\exp\left\{m'\left(\frac{\sqrt{2J_2}}{M^*_m\sigma'_m} + \ln\frac{\sigma'_m}{\sigma'_{me}} - \frac{1+e}{\lambda-\kappa}v^p\right)\right\}\left\{\frac{s_{ij}}{\sqrt{2J_2}} + \left(M^*_m - \frac{\sqrt{2J_2}}{\sigma'_m}\right)\frac{1}{3}\delta_{ij}\right\} \tag{5.19}$$

$$C = C_0\exp\left\{m'\left(-\ln\frac{\sigma'_{myi}}{\sigma'_{me}}\right)\right\} \tag{5.20}$$

When the stress ratio is zero, namely, at the end of the isotropic consolidation, from Equation (5.19), the viscoplastic volumetric strain rate is given by

$$\dot{\varepsilon}_{kk}^{vp} = C\exp\left\{m'\left(\ln\frac{\sigma'_m}{\sigma'_{me}} - \frac{1+e}{\lambda-\kappa}v^p\right)\right\}M^*_m \tag{5.21}$$

Let us explain viscoplastic parameters m' and C. All material parameters can be determined from the results of consolidation, swelling, and strain-rate controlled undrained triaxial compression tests. Secondary consolidation is the well-known time-dependent behavior of clay, and we will examine the interrelation between secondary compression rate parameter α and strain rate parameter m' in the derived constitutive equations. Under isotropic consolidation, the inelastic volumetric strain rate (secondary consolidation rate) is expressed by Equation (5.21) under the condition of $\sigma'_m = \sigma'_{me}$ (= constant) as

$$\dot{v}^p = CM^*_m\exp\left\{-m'\frac{1+e}{\lambda-\kappa}v^p\right\} \tag{5.22}$$

If we set

$$\alpha = \frac{\lambda - \kappa}{(1+e)m'} \tag{5.23}$$

then

$$\dot{v}^p = CM_m^* \exp\left\{-\frac{1}{\alpha} v^p\right\} \tag{5.24}$$

When we assume that C is constant, Equation (5.24) can be integrated under the condition of $v^p = v_0^p$ at $t = t_0$ as

$$v^p = \alpha \ln(t/t_0) + v_0^p \tag{5.25}$$

and

$$v_0^p = -\alpha \ln\{\alpha/(CM_m^* t_0)\} \tag{5.26}$$

From this explanation, m' is determined by a secondary compression rate, α, and the compression and swelling indices. On the other hand, the viscoplastic parameter, C, is related to the initial viscoplastic volumetric strain rate as

$$\dot{\varepsilon}_0^{vp} = \dot{v}_0^p = CM_m^* \tag{5.27}$$

Using secondary compression coefficient $C_a(e^p = C_a \log(t/t_0) + e_0^p)$,

$$\frac{C_\alpha}{2.303(1+e)} = \alpha \tag{5.28}$$

since $\frac{\dot{e}^p}{1+e} = \dot{\varepsilon}_{kk}^p = \dot{v}^p$.

Finally, we can obtain

$$m' = \frac{C_c - C_s}{C_\alpha} \tag{5.29}$$

where C_c and C_s are the compression index and the swelling index in terms of the void ratio. During secondary compression, the void ratio changes viscoplastically under constant stress.

When we assume that ratio k between C_c and C_s is constant, we have

$$m' = \frac{(1-k)C_c}{C_\alpha} \tag{5.30}$$

Mesri et al. (1995) experimentally found that C_c/C_α is a material constant for geomaterials. It is worth noting that, based on Mesri's work, we could say that m' is a material constant. Leroueil and Hight (2003, p. 126) have shown that, based on Mesri's work, C_c/C_α is between 17 and 35. For these values, m' is larger for inorganic clay than for organic clay. For example, Mesri et al. (1995) have experimentally obtained experimental evidence for ratio $C_\alpha/C_c = 0.03$ for Batiscan clay and Saint-Hilaire clay. Leroueil and Hight indicated that C_c/C_α remains in the narrow range as shown in Table 5.1.

The other method for determining m', based on undrained triaxial tests with different strain rates, will be explained in the following section. Viscoplastic parameter m' can be determined from undrained triaxial compression tests with different strain rates.

Since the volume change is negligible under undrained conditions, that is, $\varepsilon_{kk}=0$, we obtain the next relationship between mean effective stress, σ'_m, and inelastic volumetric strain, v^p, from Equation (5.6) as

$$d\varepsilon_{kk} = \frac{\kappa}{1+e} d\sigma'_m + dv^p = 0 \tag{5.31}$$

Integrating Equation (5.31) under the condition of $v^p = v^p_i$ at $\sigma'_m = \sigma'_{me}$ results in

$$v^p = -\frac{\kappa}{1+e} \ln(\sigma'_m/\sigma'_{me}) + v^p_i \tag{5.32}$$

where σ'_{me} denotes the stress state at the end of consolidation.

Table 5.1 Values of C_α/C_c for several geomaterials

Material	C_α/C_c
Granular soils including rockfill	0.02 ± 0.01
Shale and mudstone	0.03 ± 0.01
Inorganic clays and silts	0.04 ± 0.01
Organic clays and silts	0.05 ± 0.01
Fibrous and amorphous peats	0.06 ± 0.01

Source: Leroueil, S., and Hight, D.W., 2003, Behaviour and properties of natural soils and soft rocks, In *Characterization and Engineering Properties of Natural Soils*, T.S. Tan, K.K. Phoon, D.W. Hight, and S. Leroueil, eds., Swets and Zeitlinger, 29–254.

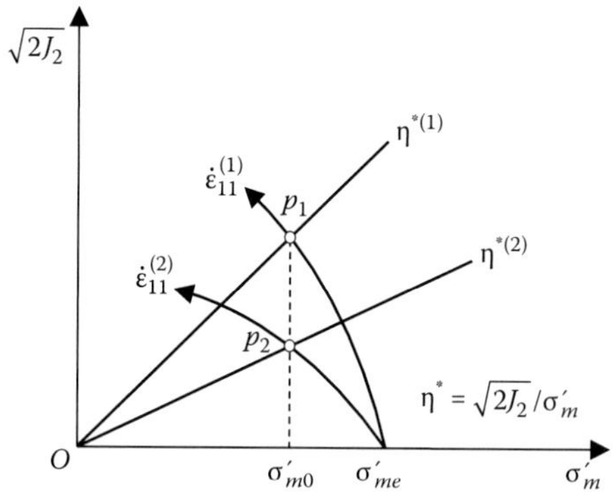

Figure 5.15 Same inelastic volumetric strain states on the different stress paths.

The meaning of Equation (5.32) is that the same inelastic volumetric strain takes place at two different stress states with different strain rates, provided the two states have the same mean effective stress and the same initial conditions.

Figure 5.15 shows this fact schematically. In other words, the inelastic volumetric strain is the same at both stress states, represented by P_1 and P_2 and lying on two different stress paths that correspond to strain rates $\dot{\varepsilon}_{11}^{(1)}$ and $\dot{\varepsilon}_{11}^{(2)}$, shown in Figure 5.15. The same inelastic volumetric strain (same strain-hardening) states on different stress paths are shown in Figure 5.15.

Under undrained conditions, total strain rate component $\dot{\varepsilon}_{ij}$ is equivalent to deviatoric strain rate component $\dot{e}_{ij}$, because Equation (5.31) is always satisfied.

We discuss the problem in a simple case of conventional axisymmetric triaxial undrained compression, $\sigma'_1 > \sigma'_2 = \sigma'_3$. Under this specific condition, the following relations are reduced:

$$s_{11} = \frac{2}{3}(\sigma_1 - \sigma_3), \quad \sqrt{2J_2} = \sqrt{2/3}(\sigma_1 - \sigma_3), \quad \varepsilon_{11} = e_{11} = \frac{2}{3}(\varepsilon_1 - \varepsilon_3) \tag{5.33}$$

Inserting these relations in Equation (5.19) results in

$$\dot{\varepsilon}_{11} = \frac{1}{2G}\dot{s}_{11} + \Phi(F)\sqrt{\frac{2}{3}} \tag{5.34}$$

$$\Phi(F) = C_0 \exp\left\{ m'\left[\frac{q}{M\sigma'_m} + \ln(\sigma'_m/\sigma'_{m0}) - \frac{1+e}{\lambda - \kappa}v^p - \ln\frac{\sigma'^{(s)}_{myi}}{\sigma'_{m0}} \right]\right\} \tag{5.35}$$

where $q = \sigma_1 - \sigma_3$, $M = \sqrt{3/2}M^*$, and G is the elastic shear modulus.

Assuming $\dot{\varepsilon}_{11} \cong \dot{\varepsilon}^p_{11}$, namely, the elastic deviatoric strain is negligible, the next relation is obtained from Equations (5.34) and (5.35) by comparing the states at P_1 and P_2, as shown in Figure 5.15.

$$\ln(\dot{\varepsilon}^{(1)}_{11}/\dot{\varepsilon}^{(2)}_{11}) = \frac{m'}{M}\left(q^{(1)}/\sigma'^{(1)}_m - q^{(2)}/\sigma'^{(2)}_m\right)$$

or

$$\frac{\dot{\varepsilon}^{(1)}_{11}}{\dot{\varepsilon}^{(2)}_{11}} = \exp\left[\frac{m'}{M}\left\{\left(\frac{q}{\sigma'_m}\right)^{(1)} - \left(\frac{q}{\sigma'_m}\right)^{(2)}\right\}\right] \tag{5.36}$$

$$m' = M\frac{\ln\dot{\varepsilon}^{(1)}_{11} - \ln\dot{\varepsilon}^{(2)}_{11}}{\left(\frac{q}{\sigma'_m}\right)^{(1)} - \left(\frac{q}{\sigma'_m}\right)^{(2)}} \tag{5.37}$$

where $q(=\sigma'_{11}-\sigma'_{33})$ is the deviator stress, σ'_m is the mean effective stress, and superscripts (1) and (2) denote two stress states on the stress paths with different strain rates and the same mean effective stress.

Figure 5.16 is prepared to evaluate the validity of Equation (5.37). It is clearly shown in the figure that a linear relationship exists between the logarithm of strain rate, $\dot{\varepsilon}_{11}$, and stress ratio, q/σ'_m, as an equi-inelastic volumetric strain line. Thus, the validity of Equation (5.34) is evaluated and the parameter can be determined from the slope of the equi-inelastic volumetric strain line.

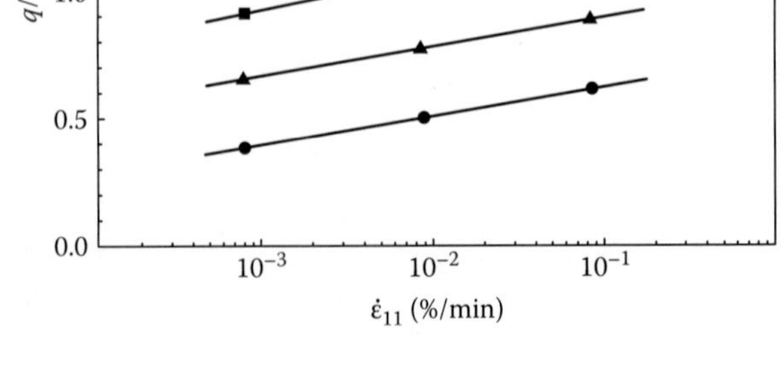

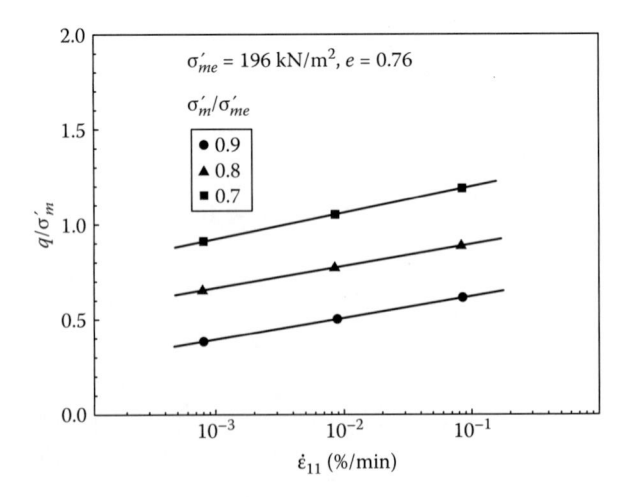

Figure 5.16 Relationship between the stress ratio and the logarithm of the strain rate.

The next task is to determine the remaining parameters, namely, C_0 and $\sigma'^{(s)}_{myi}$. Taking into account only the inelastic strain rate component in Equation (5.19), under undrained conditions, we have

$$\dot{e}_{ij}^{vp} = C \exp\left\{ m'\left(\frac{\sqrt{2J_2}}{M_m^* \sigma'_m} + \ln\frac{\sigma'_m}{\sigma'_{me}} - \frac{1+e}{\lambda - \kappa} v^p \right)\right\} \frac{s_{ij}}{\sqrt{2J_2}} \tag{5.38}$$

$$C = C_0 \exp\left\{ m'\left(-\ln\frac{\sigma'^{(s)}_{myi}}{\sigma'_{me}} \right)\right\} \tag{5.39}$$

The preceding equation indicates that it is not necessary to determine parameters C_0 and $\sigma'^{(s)}_{myi}$ separately; we only need parameter C. Substituting the strain rate, the mean effective stress, the deviator stress, and the stress ratio at the same stress point into Equation (5.38), we can determine parameter C.

5.5.1 Strain rate effect

Figures 5.17 and 5.18 show the simulated and the experimental stress–strain relations and stress paths of one-day consolidated clay under strain-rate controlled undrained triaxial conditions. The clay is reconstituted Fukakusa clay preconsolidated under a pressure of 49 kPa. The physical

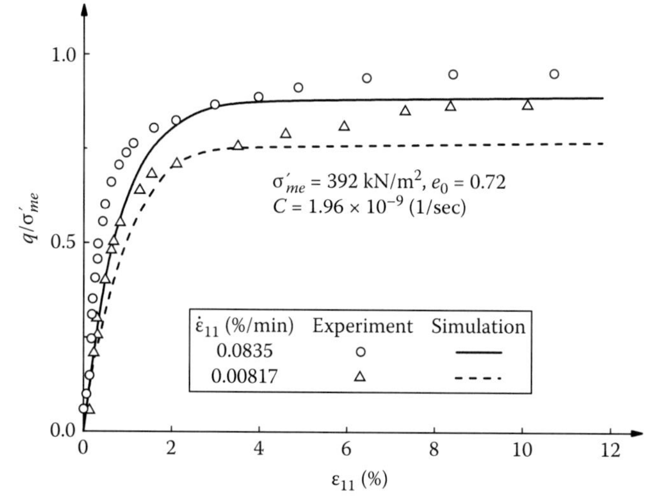

Figure 5.17 Stress–strain relations for constant strain rate triaxial tests.

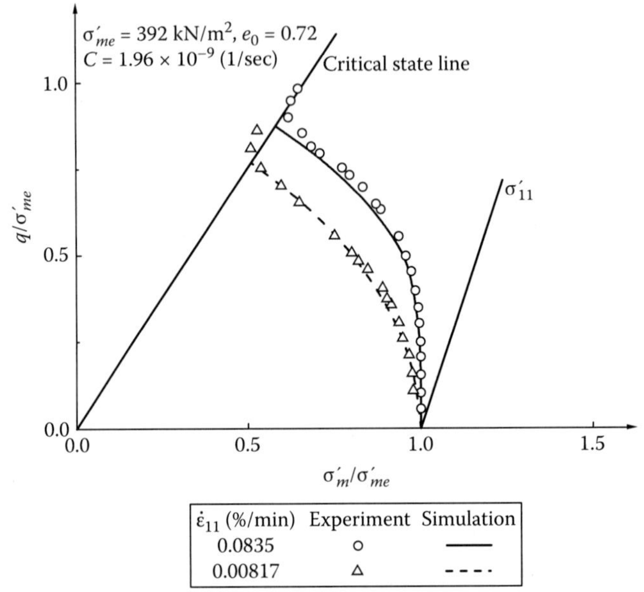

Figure 5.18 Stress paths for constant strain rate triaxial tests.

properties are as follows: specific gravity = 2.613; LL = 48.5%; PL = 26.7%; IP = 21.8%; Fines content:clay and silt content ($\leq 75\mu m$) = 67.7%; and sand content ($\geq 420\mu m$) = 5.9%. The fixed parameters used for the calculation are $\lambda = 0.1$, $\kappa = 0.02$, $M = \sqrt{\frac{3}{2}}M_m^* = 1.5$, $m' = 28.8$, and $G_0 = 363$ kPa.

The theoretical curves were obtained using Equations (5.34) and (5.35). It can be concluded that the strain-rate effect on the stress path and on the stress–strain relation is reasonably assessed by the proposed theory. With respect to elastic shear modulus, G, it is assumed that G is proportional to the square root of the mean effective stress, and it is worth noting that Young's modulus, E, is equal to three times G under undrained conditions.

5.5.2 Simulation by the Adachi and Oka's model

5.5.2.1 *Effect of secondary consolidation*

Figure 5.19 shows the calculated and the measured effective stress paths of one-day and seven-day consolidated samples, and Figure 5.20 shows the effect of secondary consolidation on the stress–strain relations. The clay sample, consolidated for a long period, increases in undrained strength and initial tangent modulus. These phenomena are in general due to "aging." The test results correspond to Shen et al. (1973).

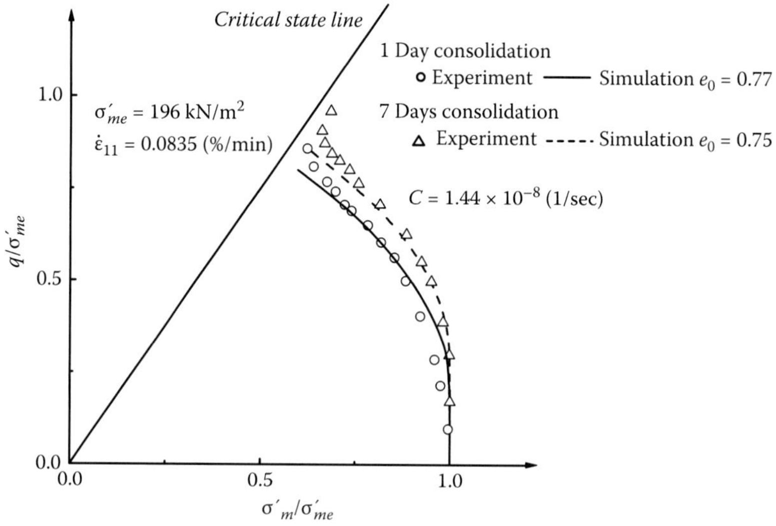

Figure 5.19 Calculated and measured effective stress paths for one-day consolidated and seven-days consolidated clays.

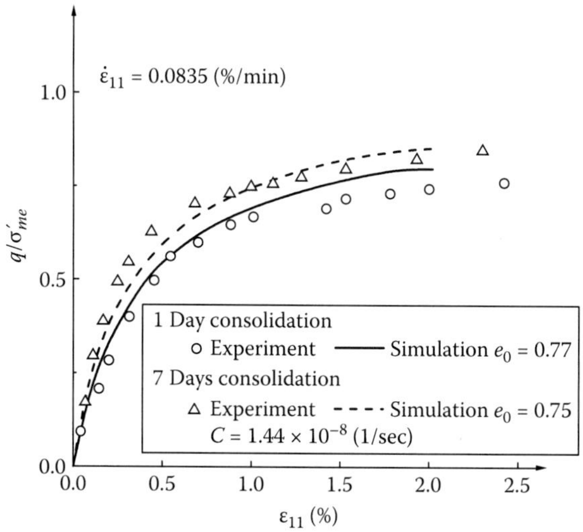

Figure 5.20 Calculated and measured stress–strain relations for one-day consolidated and seven-days consolidated clays.

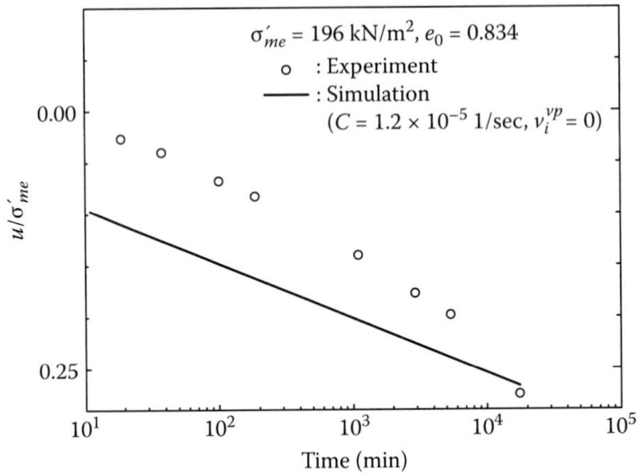

Figure 5.21 Induced pore water pressure during the undrained isotropic stress relaxation.

5.5.2.2 *Isotropic stress relaxation*

Figures 5.21 and 5.22 show the increase in pore water pressure during isotropic stress relaxation. After the first relaxation test, the sample was reconsolidated and returned to undrained conditions, namely, the second isotropic relaxation process. The results given in Figure 5.22 correspond to the second isotropic stress relaxation process, which follows the first

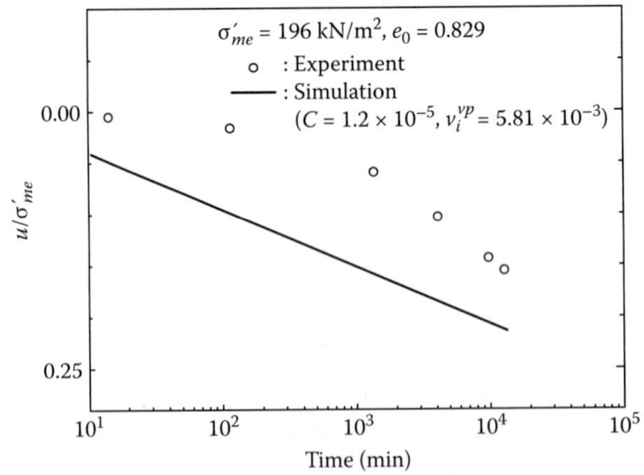

Figure 5.22 Induced pore water pressure during the undrained isotropic stress relaxation after the drainage of the pore water pressure during the stress relaxation test shown in Figure 5.21.

relaxation test results shown in Figure 5.21. Therefore, the calculated initial inelastic volumetric strain is different in the two cases. The void ratio was determined from the volume change in the clay specimen during the consolidation process. Similar experimental results have been reported by Arulanandan et al. (1971).

5.5.3 Constitutive model for anisotropic consolidated clay

For anisotropic consolidated clay, we have extended the static yield function using a second invariant of relative stress ratio, $\bar{\eta}^*_{(0)}$, as

$$f = \bar{\eta}^*_{(0)} + \tilde{M}^* \ln \frac{\sigma'_m}{\sigma'^{(s)}_{my}} = 0 \tag{5.40}$$

$$\bar{\eta}^*_{(0)} = \sqrt{\left(\eta^*_{ij} - \eta^*_{ij(0)}\right)\left(\eta^*_{ij} - \eta^*_{ij(0)}\right)} \tag{5.41}$$

where $\eta^*_{ij} = s_{ij}/\sigma'_m$ and the subscript (0) denotes the state at the end of consolidation, namely, at the initial state before deformation.

Using the extended yield function, we have obtained the viscoplastic strain rate as

$$\dot{\varepsilon}^{vp}_{ij} = C_0 \exp\left\{ m' \left(\frac{\bar{\eta}^*_{(0)}}{M^*_m} + \ln \frac{\sigma'_m}{\sigma'_{my}} \right) \right\} \left\{ \frac{\eta^*_{ij} - \eta^*_{ij(0)}}{\bar{\eta}^*_{(0)}} + \left(M^*_m - \frac{\eta^*_{kl}(\eta^*_{kl} - \eta^*_{kl(0)})}{\bar{\eta}^*_{(0)}} \right) \frac{1}{3} \delta_{ij} \right\} \tag{5.42}$$

5.6 EXTENDED VISCOPLASTIC MODEL CONSIDERING STRESS RATIO-DEPENDENT SOFTENING

As mentioned earlier, Adachi et al. (1987a) extended the original model to describe the acceleration creep behavior of clay by introducing a second material function into the model. Oka et al. (1994, 1995) studied the instability of the extended model during the undrained conventional creep process and strain localization analysis.

$$\dot{\varepsilon}^{vp}_{ij} = \gamma \langle \Phi_1(F) \rangle \Phi_2(\xi) \frac{\partial f}{\partial \sigma'_{ij}}, \quad F = \frac{f - \kappa_s}{\kappa_s} \tag{5.43}$$

where Φ_2 is the second material function.

The second material function is introduced to explain that the rate-dependency of clay vanishes at the failure state. In other words, the stress ratio at the failure state does not depend on the strain rate. In the study by Adachi et al. (1987a), the following form for the second material function is adopted:

$$\Phi_2 = 1 + \xi \tag{5.44}$$

Internal variable ξ expresses the deterioration of the materials and obeys the following evolutional equation as:

$$\dot{\xi} = \frac{M_f^{*2}}{G_2^*(M_f^* - \eta^*)^2}\dot{\eta}^* \tag{5.45}$$

where M_f^* is the value of stress ratio η^* at the failure state and G_2^* is a material parameter.

We can rewrite Equation (5.10) in an alternative form as

$$f - \kappa_s = \kappa_s\left[\Phi_1^{-1}\left\{\frac{I_2^{vp}}{\gamma\Phi_2}\left(\frac{\partial f}{\partial \sigma'_{ij}}\frac{\partial f}{\partial \sigma'_{ij}}\right)^{1/2}\right\}\right] \tag{5.46}$$

where I_2^{vp} is the second invariant of the viscoplastic strain rate tensor. From Equation (5.46), we can see that the yield function depends implicitly on both the hardening parameter and the strain rate.

5.7 ELASTOVISCOPLASTIC MODEL FOR COHESIVE SOIL CONSIDERING DEGRADATION

5.7.1 Elastoviscoplastic model considering degradation

The prediction by the extended model with the second material function (Adachi et al. 1987a) indicates that the drop in stress is rather small compared to the experimental evidence for sensitive clay and natural soil during strain softening. During the strain-softening behavior of natural clay, it is observed that strain softening follows a rather large decrease in mean effective stress. This indicates that the soil exhibits both shear and volumetric softening. In order to incorporate these features, a new model has been developed by Kimoto (2002), Kimoto et al. (2004), and Kimoto and Oka (2005), considering the degradation of soil structures and rate dependency. The model with the structural degradation can be applicable to the soft rock as well as soft and hard soils (Oka et al. 2011).

We assume an overconsolidation boundary surface that delineates the overconsolidated (OC) region $(f_b < 0)$ from the normally consolidated (NC) region $(f_b \geq 0)$, namely,

$$f_b = \bar{\eta}^*_{(0)} + M^*_m \ln \frac{\sigma'_m}{\sigma'_{mb}} = 0 \tag{5.47}$$

in which M^*_m is the value of $\eta^* = \sqrt{\eta^*_{ij}\eta^*_{ij}}$ when the volumetric strain increment changes from compression to swelling and σ'_{mb} is the hardening parameter.

Originally, the hardening rule for the σ'_{mb} surface was defined with respect to the viscoplastic volumetric strain. In order to describe the degradation of the material caused by structural changes, strain softening with viscoplastic strain is introduced in addition to strain hardening with viscoplastic volumetric strain as

$$\sigma'_{mb} = \sigma'_{ma} \exp\left(A_3 \varepsilon_v^{vp}\right), \quad A_3 = \frac{1+e_0}{\lambda - \kappa} \tag{5.48}$$

where e_0 is the initial value of the void ratio and the current void ratio can be used for the finite deformation analysis.

$$\sigma'_{ma} = \sigma'_{maf} + \left(\sigma'_{mai} - \sigma'_{maf}\right)\exp(-\beta z) \tag{5.49}$$

where σ'_{ma} is assumed to decrease with an increasing viscoplastic strain in Equations (5.48) and (5.49), and σ'_{mai} and σ'_{maf} are the initial and the final values for σ'_{ma}, respectively. σ'_{mai} corresponds to the consolidation yield stress and σ'_{maf} is determined from the difference between the peak stress and the residual stress. Material parameter, β, controls the rate of structural changes and z is an accumulation of the second invariant of the viscoplastic strain rate as

$$z = \int_0^t \dot{z}\,dt, \quad \dot{z} = \sqrt{\dot{\varepsilon}^{vp}_{ij}\dot{\varepsilon}^{vp}_{ij}}, \tag{5.50}$$

The mechanical behavior of clay at its static equilibrium state is assumed to be described by the original Cam-clay model (Adachi and Oka 1982). The following static yield function is used:

$$f_y = \bar{\eta}^*_{(0)} + \tilde{M}^* \ln \frac{\sigma'_m}{\sigma'^{(s)}_{my}} = 0 \tag{5.51}$$

In a similar way for the OC boundary surface, f_b, strain softening is defined in order to express the effect of a structural collapse through changes in $\sigma'^{(s)}_{my}$ with the viscoplastic strain, namely,

$$\sigma'^{(s)}_{my} = \{n + (1-n)\exp(-\beta z)\}\sigma'^{(s)}_{myi} \exp\left(\frac{1+e_0}{\lambda - \kappa}\varepsilon^{vp}_v\right) \tag{5.52}$$

where $n = \sigma'_{maf}/\sigma'_{mai}$ describes the degree of structure at the initial state and β is the rate of degradation.

According to the structural collapse, the decrease in $\sigma'^{(s)}_{my}$ leads to the shrinking of the static yield function.

The viscoplastic potential function is given as

$$f_p = \bar{\eta}^*_{(0)} + \tilde{M}^* \ln \frac{\sigma'_m}{\sigma'_{mp}} = 0 \tag{5.53}$$

where the dilatancy parameter $\tilde{M}^*$ is assumed to be constant in the NC region. σ'_{mp} is determined automatically from the stress state in the NC region, and it coincides with σ'_{mb} in the OC region shown in Figure 5.23. The value varies with the current stress in the OC region as

$$\tilde{M}^* = \begin{cases} M^*_m & : NC\ region \\ -\dfrac{\sqrt{\eta^*_{ij}\eta^*_{ij}}}{\ln(\sigma'_m/\sigma'_{mc})} & : OC\ region \end{cases} \tag{5.54}$$

where σ'_{mc} denotes the mean effective stress at the intersection of the overconsolidation boundary surface and the σ'_m axis as

$$\sigma'_{mc} = \sigma'_{mb} \exp\left(\frac{\sqrt{\eta^*_{ij(0)}\eta^*_{ij(0)}}}{M^*_m}\right) \tag{5.55}$$

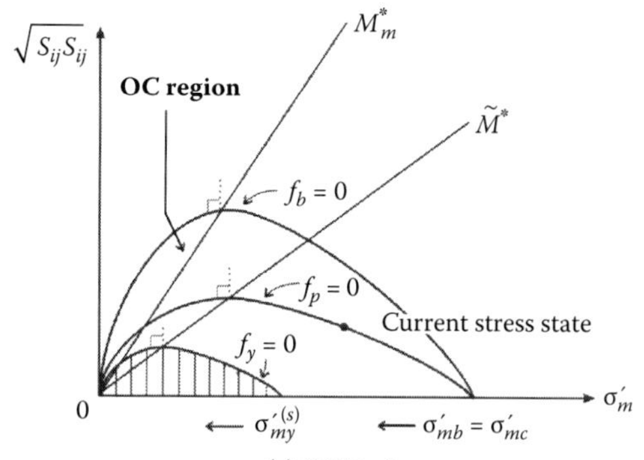

(a) OC Region

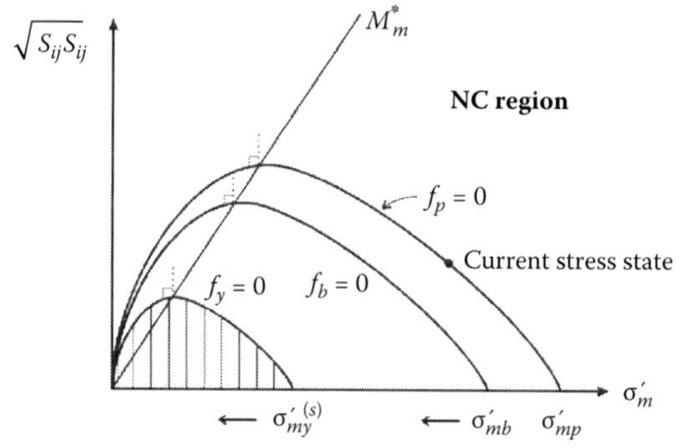

(b) NC Region

Figure 5.23 Yield surface and potential surface for isotropically consolidated clay.

According to the earlier definition, the value of dilatancy coefficient $\tilde{M}^*$ becomes zero when the stress path coincides with the mean effective stress axis during cyclic loading. Therefore, a new definition for $\tilde{M}^*$ (Kimoto et al. 2007) is introduced, as shown in Figure 5.24 which is for the isotropic consolidated one as

$$\tilde{M}^* = \begin{cases} M_m^*(\theta) & : \text{consolidated NC region} \\ (\sigma_m^*/\sigma_{mb}')M_m^*(\theta) & : \text{one as OC region} \end{cases} \tag{5.56}$$

where σ_m^* denotes the mean effective stress at the intersection of the surface, which has the same shape as f_b, and is given by

$$\sigma_m^* = \sigma_m' \exp\left(\frac{\bar{\eta}_{(0)}^*}{M_m^*(\theta)}\right) \tag{5.57}$$

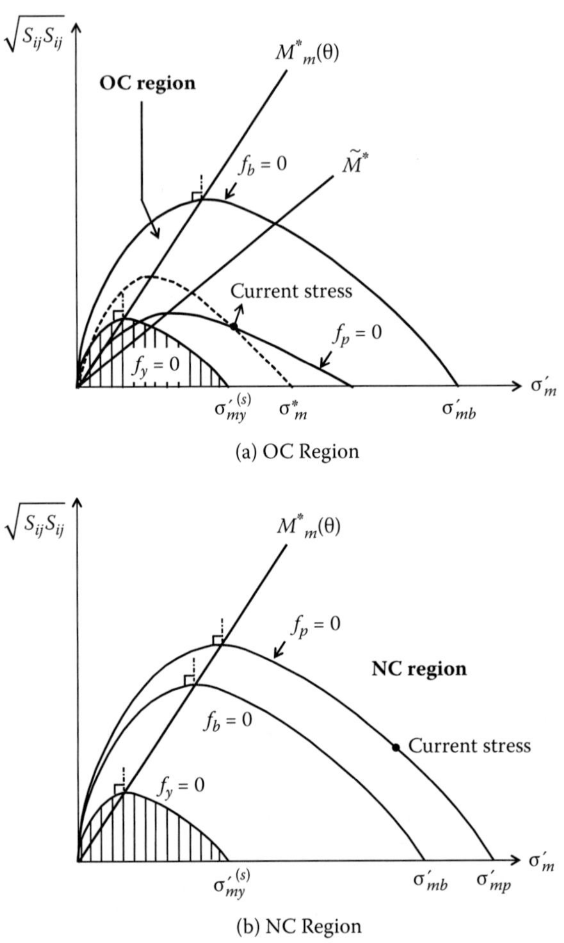

Figure 5.24 Overconsolidation boundary surface, static yield function, and viscoplastic potential function.

In Equations (5.56) and (5.57), M^*_m, in general, depends on the Lode's angle θ (Equation (4.35)). The viscoplastic strain rate tensor is given as the following equation based on the overstress type of viscoplastic theory (Perzyna 1963):

$$\dot{\varepsilon}^{vp}_{ij} = \gamma \langle \Phi_1(f_y) \rangle \frac{\partial f_p}{\partial \sigma'_{ij}} \tag{5.58}$$

where $\langle\,\rangle$ are Macaulay's brackets.

Based on the experimental results of strain–rate constant triaxial tests, material function Φ_1 is defined as

$$\begin{aligned}
\gamma\Phi_1(f_y) &= C'\sigma'_m \exp\left\{m'\left(\bar{\eta}^*_{(0)} + \tilde{M}^* \ln\frac{\sigma'_m}{\sigma'^{(s)}_{my}}\right)\right\} \\
&= C'\sigma'_m \exp\left\{m'\left(\bar{\eta}^*_{(0)} + \tilde{M}^* \ln\frac{\sigma'_m}{\sigma'^{s}_{myi}\{n+(1-n)\exp(-\beta z)\}\exp(A_3\varepsilon^{vp}_v)}\right)\right\} \\
&= C'\sigma'_m \exp\left\{m'\left(\bar{\eta}^*_{(0)} + \tilde{M}^* \ln\frac{\sigma'_m}{\frac{\sigma'^{s}_{myi}}{\sigma'_{mai}}\sigma'_{ma}\exp(A_3\varepsilon^{vp}_v)}\right)\right\} \\
&= C'\sigma'_m \exp\left\{m'\left(\bar{\eta}^*_{(0)} + \tilde{M}^* \ln\frac{\sigma'_{mai}}{\sigma'^{(s)}_{myi}}\frac{\sigma'_m}{\sigma_{mb}}\right)\right\} \\
&= C\sigma'_m \exp\left\{m'\left(\bar{\eta}^*_{(0)} + \tilde{M}^* \ln\frac{\sigma'_m}{\sigma'_{mb}}\right)\right\}
\end{aligned} \tag{5.59}$$

$$C = C'\exp\left(-m'\tilde{M}^* \ln\frac{\sigma'^{(s)}_{myi}}{\sigma'_{mai}}\right) \tag{5.60}$$

$$\begin{aligned}
\dot{\varepsilon}^{vp}_{ij} &= \gamma\langle\Phi_1(f_y)\rangle\frac{\partial f_p}{\partial\sigma'_{ij}} = C\exp\left\{m'\left(\bar{\eta}^*_{(0)} + \tilde{M}^* \ln\frac{\sigma'_m}{\sigma'_{mb}}\right)\right\} \\
&\times\left\{\frac{\eta^*_{ij} - \eta^*_{ij(0)}}{\bar{\eta}^*_{(0)}} + \left(\tilde{M}^* - \frac{\eta^*_{kl}(\eta^*_{kl} - \eta^*_{kl(0)})}{\bar{\eta}^*_{(0)}}\right)\frac{1}{3}\delta_{ij}\right\}
\end{aligned} \tag{5.61}$$

When $\sigma'_{mi} = \sigma'_{mbi} = \sigma'_{mai}$ and the shear stress is zero at the initial state after consolidation,

$$\dot{\varepsilon}^{vp}_{kki} = C\tilde{M}^* \tag{5.62}$$

where σ'_{mi}, σ'_{mai}, and σ'_{mbi} are the initial values for σ'_m, σ'_{ma}, and σ'_{mb}, respectively. C is the viscoplastic parameter corresponding to the viscoplastic volumetric strain rate at the initial stress state with zero shear stress.

In the preceding formulation, we assumed that the viscoplastic parameter, C, was scalar. However, we can generalize this assumption with a tensorial value for C such as the fourth-order isotropic tensor $C_{ijkl} = a\delta_{ij}\delta_{kl} + b(\delta_{ik}\delta_{jl} + \delta_{il}\delta_{jk})$, $C_1 = 2b, C_2 = 3a + 2b$, as has been adopted by Oka (1982, 1992) and Oka et al. (2003a).

$$\dot{\varepsilon}_{ij}^{vp} = C_{ijkl} \exp\left\{ m' \left(\bar{\eta}_{(0)}^* + \tilde{M}^* \ln \frac{\sigma'_m}{\sigma'_{mb}} \right) \right\}$$

$$\times \left\{ \frac{\eta_{kl}^* - \eta_{kl(0)}^*}{\bar{\eta}_{(0)}^*} + \left(\tilde{M}^* - \frac{\eta_{mn}^*(\eta_{mn}^* - \eta_{mn(0)}^*)}{\bar{\eta}_{(0)}^*} \right) \frac{1}{3} \delta_{kl} \right\} \tag{5.63}$$

5.7.2 Determination of the material parameters

There are ten material parameters for the proposed constitutive model. The procedure for determining these parameters is as follows. The initial void ratio, e_0, can be obtained from tests for physical properties. The compression index, λ, and swelling index, κ, are given by the slope of the isotropic consolidation and swelling tests, respectively. Compression yield stress, σ'_{mbi}, is assumed to be determined from the yield point of the isotropic consolidation tests. Elastic shear modulus, G_0, can be determined from the initial slope of the triaxial compression tests.

The stress ratio at maximum compression, M_m^*, is defined as the stress ratio whereby maximum compression occurs in the drained compression tests. For clay, however, it has been assumed to equal the stress ratio at the critical state. Herein, M_m^* is determined from the stress ratio at the residual state in the undrained triaxial compression tests.

The viscoplastic parameter, m', can be determined from undrained triaxial compression tests with different strain rates as

$$\frac{\dot{\varepsilon}_{11}^{(1)}}{\dot{\varepsilon}_{11}^{(2)}} = \exp\left[m' \sqrt{\frac{2}{3}} \left\{ \left(\frac{q}{\sigma'_m} \right)^{(1)} - \left(\frac{q}{\sigma'_m} \right)^{(2)} \right\} \right] \tag{5.64}$$

$$m' = \sqrt{\frac{3}{2}} \frac{\ln \dot{\varepsilon}_{11}^{(1)} - \ln \dot{\varepsilon}_{11}^{(2)}}{\left(\frac{q}{\sigma_m} \right)^{(1)} - \left(\frac{q}{\sigma'_m} \right)^{(2)}} \tag{5.65}$$

where $q\ (= \sigma'_{11} - \sigma'_{33})$ is the deviator stress, σ'_m is the mean effective stress, and superscripts (1) and (2) denote two stress states on the stress paths with different strain rates and the same mean effective stress.

When m' is determined, the viscoplastic parameter, C, is obtained from the deviatoric strain rate by the constitutive equation.

Alternatively, we can determine parameters m' and C through the secondary compression rate and the initial volumetric strain rate in a similar way, as shown in Equation (5.23), although the expression is a bit different since the form for the rate sensitivity function, shown in Equation (5.59), is different from that in Equation (5.13).

$$m' = \frac{\lambda - \kappa}{(1+e)\alpha \tilde{M}^*} \tag{5.66}$$

$$C = \dot{\varepsilon}^{vp}_{kk(0)} / \tilde{M}^* \tag{5.67}$$

where α is a secondary compression rate in a natural logarithm, namely,

$$v^p = \alpha \ln t / t_0 + v^p_0, \quad v^p = \varepsilon^{vp}_{kk} \tag{5.68}$$

The relationship given by Equation (5.66) can be rewritten by the secondary compression index as

$$m' = \frac{C_c - C_s}{C_\alpha \tilde{M}^*} \tag{5.69}$$

where C_c is the compression index, C_s is the swelling index, C_α is the secondary compression rate, and M^*_m is the stress ratio at the maximum compression or the stress ratio at a large strain.

Structural parameter σ'_{maf} is determined by the decrease from the peak stress to the residual stress in the undrained tests. Structural parameter β is determined by curve fitting for the strain-softening process in the undrained tests. The other possible model for the elastoviscoplastic model is a constitutive equation in which the degradation method is adopted for Equation (5.19).

5.7.3 Strain-dependent elastic shear modulus

The nonlinearity of soil stiffness has been studied extensively for materials such as sand, clay, and gravel, and has been summarized well by Ishihara (1996). For cohesive soil, several empirical equations have been proposed by considering the dependency of the shear modulus on the effective confining stress (Kokusho et al. 1982). In the original model by Adachi and Oka (1982) and Kimoto and Oka (2005), the change in the elastic shear modulus of the elastoviscoplastic model is given by the square root function of the normalized mean effective stress, namely,

$$G = G_0(e)\sqrt{\frac{\sigma'_m}{\sigma'_{m0}}} \tag{5.70}$$

in which G_0 is the value for G when $\sigma'_m = \sigma'_{m0}$ and the function of the void ratio e.

Equation (5.70) considers only the effect of the confining pressure, which can accurately approximate the variation in shear modulus at very small levels of strain. In regions with large levels of strain, however, as demonstrated by the experimental results, the strain dependency of the shear modulus should be considered as well. Various empirical formulations have been provided from the laboratory test results to express the strain dependency of the shear modulus (Hardin and Drnevich 1972; Fahey 1992; Wang and Kuwano 1999). Ogisako et al. (2007) have introduced a normalized shear modulus reduction function based on the viscoplastic shear strain in soft clay specimens and have proposed a hyperbolic equation for that expression, namely,

$$G = G_0 \frac{1}{(1+\alpha(\gamma^{vp})^r)} \tag{5.71}$$

where α and r are the experimental constants, which can be defined from the laboratory test results, and γ^{vp} is the accumulated viscoplastic shear strain given by the accumulation of the viscoplastic deviatoric strain increment, de_{ij}^{vp}, as

$$\gamma^{vp} = \int \sqrt{de_{ij}^{vp} de_{ij}^{vp}} \tag{5.72}$$

5.8 APPLICATION TO NATURAL CLAY

5.8.1 Osaka Pleistocene clay

The model is applied to Osaka Pleistocene clay, namely, Kyuhoji clay. It was sampled from the upper Pleistocene layer called Ma12, which is distributed widely in the western and eastern parts of Osaka, Japan, at a depth of 20 to 40 meters. This is marine clay containing diatoms, and it exhibits sensitive behavior due to the effect of the structures formed during the sedimentation process. Figure 5.25 compares the undrained compression test results between the undisturbed and the reconstituted samples of Kyuhoji clay (Yashima et al. 1999; Shigematsu 2002). Both the undisturbed clay and the reconstituted clay were sheared with an axial strain rate of 0.005%/min after isotropic consolidation at a confining pressure of 392 kPa, which is a little larger than the compression yield stress of 340 kPa. The initial void ratio of the undisturbed clay is larger than that of the reconstituted clay, specifically, 1.41 for the undisturbed clay and 1.02 for the reconstituted clay. The undisturbed

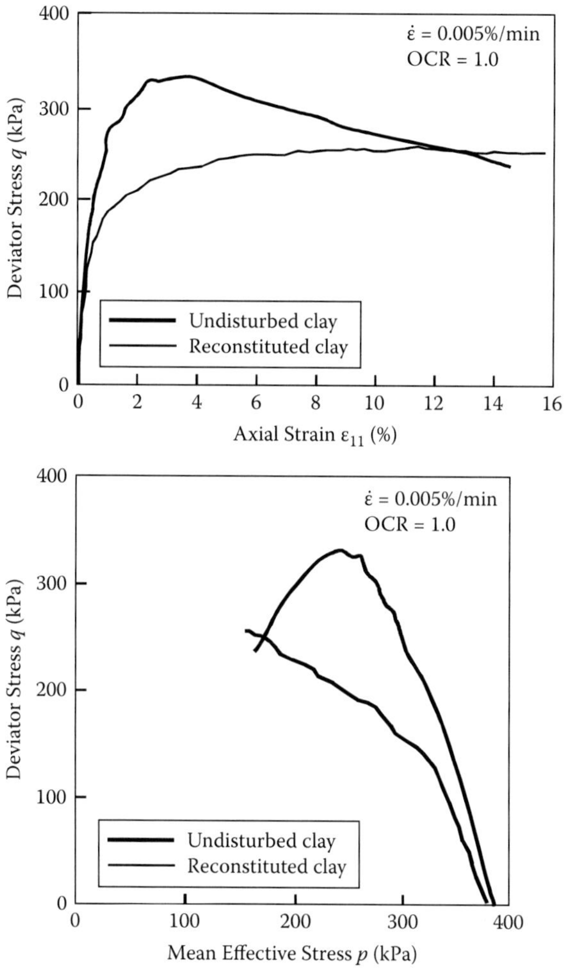

Figure 5.25 Experiments of undrained triaxial tests for Kyuhoji clay. (After Yashima, A., Shigematsu, H., Oka, F., and Nagaya, J., 1999, J. *Geotech. Eng., JSCE*, 624 (III-47): 217–229.)

clay exhibits larger strength and the deviator stress decreases after the peak stress, as seen in Figure 5.25. Figure 5.26 shows the results of simulations that give compressive strain rates under the triaxial stress state. An axial strain rate of $\dot{\varepsilon}_{11} = 0.005\%/\text{min}$ is provided for the calculations. The material parameters used in the simulations are shown in Table 5.2. The structural parameter, σ'_{maf}, is set to be 280 kPa for the undisturbed clay, and β is set to be 10 for the undisturbed clay and 0 for reconstituted clay. $\beta = 0$ provides the original model, which does not describe structural changes. The values for

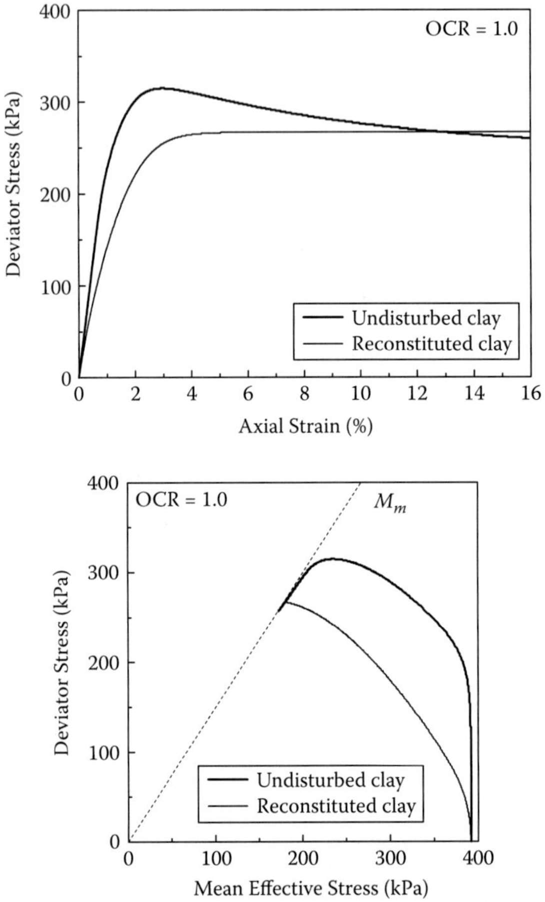

Figure 5.26 Simulations of undrained triaxial tests for Kyuhoji clay.

C contain $\sigma_{myi}^{(s)}$ concerning the degree of initial structures in the derivation. Since the degree of the initial structure of the reconstituted clay is thought to be lower than that of the undisturbed clay, a larger value for C is given for the reconstituted clay. Figures 5.25 and 5.26 confirm that the proposed model can describe the difference in behavior between the highly structured soil and the lowly structured soil.

5.8.2 Osaka Holocene clay

The model of Equation (5.61) has been used for various types of clay. Figures 5.27 and 5.28 show the stress–strain relations and the stress paths of Osaka Holocene clay in triaxial compression tests under undrained

Table 5.2 Material parameters for Kyuhoji clay

	Undisturbed	*Reconstituted*
Elastic shear modulus G_0	8333 (kPa)	6330 (kPa)
Compression index λ		0.327
Swelling index κ		0.028
Initial void ratio e_0	1.41	1.02
Compression yield stress	$\sigma'_{mbi} = \sigma'_{mai}$ = 392 (kPa)	
Stress ratio at failure M^*_m		1.22
Viscoplastic parameter m'		21.5
Viscoplastic parameter C	4.5×10^{-11} (1/s)	2.5×10^{-8} (1/s)
Structural parameter σ'_{maf}	280 (MPa)	392 (MPa)
Structural parameter β	10.0	0

conditions (Mirjalili 2010; Mirjalili et al. 2011). The material parameters used in the analysis are listed in Table 5.3. The figures indicate that the behavior is rate dependent and the stress–strain relations exhibit extensive strain softening in the overconsolidated range. Figures 5.27 and 5.28 illustrate a comparison between the simulated results by the proposed model and the experimental results. From the figures, it is seen that the model can well simulate the rate-dependent behavior of soft clay as well as strain softening.

5.8.3 Elastoviscoplastic model based on modified Cam-clay model

When we use the modified Cam-clay model (Roscoe and Burland 1968), the following yield and plastic potential functions are recommended, since if we use the original yield functions by Roscoe and Burland, we encounter complexity in determining the viscoplastic parameter, m' (Karunawardena 2007; Kimoto et al. 2011):

$$f_y = \ln\left(\frac{\bar{\eta}_\chi^{*2} + M^{*2} - \chi^{*2}}{M^{*2} - \chi^{*2}}\right) + \ln\left(\frac{\sigma'_m}{\sigma'^{(s)}_{my}}\right) = 0 \tag{5.73}$$

$$f_p = \ln\left(\frac{\bar{\eta}_\chi^{*2} + M^{*2} - \chi^{*2}}{M^{*2} - \chi^{*2}}\right) + \ln\left(\frac{\sigma'_m}{\sigma'_{mp}}\right) = 0 \tag{5.74}$$

$$\bar{\eta}^*_x = \sqrt{(\eta^*_{ij} - \chi^*_{ij})(\eta^*_{ij} - \chi^*_{ij})}, \quad \chi^* = \sqrt{\chi^*_{ij}\chi^*_{ij}}, \tag{5.75}$$

where χ^*_{ij} is a kinematical hardening parameter.

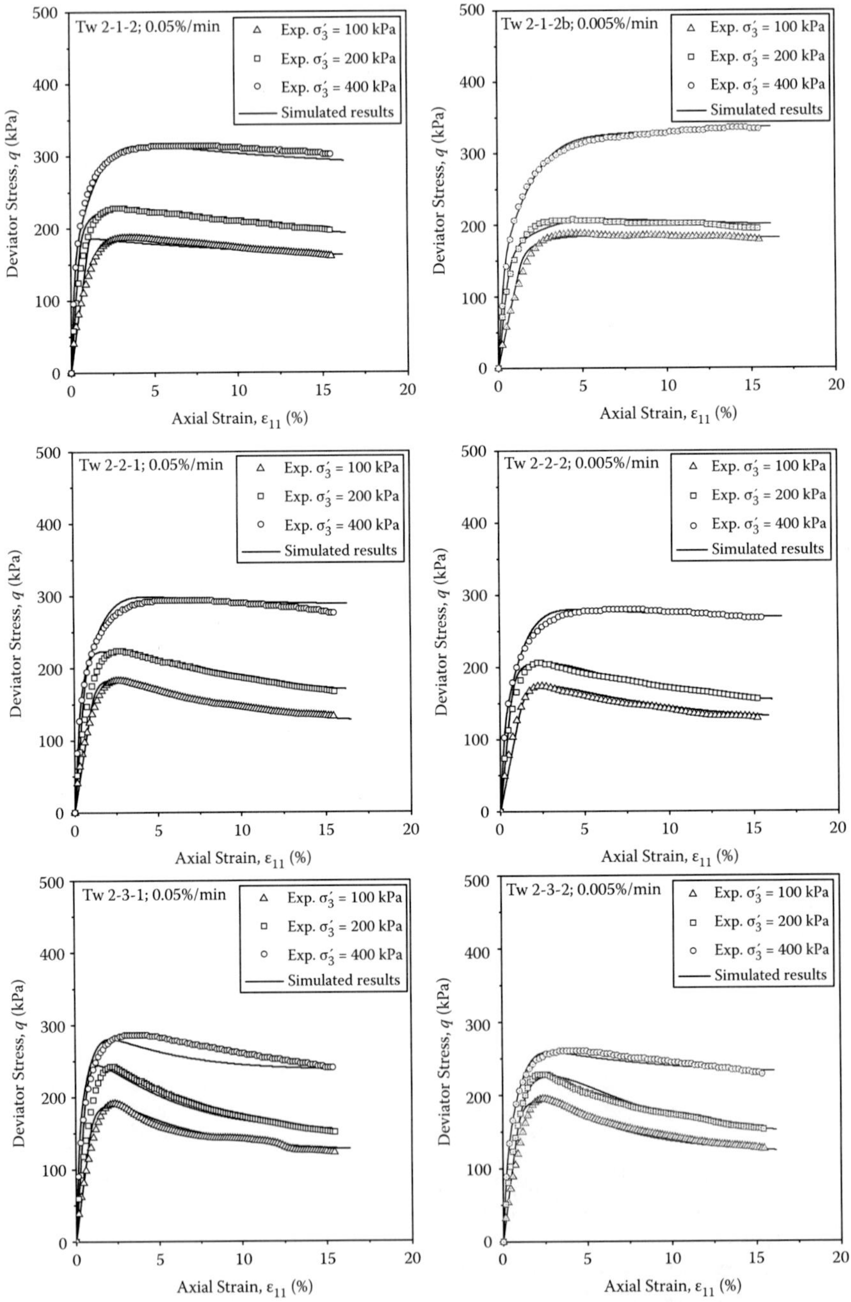

Figure 5.27 Experimental and simulated stress–strain relations of Torishima clay.

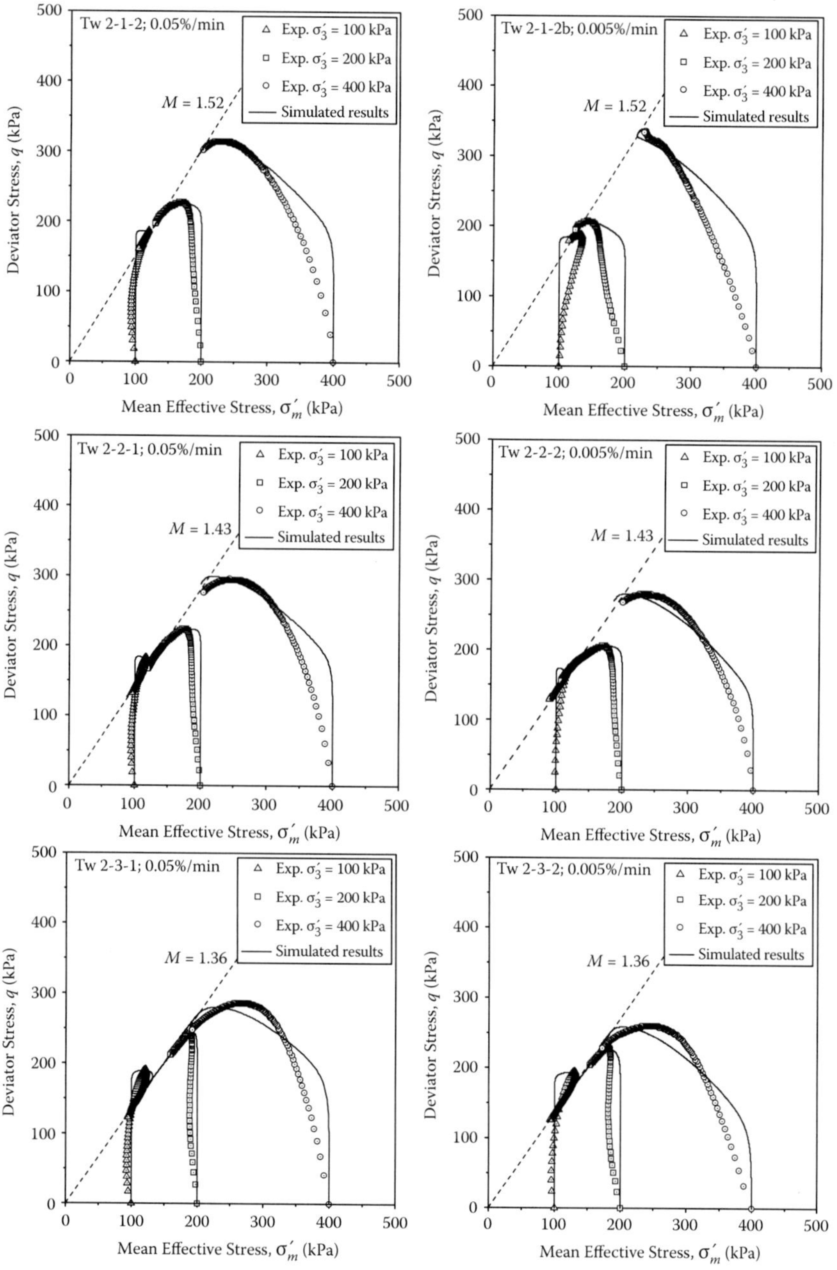

Figure 5.28 Experimental and simulated stress paths of Torishima clay.

Table 5.3 Material parameters of soft clay layer Ac2

	Ac2-U	*Ac2-M*	*Ac2-L*
Depth (m)	11.0 – 15.0	15.0 – 19.0	19.0 – 23.0
	Tw 2-1-2	Tw 2-2-1	Tw 2-3-1
Test no.	Tw 2-1-2b	Tw 2-2-2	Tw 2-3-2
Initial void ratio, e_0	1.25	1.65	1.42
Poisson's ratio, ν	0.3	0.3	0.3
Initial elastic shear modulus[(1)] G_0 *(kPa)*	3759	3927	5993
Compression index, λ	0.341	0.593	0.652
Swelling index, κ	0.019	0.027	0.014
Stress ratio at failure, M^*_{mc}	1.24	1.18	1.12
Viscoplastic parameter, m'	24.68	28.2	21.15
Viscoplastic parameter, C_l *(1/s)*	3.83×10^{-11}	1.85×10^{-11}	8.99×10^{-11}
Viscoplastic parameter, C_2 *(1/s)*	3.83×10^{-11}	1.85×10^{-11}	8.99×10^{-11}
Structural parameter, $n = \sigma'_{maf}/\sigma'_{mai}$	0.83	0.67	0.60
Structural parameter, β	10	15	20
Strain dependent parameter, a	20	10	10
Strain dependent parameter, r	0.4	0.4	0.4

[(1)] Based on the mean effective stress at the depth of the specimen.

It is interesting to note that the shape of the yield surface and the viscoplastic potential are the same as the original modified Cam-clay model.

5.9 CYCLIC ELASTOVISCOPLASTIC MODEL

Several constitutive models have been proposed to describe the rheological behavior of clay under static loading conditions, as mentioned in Section 5.3. However, few viscoplastic constitutive models are available for analysis under dynamic loading conditions (e.g., Oka 1992; Modaressi and Laloui 1997; Oka et al. 2004; Maleki and Cambou 2009). Oka (1992) developed a cyclic elastoviscoplastic constitutive model for clay based on nonlinear kinematic hardening rules (Armstrong and Frederick 1966; Chaboche and Rousselier 1983). Later, Oka, Kodaka, et al. (2003) proposed a cyclic viscoelastic–viscoplastic model by incorporating the viscoelastic feature into the constitutive equations, in which the behavior of clay can be described not only in the range of middle to high levels of strain, but also in the range of low levels of strain. The models have been successively applied to the dynamic analysis of the ground during earthquakes, considering liquefaction (Oka, Uzuoka, et al. 2003), as shown in Chapter 9.

5.9.1 Cyclic elastoviscoplastic model based on nonlinear kinematical hardening rule

Oka (1992) proposed a cyclic elastoviscoplastic model by adopting the nonlinear kinematical hardening rule by Armstrong and Frederick (1966).

The static yield function is given by

$$f_y = \bar{\eta}^*_{(x)} + \tilde{M}^* \ln \frac{\sigma'_m}{\sigma'^{(s)}_{my}} = 0 \tag{5.76}$$

$$\bar{\eta}^*_{(x)} = \sqrt{(\eta^*_{ij} - x^*_{ij})(\eta^*_{ij} - x^*_{ij})} \tag{5.77}$$

where χ^*_{ij} is a hardening parameter described by

$$dx^*_{ij} = B^*(A^* de^{vp}_{ij} - x^*_{ij} d\gamma^{vp}) \quad d\gamma^{vp} = \sqrt{de^{vp}_{ij} de^{vp}_{ij}} \tag{5.78}$$

in which B^* and $A^* = M^*_f$ are kinematical hardening parameters.

The plastic potential function is given by

$$f_p = \bar{\eta}^*_{(x)} + \tilde{M}^* \ln \frac{\sigma'_m}{\sigma'_{mp}} = 0 \tag{5.79}$$

where $\tilde{M}^*$ is constant in normally consolidated clay and varies in the overconsolidated region as

$$\tilde{M}^* = \begin{cases} M^*_m & : \quad f_b \geq 0 \\ -\dfrac{\sqrt{\eta^*_{ij}\eta^*_{ij}}}{\ln(\sigma'_m/\sigma'_{mc})} & : \quad f_b < 0 \end{cases} \tag{5.80}$$

in which σ'_{mc} is the value at the intersection point between the overconsolidated boundary surface and mean effective stress axis.

$$\sigma'_{mc} = \sigma'_{mb} \exp \frac{\sqrt{\eta^*_{ij(0)}\eta^*_{ij(0)}}}{M^*_m} \tag{5.81}$$

For the isotropic stress conditions, σ'_{mc} corresponds to σ'_{mb}.

The flow rule is now generalized as

$$\dot{\varepsilon}^{vp}_{ij} = C_{ijkl} \langle \Phi_1(f_y) \rangle \frac{\partial f_p}{\partial \sigma'_{kl}} \tag{5.82}$$

$$C_{ijkl} = a\delta_{ij}\delta_{kl} + b(\delta_{ik}\delta_{jl} + \delta_{jk}\delta_{il})$$

Finally, the deviatoric viscoplastic strain rate tensor, $\dot{e}_{ij}^{vp}$, and the viscoplastic volumetric strain rate, $\dot{\varepsilon}_{kk}^{vp}$, are obtained as

$$\dot{e}_{ij}^{vp} = C_{01} \exp(m_0' \bar{\eta}_\chi) \frac{\eta_{ij}^* - \chi_{ij}^*}{\bar{\eta}_\chi} \tag{5.83}$$

$$\dot{\varepsilon}_{kk}^{vp} = C_{02} \exp(m_0' \bar{\eta}_\chi) \left\{ \tilde{M}^* - \frac{\eta_{mn}^* (\eta_{mn}^* - \chi_{mn}^*)}{\bar{\eta}_\chi} \right\} \tag{5.84}$$

$$C_{01} = 2b, \; C_{02} = 3a + 2b$$

5.9.2 Cyclic elastoviscoplastic model considering structural degradation

Despite the ability of the above model (Oka, 1992) to explain the deformation characteristics under cyclic loading conditions, the effect of the structural degradation of clay particles was disregarded. Taking into account structural degradation and microstructural changes, a cyclic elastoviscoplastic model is developed based on the nonlinear kinematic hardening rules for changes in both the stress ratio and the mean effective stress. In addition, the kinematic hardening rule for changes in viscoplastic volumetric strain is generalized to predict the behavior during the cyclic loading process (Shahbodagh 2011).

5.9.2.1 *Static yield function*

The static yield function is obtained by considering the nonlinear kinematic hardening rule for changes in the stress ratio, the mean effective stress, and the viscoplastic volumetric strain, as

$$f_y = \bar{\eta}_\chi^* + \tilde{M}^* \left(\ln \frac{\sigma'_{mk}}{\sigma'^{(s)}_{my}} + \left| \ln \frac{\sigma'_m}{\sigma'_{mk}} - y_m^* \right| \right) = 0 \tag{5.85}$$

$$\bar{\eta}_\chi^* = \left\{ \left(\eta_{ij}^* - \chi_{ij}^* \right) \left(\eta_{ij}^* - \chi_{ij}^* \right) \right\}^{\frac{1}{2}} \tag{5.86}$$

in which σ'_{mk} is the unit value of the mean effective stress, y_m^* is the scalar kinematic hardening parameter, and $\sigma'^{(s)}_{my}$ denotes the static hardening parameter. χ_{ij}^* is the so-called back stress parameter, which has the same dimensions as the stress ratio, η_{ij}^*.

Incorporating strain softening into the structural degradation, the hardening rule for $\sigma'^{(s)}_{my}$ can be expressed as

$$\sigma'^{(s)}_{my} = \frac{\sigma'_{maf} + (\sigma'_{mai} - \sigma'_{maf})\exp(-\beta z)}{\sigma'_{mai}} \sigma'^{(s)}_{myi} \tag{5.87}$$

5.9.2.2 *Viscoplastic potential function*

In the same manner as for the static yield function, the viscoplastic potential function, f_p, is given by

$$f_p = \bar{\eta}^*_\chi + \tilde{M}^* \left(\ln \frac{\sigma'_{mk}}{\sigma'_{mp}} + \left| \ln \frac{\sigma'_m}{\sigma'_{mk}} - y^*_m \right| \right) = 0 \tag{5.88}$$

The dilatancy coefficient, $\tilde{M}^*$, is defined separately for the NC region and the OC region as

$$\tilde{M}^* = \begin{cases} M^*_m & : \text{NC region} \\ (\sigma^*_m / \sigma'_{mb}) \ M^*_m & : \text{OC region} \end{cases} \tag{5.89}$$

where σ'_m is the mean effective stress at the intersection of the surface and the axis of anisotropic consolidation

$$\bar{\eta}^*_{(0)} + M^*_m \ln \frac{\sigma'_m}{\sigma^*_m} = 0 \tag{5.90}$$

and σ^*_m is given by

$$\sigma^*_m = \sigma'_m \ \exp\left(\frac{\bar{\eta}^*_{(0)}}{M^*_m} \right) \tag{5.91}$$

Figure 5.29 illustrates the overconsolidation boundary surface ($f_b = 0$), the static yield surface ($f_y = 0$), and the viscoplastic potential function ($f_p = 0$) for the isotropically consolidated soil in the OC and NC regions, respectively.

5.9.2.3 *Kinematic hardening rules*

The evolution equation for the nonlinear kinematic hardening parameter, χ^*_{ij}, is given by

$$d\chi^*_{ij} = B^* \left(A^* de^{vp}_{ij} - \chi^*_{ij} d\gamma^{vp} \right) \tag{5.92}$$

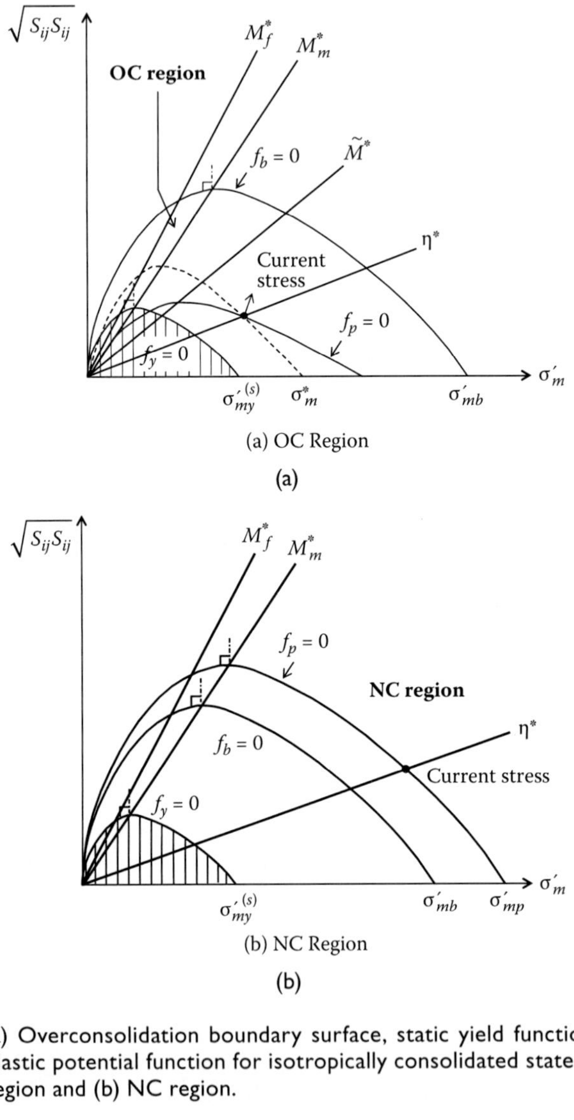

(a) OC Region

(a)

(b) NC Region

(b)

Figure 5.29 (a) Overconsolidation boundary surface, static yield function, and viscoplastic potential function for isotropically consolidated state in the (a) OC region and (b) NC region.

where A^* and B^* are material parameters, de_{ij}^{vp} is the viscoplastic deviatoric strain increment tensor, and $d\gamma^{vp} = \sqrt{de_{ij}^{vp} de_{ij}^{vp}}$ is the viscoplastic shear strain increment tensor. A^* is related to the stress ratio at failure, namely, $A^* = M_f^*$, and B^* is proposed to be dependent on the viscoplastic shear strain as

$$B^* = (B_{\max}^* - B_1^*)\exp\left(-C_f \gamma_{(n)}^{vp^*}\right) + B_1^* \tag{5.93}$$

in which B_1^* is the lower boundary of B^*, C_f is the parameter controlling the amount of reduction, and $\gamma_{(n)}^{vp^*}$ is the accumulated value of viscoplastic shear strain between two sequential stress reversal points in the previous cycle. $B_{\max}^*$ is the maximum value of parameter B^*, which is defined following the proposed method by Oka et al. (1999) as

$$B_{\max}^* = \begin{cases} B_0^* & \text{:Before reaching failure line} \\ \dfrac{B_0^*}{1+\gamma_{(n)\max}^{vp^*}/\gamma_{(n)r}^{vp^*}} & \text{:After reaching failure line} \end{cases} \tag{5.94}$$

where B_0^* is the initial value of B^*, $\gamma_{(n)\max}^{vp^*}$ is the maximum value of $\gamma_{(n)}^{vp^*}$ in past cycles, and $\gamma_{(n)r}^{vp^*}$ is the viscoplastic reference strain.

In order to improve the predicted results under cyclic loading conditions, a scalar nonlinear kinematic hardening parameter, y_m^*, can be decomposed into two parts as

$$y_m^* = y_{m1}^* + y_{m2}^* \tag{5.95}$$

Then, two hardening parameters follow the evolutional equations as

$$dy_{m1}^* = B_2^*\left(A_2^* d\varepsilon_v^{vp} - y_{m1}^*\left|d\varepsilon_v^{vp}\right|\right) \tag{5.96}$$

$$y_{m2}^* = B_3^* d\varepsilon_v^{vp} \tag{5.97}$$

where A_2^*, B_2^* and B_3^* are material parameters, and $d\varepsilon_v^{vp}$ is the increment in the viscoplastic volumetric strain tensor. The values for A_2^*, B_2^* and B_3^* are determined by a data-adjusting method from the laboratory test data.

5.9.2.4 *Strain-dependent shear modulus*

The degradation of the elastic shear modulus from the beginning of loading can be expressed by its dependency on accumulated viscoplastic shear strain γ^{vp} as

$$G = \frac{G_0}{\left(1+\alpha\left(\gamma^{vp}\right)^r\right)}\sqrt{\frac{\sigma_m'}{\sigma_{m0}'}} \tag{5.98}$$

where r and α are the strain-dependent parameters, which can be determined from the laboratory test results. In this study, based on the experimental results, $r = 0.4$ is chosen, and $\gamma_{(n)}^{vp^*} = \int_{(n)} d\gamma^{vp^*}$ is the accumulated plastic shear strain between two sequential stress reversal points in the nth cycle.

5.9.2.5 Viscoplastic flow rule

Based on the overstress type of viscoplastic theory first adopted by Perzyna (1963), viscoplastic strain rate tensor $\dot{\varepsilon}_{ij}^{vp}$ is defined as

$$\dot{\varepsilon}_{ij}^{vp} = C'_{ijkl}\langle\Phi(f_y)\rangle\frac{\partial f_p}{\partial\sigma'_{kl}} \tag{5.99}$$

$$\langle\Phi\, f_y)\rangle = \begin{cases} (f_y) & : f_y > 0 \\ 0 & : f_y \leq 0 \end{cases} \tag{5.100}$$

where $\langle\rangle$ are Macaulay's brackets, and $\Phi\,(f_y)$ is the rate-sensitive material function and is determined as

$$C'_{ijkl}\Phi\,(f_y) = C'_{ijkl}\sigma'_m\ \exp\left\{m'\tilde{M}^*\ln\frac{\sigma'_{mai}}{\sigma'^{(s)}_{myi}}\right\}$$

$$\times\exp\left\{m'\left(\bar{\eta}^*_\chi + \tilde{M}^*\left(\ln\frac{\sigma'_{mk}}{\sigma'_{ma}} + \left|\ln\frac{\sigma'_m}{\sigma'_{mk}} - y^*_{m1}\right|\right)\right)\right\}$$

$$= C_{ijkl}\sigma'_m\ \exp\left\{m'\left(\bar{\eta}^*_\chi + \tilde{M}^*\left(\ln\frac{\sigma'_{mk}}{\sigma'_{ma}} + \left|\ln\frac{\sigma'_m}{\sigma'_{mk}} - y^*_{m1}\right|\right)\right)\right\} \tag{5.101}$$

in which m' is the viscoplastic parameter and

$$C_{ijkl} = a\delta_{ij}\delta_{kl} + b(\delta_{ik}\delta_{jl} + \delta_{il}\delta_{jk}) \tag{5.102}$$

where C'_{ijkl} is a fourth-order isotropic tensor. a and b in Equation (5.99) are material constants.

Finally, by combining Equations (5.99), (5.101), and (5.102), the viscoplastic deviatoric strain rate, $\dot{e}_{ij}^{vp}$, and the viscoplastic volumetric strain rate, $\dot{\varepsilon}_{ij}^{vp}$, can be expressed as

$$\dot{e}_{ij}^{vp} = C_1\ \exp\left\{m'\left(\bar{\eta}^*_\chi + \tilde{M}^*\left(\ln\frac{\sigma'_{mk}}{\sigma'_{ma}} + \left|\ln\frac{\sigma'_m}{\sigma'_{mk}} - y^*_{m1}\right|\right)\right)\right\}\frac{\eta^*_{ij} - \chi^*_{ij}}{\bar{\eta}^*_\chi} \tag{5.103}$$

$$\dot{\varepsilon}_{kk}^{vp} = C_2 \exp\left\{m'\left(\bar{\eta}_\chi^* + \tilde{M}^*\left(\ln\frac{\sigma'_{mk}}{\sigma'_{ma}} + \left|\ln\frac{\sigma'_m}{\sigma'_{mk}} - y_{m1}^*\right|\right)\right)\right\}$$
$$\times\left\{\tilde{M}^* \frac{\ln\frac{\sigma'_m}{\sigma'_{mk}} - y_{m1}^*}{\left|\ln\frac{\sigma'_m}{\sigma'_{mk}} - y_{m1}^*\right|} - \frac{\eta_{mn}^*(\eta_{mn}^* - \chi_{mn}^*)}{\bar{\eta}_\chi^*}\right\} \tag{5.104}$$

where $C_1 = 2b\exp\{m'\tilde{M}^* \ln\frac{\sigma'_{mai}}{\sigma'^{(s)}_{myi}}\}$ and $C_2 = (3a+2b)\exp\{m'\tilde{M}^* \ln\frac{\sigma'_{mai}}{\sigma'^{(s)}_{myi}}\}$ are the viscoplastic parameters for the deviatoric and the volumetric strain components, respectively.

To verify the performance of the proposed model, the modeling of the soft clay samples is carried out by integration of the constitutive equations under undrained triaxial conditions. Figure 5.30 shows the simulated results for Nakanoshima soft clay from Osaka, Japan in which B_3^* is zero. The material parameters are listed in Table 5.4. It is seen that the mean effective stress decreases with an increase in loading cycles.

As for the shape of cyclic stress–strain curve, we can improve it if we use the following formula for the elastic shear modulus:

$$G = \frac{G_0(e)}{\left(1+\alpha\left(\gamma_{(n)}^{vp*}\right)^r\right)}\sqrt{\frac{\sigma'_m}{\sigma'_{m0}}} \tag{5.105}$$

where $\gamma_{(n)}^{vp*} = \int_{(n)} d\gamma^{vp*}$ is the accumulated plastic shear strain between two sequential stress reversal points in the nth cycle.

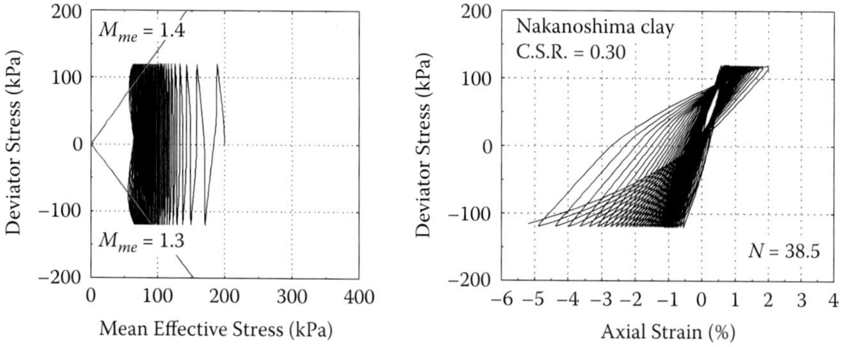

Figure 5.30 Simulated results of cyclic behavior of an undrained triaxial test.

Table 5.4 Material parameters of Nakanoshima clay

Initial void ratio, e_0	1.373	1.573
Compression index, λ	0.2173	
Swelling index, κ	0.0344	
Initial elastic shear modulus (kPa), G_0	22670	
Stress ratio at compression, M^*_{mc}	1.143	
Stress ratio at extension, M^*_{me}	1.061	
Viscoplastic parameter, m'	22.7	
Viscoplastic parameter (1/s), C_1	1.00×10^{-5}	
Viscoplastic parameter (1/s), C_2	3.30×10^{-6}	
Structural parameter, $n = \sigma'_{maf} / \sigma'_{mai}$ [1]	0.325	
Structural parameter, β	3.7	5.7
Hardening parameter, B^*_0	105	
Hardening parameter, B^*_1	1.0	
Hardening parameter, C_f	5	75
Reference value of viscoplastic strain (%), $\gamma^{vp*}_{(n)r}$	3.5	
Strain-dependent parameter, a	10	1
Scalar hardening parameter, A^*_2	5.1	
Scalar hardening parameter, B^*_2	2.6	
Overconsolidation ratio, OCR	1.0	

[1] $\sigma'_{mai} = \sigma'_{m0} \times OCR$

REFERENCES

Aboshi, H. 1973. An experimental investigation on the similitude in the consolidation of a soft clay including the secondary creep settlement, 8th ICSMFE, Moscow, 4(3):88.

Adachi, T., and Oka, F. 1982. Constitutive equations for normally consolidated clay based on elasto-viscoplasticity, *Soils and Foundations*, 22 (4):55–70.

Adachi, T., and Oka, F. 1984. Constitutive equations for sands and overconsolidated clays and assigned works for sand, Constitutive Relations for soils, Results of the International Workshop on Constitutive Relations for Soils, G. Gudehus, F. Darve, and I. Vardoulakis, eds., 6–8 September 1982, Grenoble, Balkema, 141–157.

Adachi, T., and Oka, F. 1995. An elasto-plastic constitutive model for soft rock with strain softening, *Int. J. Numerical Anal. Methods in Geomechanics*, 19: 233–247.

Adachi, T., Oka, F., Hirata, T., Hashimoto, T., Nagaya, J., Mimura, M., and Pradhan, T.B.S. 1995. Stress-strain and yielding characteristics of eastern Osaka clay, *Soils and Foundations*, 35(3):1–13.

Adachi, T., and Oka, F., and Koike, M. 2003. An elasto-viscoplastic constitutive model with strain softening for a sedimentary soft rock, Proceedings of International Workshop on Prediction and Simulation Methods in Geomechanics, Athens, F. Oka, I. Vardoulakis, A. Murakami, and T. Kodaka eds., JGS, 5–8.

Adachi, T., Oka, F., and Koike, M. 2005. An elasto-viscoplastic constitutive model with strain-softening for soft sedimentary rocks, *Soils and Foundations*, 45(2):125–133.

Adachi, T., Oka, F., and Mimura, M. 1987a. An elasto-viscoplastic theory for clay failure, Proc. 8th Asian Reg. Conf. Soil Mech. Found. Eng., Kyoto, 1, JSSMFE, 5–8.

Adachi, T., Oka, F., and Mimura, M. 1987b. Mathematical structure of an overstress elasto-viscoplastic model for clay, *Soils and Foundations*, 27(3):31–42.

Adachi, T., Oka, F., and Mimura, M. 1996. Modeling aspects associated with time dependent behavior of soils, Measuring and modeling time dependent soil behavior, Geotechnical Special Publication No. 61 ASCE, T.C. Sheahan and V. Kaliakin, eds., 61–95.

Adachi, T., Oka, F., and Poorooshasb, H.B. 1990. A constitutive model for frozen sand, *J. Energy Resour. Tech., ASME*, 112:208–212.

Adachi, T., and Okano, M. 1974. A constitutive equation for normally consolidated clay, *Soils and Foundations*, 14(4):55–73.

Akai, K., Adachi, T., and Ando, N. 1974. Stress-strain-time relationship of saturated clay, Proc. JSCE, 225:53–61 (In Japanese).

Armstrong, P.J., and Frederick, C.O. 1966. A mathematical representation of the multiaxial Bauschinger effect, C. E. G. B. Report RD/B/N 731.

Arulanandan, K.-Shen, C.-K., and Young, R. B. 1971. Undrained creep behavior of a coastal organic clay, *Géotechnique*, 21(4): 359–375.

Asaoka, A., Nakano, M., Noda, T., and Kaneda, K. 2000. Delayed compression/consolidation of natural clay due to degradation of soil structure, *Soils and Foundations*, 40(3):75–85.

Aubry, D., Kodaissi, E., and Meimon, Y. 1985. A viscoplastic constitutive equation for clays including a damage law, Proceedings 5th International Conference on Numerical Methods in Geomechanics, Nagoya, T. Kawamoto and Y. Ichikawa, eds., Balkema, 1:421–428.

Baladi, G.Y., and Rohani, B. 1984. Development of an elasto-viscoplastic constitutive relationship for earth materials, In *Mechanics of Engineering*, C.S. Desai and R.H. Gallerger, eds., John Wiley & Sons, New York, 23–43.

Bjerrum, L. 1967. Engineering geology of Norwegian normally consolidated marine clays as related to settlements of buildings, *Géotechnique*, 17(2):81–118.

Bjerrum, L. 1973. Problems of soil mechanics and construction on soft clays, state-of-the-art report to Session IV, Proc. 8th ICSMFE, Moscow, 3:111–159.

Chaboche, J.L., and Rousselier, G. 1983. On the plastic and viscoplastic constitutive equations Part I and Part II. *J. Pressure Vessel Technol., Trans. ASME*, 105:153–164.

Dafalias, Y. 1982. Bounding surface elasto-viscoplasticity for particulate cohesive media, Proceedings. of IUTAM Conference on Deformation and Failure of Granular Materials, P.A. Vermeer and H.J. Luger, eds., Balkema, Rotterdam, 97–107.

Di Benedetto, H., Tatsuoka, F. and Ishihara, M. 1997. Time-dependent deformation characteristics of sand and their constitutive modeling, *Soils and Foundations*, 42(2):1–22.

Dragon, A. and Mroz, Z. 1979. A model for plastic creep of rock-like materials accounting for the kinematics of fracture, *Int. J. Rock Mech. Min. Sci. Geomech.*, Abstr., 16:253–259.

Drucker, D.C. 1959. A definition of stable inelastic materials, *J. Appl. Mech., Trans. ASME*, 26:101–106.

Duvaut, G., and Lions, J.L. 1976. *Inequalities in mechanics and physics*, Springer-Verlag, Berlin (Les inequations en mechanique et en physique, Dunod, Paris).

Eyring, H. 1936. Viscosity, plasticity, and diffusion as example of absolute reaction rates, *J. Chem. Phys.*, 4(4):283–291.

Fahey, M. 1992. Shear modulus of cohesionless soil:variation with stress-strain level, *Can. Geotech.* J., 29(1):157–161.

Febro-Cordero, E., and Mesri, G. 1974. Influence of testing conditions on creep behavior of clay, I Report No. FRA-ORD&D-75-29, UILU-ENG-74-2031.

Finn, L., and Shead, D. 1973. Creep and creep rupture of a sensitive clay, 8th ICSMFE, Moscow, 1(1):135–142.

Graham, J., Crooks, J.H.A., and Bell, A.L. 1983. Time effects on the stress-strain behavior of soft marine clays, *Géotechnique*, 33(3):327–340.

Hardin, B.O., and Drnevich, V.P. 1972. Shear modulus and damping in soils: design equation and curves. J. Soil Mech. Foundations, ASCE, 98(SM7):667–692.

Hirst, T.J. 1968. The influences of compositional factors on the stress-strain-time behavior of soils, PhD thesis, University of California, Berkeley.

Hohenemser, K., and Prager, W. 1932. Über die ansätze der mechanik isotroper kontinua, ZAMM, 12:216–226.

Hori, M. 1974. Fundamental study of wave propagation characteristics through soils, PhD thesis, Kyoto University.

Ishihara, K. 1996. *Soil Behavior in Earthquake Geotechnics.* Oxford University Press, 85–152.

Kabbaj, M., Oka, F., Leroueil, S., and Tavenas, F. 1985. Consolidation of natural clays and laboratory testing, Consolidation of Soils: Testing and Evaluation, ASTM Special Technical Publication 892, R.N. Yong and F.C. Townsend, eds., 378–404.

Kabbaj, M., Tavenas, F., and Leroueil, S. 1988. In situ and laboratory stress-strain relationships, *Géotechnique*, 38(1):83–100.

Karunawardena, A. 2007. Consolidation analysis of Sri Lankan peaty clay using elasto-viscoplastic theory, PhD thesis, Kyoto University.

Katona, M.G. 1984. Evaluation of viscoplastic cap model, *J. Geotech. Eng., ASCE*, 110(8):1106–1125.

Kimoto, S. 2002. Constitutive models for geomaterials considering structural changes and anisotoropy, PhD thesis, Kyoto University, Japan.

Kimoto, S., and Oka, F. 2005. An elasto-viscoplastic model for clay considering destructuralization and consolidation analysis of unstable behavior, *Soils and Foundations*, 45(2):29–42.

Kimoto, S., Oka, F., and Higo, Y. 2004. Strain localization analysis of elasto-viscoplastic soil considering structural degradation, *Comput. Meth. Appl. Mech. Eng.*, 193:2845–2866.

Kimoto, S., Oka, F., and Karunawardena, A. 2011. An elasto-viscoplastic model based on the modified Cam-clay model, Proceedings of Annual Meeting of JSCE, 307–308 (In Japanese).

Kimoto, S., Oka, F., Watanabe, T., and Sawada, M. 2007. Improvement of the visco-plastic potential and static yield function for the elasto-viscoplastic constitutive model. Proceedings of 62nd Annual Meeting of JSCE, Hiroshima, 637–638.

Kokusho, T., Yoshida, Y., and Esashi, Y. 1982. Dynamic properties of soft clays for wide strain range, *Soils and Foundations*, 22:1–18.

Kondner, R.L., and Ho, M.M.K. 1965. Viscoelastic response of a cohesive soil in the frequency domain, *Trans. Soc. Rheology*, 9(2):329–342.

Kovacs, W.D., Seed, H.B., and Chan, C.K. 1971. Dynamic modulus and damping ratio for a soft clay, *J. Soil Mech. Foundations, ASCE*, 97(SM1):59–75.

Kutter, B.L., and Sathialinggam, N. 1992. Elastic-viscoplastic modeling of the rate-dependent behaviour of clays, *Géotechnique*, 42(3):427–441.

Leroueil, S. 1988. Recent developments in consolidation of natural clays, Tenth Canadian Geotechnical Colloquium, *Can. Geotech. J.*, 25:85–107.

Leroueil, S., and Hight, D.W. 2003. Behaviour and properties of natural soils and soft rocks, In *Characterization and Engineering Properties of Natural Soils*, T.S. Tan, K.K. Phoon, D.W. Hight, and S. Leroeuil, eds., Swets and Zeitlinger, Lisse, 29–254.

Leroueil, S., Kabbaj, M., Tavenas, F., and Bouchard, R. 1985. Stress-strain-strain rate relation for the compressibility of sensitive natural clays, *Géotechnique*, 36(2):288–290.

Leroueil, S., Samson, L., and Bozozuk, M. 1983. Laboratory and field determination of preconsolidation pressures at Gloucester, *Can. Geotech. J.*, 20(3):477–490.

Leroueil, S., Tavenas, F., Brucy, F., La Rochelle, P., and Roy, M. 1979. Behavior of destructed natural clays, J. Geotech. Eng., ASCE, 115(6):759–778.

Maleki, M. and Cambou, B., 2009. A cyclic elastoplastic-viscoplastic constitutive model for soils, *Geomech. Geoeng.: Int. J.*, 4(3):209–220.

Matsui, T., and Abe, N. 1985. Elasto/viscoplastic constitutive equation of normally consolidated clays based on flow surface theory, Proceedings of 5th International Conference on Numerical Methods in Geomechanics, Nagoya, T. Kawamoto and Y. Ichikawa, eds., Balkema, Rotterdam, 1:407–413.

Mesri, G., and Choi, Y.K. 1979. Excess pore water pressure during consolidation, Proc. 6th Asian Regional Conf. on SMFE, 1:151–154.

Mesri, G., Shahien, M., and Feng, T.W. 1995. Compressiblity parameters during primary consolidation, Proceedings International Symposium on Compression and Consolidation of Clayey Soils, IS-Hiroshima, H. Yoshikuni and O. Kusakabe, eds., Balkema, Rotterdam, 2:1021–1037.

Mirjalili, M. 2010. Numerical analysis of a large-scale levee on soft soil deposits using two-phase finite deformation theory, PhD thesis, Kyoto University.

Mirjalili, M., Shahbodagh Khan, B., Oka, F. and Kimoto, S. 2011. Dynamic Strain Localization Analysis of Elasto-Viscoplastic Soil Using Updated Lagrangian Finite Element Formulation, Proceedings of the 2nd International Symposium on Comutational Geomechanics, S. Pietruzczak and G.N. Pande eds., International Center for Computational Engineering, 203–210.

Mitchell, J.K. 1986. Practical problems from surprising soil behavior, 20th Terzaghi Lecture, *J. Geotech. Eng., ASCE*, 112(3):259–289.

Mitchell, J.K. Campanella, R.G., and Singh, A. 1968. Soil creep as a rate process, *ASCE J. SMFD*, 94(1):231–254.

Modaressi, H., and Laloui, L., 1997. A thermo-viscoplastic constitutive model for clay, *Int. J. Numer. Anal. Meth. Geomech.*, 21:313–335.

Murayama, S., and Shibata, T. 1964. *Flow and stress relaxation of clays,* Proceedings of IUTAM Symposium on Rheology and Soil Mech., J. Kravtchenko and P.M. Sirieys, eds., Springer-Verlag, Berlin, 99–129.

Murayama, S. 1983. Formation of stress-strain-time behavior of soils under deviatoric stress condition, *Soils and Foundations*, 23(2): 43–57.

Nova, R. 1982. A viscoplastic constitutive model for normally consolidated clay. Proceedings of IUTAM Conference on Deformation and Failure of Granular Materials, P.A. Vermeer and H.J. Luger, eds., Blakema, Rotterdam, 287–295.

Ogisako, E., Nishio, S., Denda, A., Oka, F., and Kimoto, S. 2007. Simulation of triaxial compression tests on soil samples obtained from seabed ground in deep sea by elasto-viscoplastic constitutive equation, Proceedings of the Seventh ISOPE Ocean Mining & Gas Hydrates Symposium, Lisbon, Portugal, Chung and Komai, eds., ISOPE, 63–68.

Oka, F. 1979. Constitutive theory for solid-fluid mixture and its application to stress wave propagation through cohesive soil, *Trans. JSCE*, 272:117–130.

Oka, F. 1981. Prediction of time-dependent behaviour of clay, Proceedings of 10th International Conference on Soil Mechanics and Foundation Engineering, Stockholm, Balkema, 1:215–218.

Oka, F. 1982. Constitutive equations for granular materials in cyclic loadings, Proceedings of IUTAM Conference on Deformation and Failure of Granular Materials, Delft, P.A. Vermeer and H.J. Luger eds., Balkema, 297–306.

Oka, F. 1985. Elasto/viscoplastic constitutive equations with memory and internal variables, *Computers and Geotechnics*, 1:59–69.

Oka, F. 1992. A cyclic elasto-viscoplastic constitutive model for clay based on the non-linear-hardening rule, Proceedings of 4th International Symposium on Numerical Models in Geomechanics, G.N. Pande and S. Pietruszczak, eds., Swansea, UK, Balkema, Rotterdam, 1:105–114.

Oka, F., and Adachi, T. 1985. A constitutive equation of geologic materials with memory, Proceedings of 5th International Conference on Numerical Methods in Geomechanics, Balkema, 1:293–300.

Oka, F., Adachi, T., and Yashima, A. 1994. Instability of an elasto-viscoplastic constitutive model for clay and strain localization, *Mech. Mater.*, 18:119–129.

Oka, F., Adachi, T., and Yashima, A. 1995. A strain localization analysis of clay using a strain softening viscoplastic model, *Int. J. Plasticity*, 11, 5:523–545.

Oka, F., Feng, H., Kimoto, S., Kodaka, T., and Suzuki, H. 2008. A numerical simulation of triaxial test of unsaturated soil at constant water and air content by using an elasto-viscoplastic model, In Unsaturated Soils: Advances in Geo-Engineering, D.G. Toll, C.E. Augrade, D. Gallipoli, and S.J. Wheeler, eds., Taylor and Francis Group, London, 735–741.

Oka, F., Higo, Y., and Kimoto, S. 2002. Effect of dilatancy on the strain localization of water-saturated elasto-viscoplastic soil, *Int. J. Solids Struct.*, 39:3625–3647.

Oka, F., Kimoto, S., and Nguyen, Q.H. 2011. Elasto-viscoplastic constitutive modeling of soft sedimentary rocks, Proceedings of 15th European Conference on Soil Mechanics and Geotechnical Engineering, A. Anagnostopoulos, M. Pachakis, and C. Tsatsanifos, eds., IOS press, Amsterdam, 1:563–568.

Oka, F., Kodaka, T., and Kim, Y.-S. 2003a. A cyclic viscoelastic-viscoplastic constitutive model for clay and liquefaction analysis of multi-layered ground, *Int. J. Numer. Analy. Meth. Geomech.*, 28:131–179.

Oka, F., Kodaka, T., and Kim, Y.-S. 2004. A cyclic viscoelastic-viscoplastic constitutive model for clay and liquefaction analysis of multi-layered ground, *Int. J. Numer. Anal. Meth. Geomech.*, 28(2):131–179.

Oka, F., Kodaka, T., Kimoto, S., Ichinose, T., and Higo, Y. 2005. Strain localization of rectangular clay specimens under undrained triaxial compression conditions. Proc. 16th ICSMGE, 2:841–844.

Oka, F., Tavenas, F., and Leroueil, S. 1991. An elasto-viscoplastic FEM analysis of sensitive clay foundation beneath embankment, Proceedings of 7th International Conference on Computer Methods and Advanced in Geomechanics, Cairns, G. Beer, J.R. Booker, and J.P. Carter, eds., Balkema, 2:1023–1028.

Oka, F., Uzuoka, R. Tateishi, A., and Yashima, A. 2003b. A cyclic elasto-plastic model for sand and its application to liquefaction analysis, Constitutive Modeling of Geomaterials, selected contributions from the Frank L. DiMaggio Symposium, Inelastic Behavior Committee Engineering Mechanics Division, ASCE, CRC Press, 75–99.

Oka, F., Yashima, A., Tateishi, A., Taguchi, Y., and Yamashita, S. 1999. A cyclic elasto-plastic constitutive model for sand considering a plastic-strain dependence of the shear modulus, *Géotechnique*, 49(5):661–680.

Peirce, D., Shih, C.F., and Needleman, A. 1984. A tangent modulus method for rate dependent solids, *Comput. Struct.*, 18(5):845–887.

Perzyna, P. 1963. The constitutive equations for work hardening and rate sensitive plastic materials, Proc. Vibrational Problems, Warsaw, 4(3):281–290.

Phillips, A., and Wu, H.-C. 1973. A theory of viscoplasticity, Int. *J. Solids Struct.*, 9:15–30.

Potts, D.M., and Zdravkovic, L. 1999. *Finite Element Analysis in Geotechnical Engineering: Theory*, Thomas Telford, London.

Richardson, A.M. Jr., and Whitman, R. 1963. Effect of the strain rate upon undrained shear resistance, *Géotechnique*, 13(4):310–324.

Roscoe, K.H., and Burland, J.B. 1968. On the generalised stress-strain behaviour of "wet" clay, In *Engineering Plasticity*, J. Heyman and F.A. Leckie, eds., Cambridge University Press, 535–609.

Roscoe, K.H., Schofield, A.N., and Thurairajah, A. 1963. Yielding of clays in states wetter than critical, *Géotechnique*, 13(3):211–240.

Saito, M. 1992. Jissho Doshitsu Kogaku, Gihodo Syuppan Co. Ltd., 156 (In Japanese).

Saito, M., and Uezawa, H. 1961. Failure of soil due to creep, 5th ICSMFE, Paris, 1:315–318.

Sällfors, G. 1975. Preconsolidation pressure of soft high-plastic clays, PhD thesis, Chalmers University of Technology, Gothenburg, Sweden.

Sawada, K., Yashima, A., and Oka, F. 2001. Numerical analysis of deformation of saturated clay based on Cosserat type elasto-viscoplastic model, *J. Soc. Mater. Sci. Japan*, 50(6):585–592 (In Japanese).

Seed, H.B., Wong, R.T., Idriss, I.M., and Tokimatsu, K. 1986. Moduli and damping factors for dynamics analyses of cohesionless soils, *J. Geotech. Eng.*, 112(11):1016–1032.

Sekiguchi, H. 1977. Rheological characteristics of clays, Proc. 9th ICSMFE, Tokyo, 1:289–292.

Sekiguchi, H., and Ohta, H. 1977. Induced anisotropy and time dependency in Clays, Proc. Speciality Session 9 9th ICSMFE, Tokyo, JSSMFE: 229–238.

Shahbodagh, K. B. 2011. Large deformation dynamic analysis method for partially saturated elasto-viscoplastic soils, PhD thesis, Kyoto University.

Shen, C.-K., Arulanandan, K., and Smith, W.S. 1973. Secondary consolidation and strength of a clay, *J. Soil Mech. Foundation Div.*, Proc. ASCE, 99 (SM 1):95–110.

Shigematsu, H. 2002. Study on the microstructure and mechanical behavior of natural sedimentary soils with aging effect, Doctoral thesis, Gifu University, Japan (In Japanese).

Singh, A., and Mitchell, J.K. 1968. General stress-strain-time function for soils, *J. Soil Mech. Foundat. Eng., ASCE*, 94, SM1:21–46.

Singh, A., and Mitchell, J.K. 1969. Creep potential and creep rupture of soils, Proc. 7th ICSMFE, Mexico, 379–384.

Suklje, L. 1957. The analysis of the consolidation process by the isotaches method, Proc. 4th ICSMFE, London, 1:200–206.

Suklje, L. 1969. *Rheological Aspects of Soil Mechanics*, Wiley-Interscience, London.

Terzaghi, K. 1944. Ends and means in soil mechanics, *Eng. J. Can.*, 27:608–613.

Vaid, Y. P., and Campanella, M. 1977. Time-dependent behavior of undisturbed clay, J. Geotech. Eng. Div., Proc. ASCE, 103 (GT7):693–709.

Walker, L.K. 1969. Secondary compression in the shear of clays, J. Soil Mech. Foundat. Div., Proc. ASCE, 95 (SM1):167–188.

Wang, G.X., and Kuwano, J. 1999. Modeling of strain dependency of shear modulus and damping of clayey sand, *Soil Dynamics Earthquake Eng.*, 18:463–471.

Yashima, A., Shigematsu, H., Oka, F., and Nagaya, J. 1999. Mechanical behavior and micro-structure of Osaka upper-most Pleistocene marine clay, *J. Geotech.* Eng., JSCE, 624 (III-47):217–229 (In Japanese).

Yin, J.-H., and Graham, J. 1999. Elasto visco-plastic modelling of the time-dependent stress-strain behavior of soils. *Can. Geotech. J.*, 36(4):736–745.

Yin, Z.Y., and Karstunen, M. 2008. Influence of anisotropy, destructuration and viscosity on the behavior of an embankment on soft clay, Proceedings of the 12th IACMAG, 1Goa, India, D.N. Singh, ed., 4729–4735.

Yong, R.N., and Japp, R.D. 1969. Stress-strain behavior of clays in dynamic compression, Vibrational effects on earthquakes on soil and foundations, ASTM, STP, 450:233–262.

Zienkiewicz, O.C., Humpheson, C., and Lewis, R.W. 1975. Associated and non-associated viscoplasticity and plasticity in soil mechanics, *Géotechnique*, 25(4):671–689.

Chapter 6

Virtual work theorem and finite element method

In the present chapter, the virtual work theorem is reviewed and the finite element method for multiphase geomaterials is presented.

6.1 VIRTUAL WORK THEOREM

6.1.1 Boundary value problem

Finding the displacement and the stress fields that satisfy the constitutive equations, such as the elastoplastic model, the strain–displacement relations, the equations of motion, and the initial and the boundary conditions, is called an elastic–plastic boundary value problem.

When governing equations such as the equations of motion are formulated in an incremental form, the boundary value problem is called an incremental boundary value problem and the incremental formulation is usually applied to nonlinear problems.

- Equilibrium equations—In body B, neglecting the acceleration term, the following equations hold:

$$\sigma_{ji,j} + \bar{F}_i = 0 \tag{6.1}$$

where $\sigma_{ij}(i, j = 1,2,3)$ are the components of the symmetric stress tensor assuming the equilibrium of the moment $(\sigma_{ij} = \sigma_{ji})$. $\bar{F}_i$ is a body force and subscript i denotes $\frac{\partial}{\partial x_i}$.

- Compatibility conditions—In body B,

$$\varepsilon_{ij} = \frac{1}{2}(u_{i,j} + u_{j,i}) \tag{6.2}$$

where ε_{ij} is the strain tensor and u_i is the displacement vector.

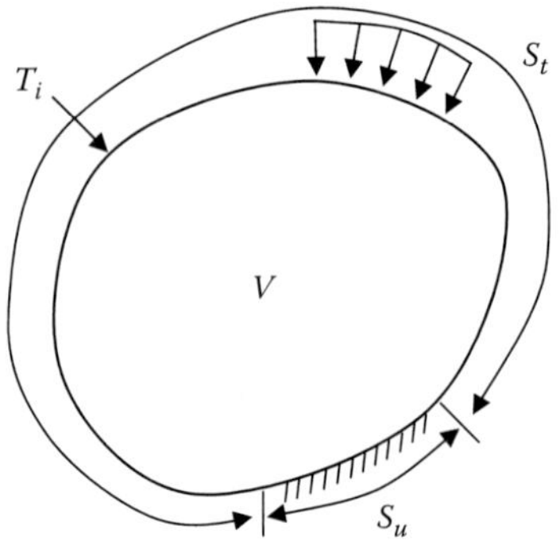

Figure 6.1 Body B and boundary conditions.

- Boundary conditions—As shown in Figure 6.1, boundary S can be divided into two parts, namely, stress traction boundary, S_t; and displacement boundary, S_u.

$$S = S_t + S_u \tag{6.3}$$

  - Stress traction boundary conditions—For S_t

$$T_i = \sigma_{ji} n_j = \overline{T}_i \tag{6.4}$$

where n_i is a component of the unit outward normal vector on the boundary and T_i is the component of the stress traction vector.

  - Displacement boundary conditions—For S_u

$$u_i = \overline{u}_i \tag{6.5}$$

- Constitutive equations (stress-strain relations)—For linear elastic bodies, the model can be expressed as

$$\sigma_{ij} = E_{ijkl} \varepsilon_{kl} \tag{6.6}$$

where E_{ijkl} is a fourth-order elastic coefficient tensor.

The displacements and the displacement rates that satisfy the boundary conditions are called kinematically admissible displacements or displacement rate fields, and the strain and the strain rate fields that satisfy the compatibility conditions and the aforementioned boundary conditions are called kinematically admissible strain and strain rate fields. The stress fields that satisfy the stress traction boundary conditions are called statically admissible.

When the equilibrium Equation (6.1) at time t holds and Equations (6.1) to (6.4) hold between time t and time $t + dt$, the boundary value problem is called an incremental boundary problem.

For incremental boundary value problems, Equations (6.1) to (6.6) become as follows:

$$\dot{\sigma}_{ji,j} + \dot{\bar{F}}_i = 0 \tag{6.7}$$

where the super dot denotes the time differentiation and $\dot{\sigma}_{ij}(i, j = 1,2,3)$ is the stress rate tensor.

$$\dot{\varepsilon}_{ij} = \frac{1}{2}(\dot{u}_{i,j} + \dot{u}_{j,i}) \tag{6.8}$$

where $\dot{\varepsilon}_{ij}$ is the stain rate tensor and $\dot{u}_i$ is the displacement rate vector.

On stress traction boundary, S_t,

$$\dot{\sigma}_{ji} n_j = \dot{\bar{T}}_i \tag{6.9}$$

where n_i is the unit outward normal vector on the boundary and $\dot{T}_i$ is the stress rate vector.

On displacement rate boundary, S_u,

$$\dot{u}_i = \dot{\bar{u}}_i \tag{6.10}$$

$$\dot{\sigma}_{ij} = E_{ijkl}\dot{\varepsilon}_{kl} \tag{6.11}$$

6.1.2 Virtual work theorem

The weak form of the equilibrium equations, with respect to the kinematically admissible displacements and strain fields for continuously differentiable body B, are derived next. On displacement boundary, S_u, the displacement is set to be $\bar{u}_i$. Herein, for such arbitrary kinematically admissible strain and displacement fields as ε^{**} and u_i^{**}, respectively, we consider the weak form and assume σ_{ij} and u_i^{**} are single-valued and continuous in B.

Since boundary S is divided into two parts, namely, stress boundary, S_p, and displacement boundary, S_u, we have

$$S = S_u + S_p$$
$$\int_V (\sigma_{ij,j} + \bar{F}_i) u_i^{**} dv + \lambda \int_{Sp} (\bar{T}_i - \sigma_{ij} n_j) u_i^{**} ds = 0 \tag{6.12}$$

where λ is an arbitrary scalar.

Using the Gauss theorem, we obtain

$$\int_V \sigma_{ij,i} u_j^{**} dv = \int_S \sigma_{ij} u_j^{**} n_i ds - \int_v \sigma_{ij} u_{j,i}^{**} dv \tag{6.13}$$

Considering the symmetry of the stress tensor,

$$\int_V \bar{F}_i u_i^{**} dv + \int_{Su} \sigma_{ij} u_j^{**} n_i ds - \int_V \sigma_{ij} u_{i,j}^{**} dv + \lambda \int_{Sp} \bar{T}_j u_j^{**} ds$$
$$+ (1-\lambda) \int_{St} \sigma_{ij} n_j u_i^{**} ds = 0 \tag{6.14}$$

If we set $\lambda = 1$ and $u_j^{**} = \bar{u}_j$ on S_u, Equation (6.14) becomes

$$\int_v \sigma_{ij} \varepsilon_{ij}^{**} dv = \int_{St} \bar{T}_i u_i^{**} ds + \int_{Su} \sigma_{ij} \bar{u}_j n_i ds + \int_v \bar{F}_i u_i^{**} dv \tag{6.15}$$

Conversely, if Equation (6.15) holds, the boundary conditions and the equilibrium equations can be derived. Hence, we can solve Equation (6.15) instead of solving the equilibrium equations with the boundary conditions.

Now for the other kinematically admissible strain field, ε^{***}, and the displacement field, u_i^{***}, we can consider a similar weak form. Taking the difference in the weak form and Equation (6.15), we get a new weak form, namely,

$$\int_v \sigma_{ij} \hat{\varepsilon}_{ij}^{**} dv = \int_{St} \bar{T}_i \hat{u}_i^{**} ds + \int_v \bar{F}_i \hat{u}_i^{**} dv \tag{6.16}$$

where $\hat{u}_i^{**} = u_i^{**} - u_i^{***}$ and $u_i = \bar{u}_i$ on S_u are the additional conditions. Equation (6.1) includes the first-order spatial derivative of the stress, but Equation (6.15) does not.

As mentioned earlier, Equation (6.10) is called a weak form since the order of the differentiability required has been weakened. It is worth noting that the virtual work theorem does not depend on the constitutive model.

When the displacement rate may be discontinuous on Γ, the virtual work theorem for incremental boundary value problems is expressed by

$$\int_{v-\Gamma} \dot{\sigma}_{ij}\dot{\varepsilon}_{ij}^{**}dv = \int_{v-\Gamma} \dot{\bar{F}}_i\dot{u}_i^{**}dv + \int_{St} \dot{\bar{T}}_i\dot{u}_i^{**}ds + \int_{Su} \dot{\sigma}_{ij}\bar{\dot{u}}_i n_j ds + \int_{\Gamma} \dot{\sigma}_{ij}n_j[\dot{u}_j^{**}]ds \quad (6.17)$$

where $\dot{\varepsilon}_{ij}^{**}$ is the strain rate tensor and $\dot{u}_i^{**}$ is the velocity. The brackets [ ] denote the discontinuity across Γ.

The equation equivalent to Equation (6.16) is given by

$$\int_{v-\Gamma} \dot{\sigma}_{ij}\hat{\dot{\varepsilon}}_{ij}^{**}dv = \int_{v-\Gamma} \dot{\bar{F}}_i\hat{\dot{u}}_i^{**}dv + \int_{St} \dot{\bar{T}}_i\hat{\dot{u}}_i^{**}ds + \int_{\Gamma} \dot{\sigma}_{ij}n_j[\hat{\dot{u}}_j^{**}] \quad (6.18)$$

When discontinuity Γ does not exist, we have

$$\int_{v} \dot{\sigma}_{ij}\hat{\dot{\varepsilon}}_{ij}^{**}dv = \int_{St} \dot{\bar{T}}_i\hat{\dot{u}}_i^{**}ds + \int_{v} \dot{\bar{F}}_i\hat{\dot{u}}_i^{**} \quad (6.19)$$

where $\hat{\dot{u}}_i^{**} = \dot{u}_i^{**} - \dot{u}_i^{**'}$ and the supplementary conditions are $\dot{u}_i = \bar{\dot{u}}_i$ on S_u.

6.2 FINITE ELEMENT METHOD

6.2.1 Discretization of equilibrium equation

The finite element method is an approximation analysis method used to solve boundary value problems based on variational principles, the weighted residual method, the Galerkin method, and so on. In the finite element method, the body is divided into a finite number of elements. If the compatibility conditions and the completeness conditions are satisfied in the formulation of the finite element method, the solution converges to an accurate solution when the size of the element becomes small.

The formulation given later is for a finite element method; it is based on the virtual work theorem and is called the displacement method, in which the displacements are taken as unknowns. In the displacement method, the displacements or displacement increments in the element are interpolated by the values at nodal points. By solving the equations for all of the elements under the boundary conditions, the displacement increments at the nodal points are obtained for the incremental problems. The stain increments are then calculated by the displacement increments, and the stress is calculated using constitutive equations. Thereafter, the calculation moves to the next step.

Let $\Delta\varepsilon_{ij}$ denote the strain increment tensor, Δu_i the displacement increment vector, $\Delta\hat{\varepsilon}_{ij}$ the virtual strain increment tensor, $\Delta\hat{u}_i$ the virtual displacement increment vector, Δu_w the pore water pressure in the element, and $\Delta\sigma'$ the effective stress increment tensor.

The virtual work theorem for a rate type of formulation is expressed in matrix form as

$$\int_V \{\Delta\hat{\varepsilon}\}^T\{\Delta\sigma\}dV (= \int_V \{\Delta\hat{\varepsilon}\}^T\{\Delta\sigma'\}dV + \int_V \{\Delta\hat{\varepsilon}\}^T\{\Delta u_W\}dV)$$

$$= \int_V \{\Delta\hat{u}\}^T\{\Delta F_b\}dV + \int_S \{\Delta\hat{u}\}^T\{\Delta T_s\}dS \tag{6.20}$$

In the preceding equation, the strain increment vector, the total stress vector, the effective stress vector, the displacement increment vector, the pore pressure vector, the body force vector, and the surface force vector for two-dimensional problems are as follows:

Total increment stress vector: $\{\Delta\sigma\}$; $\{\Delta\sigma\}^T = \{\Delta\sigma_{xx}, \Delta\sigma_{yy}, \Delta\sigma_{xy}\}$
Effective increment stress vector: $\{\Delta\sigma'\}$; $\{\Delta\sigma'\}^T = \{\Delta\sigma'_{xx}, \Delta\sigma'_{yy}, \Delta\sigma'_{xy}\}$
Displacement increment vector: $\{\Delta u\}$; $\{\Delta u\}^T = \{\Delta u_x, \Delta u_y\}$, $\{\Delta\hat{u}\}$; $\{\Delta\hat{u}\}^T = \{\Delta\hat{u}_x, \Delta\hat{u}_y\}$
Strain increment vector: $\{\Delta\varepsilon\}$; $\{\Delta\varepsilon\}^T = \{\Delta\varepsilon_{xx}, \Delta\varepsilon_{yy}, \Delta\gamma_{xy}\}$
Pore pressure increment vector: $\{\Delta u_W\}$; $\{\Delta u_W\}^T = \{\Delta u_W, \Delta u_W, 0\}$
Body force increment vector: $\{\Delta F_b\}$
Surface force increment vector: $\{\Delta T_s\}$

The displacement increment $\{\Delta u\}$ at a point in the element is approximated by the interpolation function as

$$\{\Delta u\} = [N]\{\Delta u^*\}, \{\Delta\hat{u}\} = [N]\{\Delta\hat{u}^*\} \tag{6.21}$$

where $[N]$ is the interpolation function, and $\{\Delta\hat{u}\} = [N]\{\Delta\hat{u}^*\}$ and $\{\Delta u^*\}$ are the nodal increment displacement vectors.

The strain increment vector, $\{\Delta\varepsilon\}$, and volumetric strain, $\Delta\varepsilon_v$, of the element are given by

$$\{\Delta\varepsilon\} = [B]\{\Delta u^*\}, \quad \{\Delta\hat{\varepsilon}\} = [B]\{\Delta\hat{u}^*\} \tag{6.22}$$

$$\{\Delta\hat{\varepsilon}_v\} = [B_v]^T\{\Delta\hat{u}^*\}, \quad \Delta\hat{\varepsilon}_v = [B_v]^T\{\Delta\hat{u}^*\} \tag{6.23}$$

where $[B]$ is a strain-nodal displacement matrix and $[B_v]^T$ is a volumetric strain-nodal displacement matrix.

Using Equations (6.21) and (6.22), Equation (6.20) becomes

$$\int_V \{\Delta\hat{u}^*\}^T[B]^T\{\Delta\sigma'\}dV + \int_V \{\Delta\hat{\varepsilon}\}^T\{\Delta u_W\}dV$$
$$= \int_V \{\Delta\hat{u}^*\}^T[N]^T\{\Delta F_b\}dV + \int_S \{\Delta\hat{u}^*\}^T[N]^T\{\Delta T_s\}dS \tag{6.24}$$

In view of Equation (6.23), the second term on the right-hand side of the preceding equation becomes

$$\int_V \{\Delta\hat{\varepsilon}\}^T\{\Delta u_W\}dV = \int_V \{\Delta\hat{\varepsilon}_{xx}, \Delta\hat{\varepsilon}_{yy}, \Delta\hat{\gamma}_{xy}\}\begin{Bmatrix} \Delta u_W \\ \Delta u_W \\ 0 \end{Bmatrix} dV$$
$$= \Delta u_W \int_V \{\Delta\hat{\varepsilon}_{xx} + \Delta\hat{\varepsilon}_{yy}\}dV = \Delta u_w \int_V \Delta\hat{\varepsilon}_v dV \tag{6.25}$$
$$= \Delta u_w \int_V [B_v]^T\{\Delta\hat{u}^*\}dV$$

Let the constitutive model be written as

$$\{\Delta\sigma'\} = [D]\{\Delta\varepsilon\} - \{\Delta\sigma^*\}, \qquad \{\Delta\sigma^*\} = [D]\{\Delta\varepsilon^{vp}\} \tag{6.26}$$

where $[D]$ is the elastic coefficient matrix, $\{\Delta\sigma^*\}$ is the viscoplastic relaxation stress vector, and $\{\Delta\varepsilon^{vp}\}$ is the viscoplastic strain increment vector.

As to the estimation method for the viscoplastic increment vector, there are two methods, namely, the explicit scheme and the implicit scheme.

Substituting Equation (6.26) into Equation (6.24) yields

$$\int_V \{\Delta\hat{u}^*\}^T [B]^T ([D]\{\Delta\varepsilon\} - \{\Delta\sigma^*\}) dV + \Delta u_W \int_V \{\Delta\hat{u}^*\}^T [B_v] dV$$
$$= \int_V \{\Delta\hat{u}^*\}^T [N]^T \{\Delta F_b\} dV + \int_S \{\Delta\hat{u}^*\}^T [N]^T \{\Delta T_S\} dS \tag{6.27}$$

Since $\{\Delta\hat{u}^*\}^T$ is arbitrary, except for the boundary where the displacements are prescribed, by substituting Equation (6.22), we have

$$\int_V [B]^T [D][B]\{\Delta u^*\} dV + \{K_V\}\Delta u_W = \int_V [B]^T \{\Delta\sigma^*\} dV$$
$$+ \int_V [N]^T \{\Delta F_b\} dV + \int_S [N]^T \{\Delta T_s\} dS$$

$$\{K_v\} = \int_V [B_v] dV \tag{6.29}$$

For solving Equation (6.28), we need to discretize the equations of motion for the liquid phase.

For simplicity, let us assume zero pore water pressure, $\Delta u_W = 0$. Thus, we have

$$[K]\{\Delta u^*\} = \{\Delta Q\} \tag{6.30}$$

$$[K] = \int_V [B]^T [D][B] dV \tag{6.31}$$

$$\{\Delta Q\} = \int_V [N]^T \{\Delta F_b\} dV + \int_S [N]^T \{\Delta T_s\} dS + \int_V [B]^T \{\Delta\sigma^*\} dV \tag{6.32}$$

where $\{\Delta Q\}$ is the force vector and $[K]$ is the stiffness matrix. $[K]$ is in general nonlinear, and Equation (6.30) has been solved by an iteration method such as Newton–Raphson method (Zienkiewicz 1977).

6.2.2 Discretization of continuity equation

When we solve the solid–water saturated problem, $\Delta u_w \neq 0$ in Equation (6.28), we need to simultaneously solve the equilibrium equations for the pore water. In the following section, the standard formulation for the continuity equation will be presented based on the weighted residual method, subdomain collocation method, and Galerkin method. Herein, the simplest method based on the finite difference method is presented using a finite element grid.

As is described by Equation (2.45), a continuity equation can be derived from the equation of motion disregarding the acceleration term, the compressibility of water, and the mass conservation law as

$$-\frac{k}{\gamma_w}\frac{\partial^2 u_w}{\partial x_i^2} = \frac{\partial \varepsilon_{kk}}{\partial t} \tag{6.33}$$

where k is the permeability coefficient and γ_w is the unit weight of water $\rho^f g$.

In view of Figure 6.2, discretizing Equation (6.33) gives

$$\{K_v\}^T\{\Delta u^*\} = -\beta u_w(t+\Delta t) + \sum_{i=1}^{4} \beta_i u_{wi}(t+\Delta t) \tag{6.34}$$

$$\beta_i = -\frac{k\Delta t b_i}{\gamma_w s_i}, \qquad \beta = \sum_{i=1}^{4} \beta_i \tag{6.35}$$

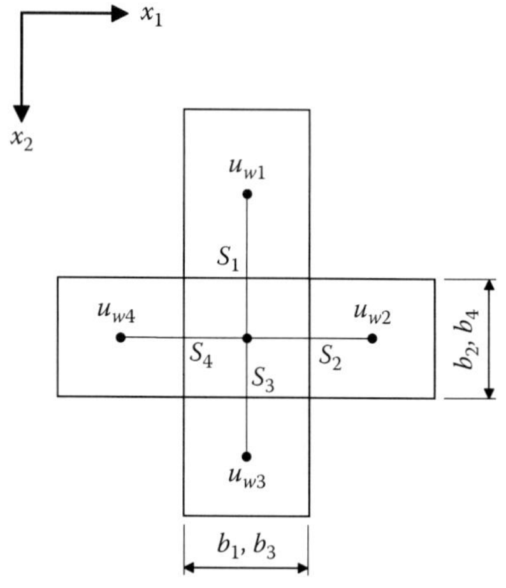

Figure 6.2 Discretization of pore water pressure.

In the derivation of Equation (6.34) from Equation (6.33), it is worth noting that the left-hand side of Equation (6.34) is a change in volume of the element.

Equations (6.28) and (6.34) can be written in matrix form as

$$\begin{bmatrix} [K], & \{K_v\} \\ \{K_v\}^T, & \beta \end{bmatrix} \begin{Bmatrix} \{\Delta u^*\} \\ u_w(t+\Delta t) \end{Bmatrix} = \begin{Bmatrix} \{K_v\}u_w(t)+\{\Delta Q\} \\ \sum_{i=1}^{4} \beta_i u_{wi}(t+\Delta t) \end{Bmatrix} \tag{6.36}$$

where $\Delta u^* = u^*(t+dt) - u^*(t)$.

Solving the aforementioned equations under the appropriate boundary and initial conditions yields the solution of consolidation problems.

6.2.3 Interpolation function

In the finite element method, the body is divided into elements, as shown in Figure 6.3, in which triangular elements are adopted.

The state variable Q in the element is interpolated by the value of Q_i (i is number of nodes) at the nodes of the element:

$$Q = \sum_{i=1}^{n} Q_i N_i \tag{6.37}$$

When the interpolation function of the element is continuous and differentiable to the necessary degree, the element is compatible and this condition is called compatibility. The compatibility condition ensures that the displacement in the elements and the displacements along the edge of the elements are continuous (Bathe 1996; Belytschko et al. 2000).

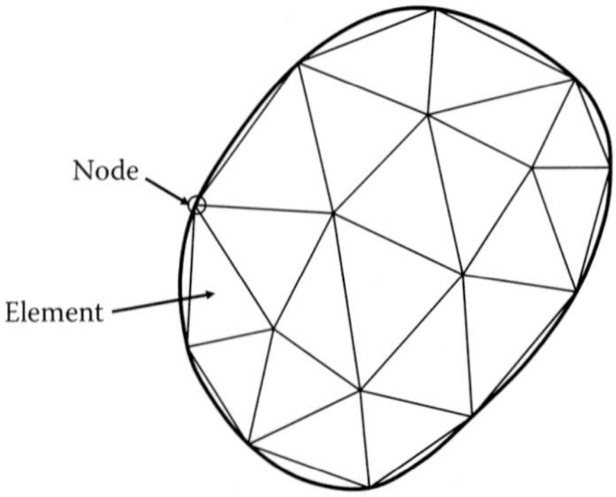

Figure 6.3 Division of the body into finite elements.

The finite element solution converges to the exact solution if the finite element and the mesh are compatible and complete. The completeness is defined by the basis functions, that is, the interpolation functions are complete in the solution space and the reproduction of the constant strain and the rigid body motions. For example, the displacement in the element is approximated by

$$u = \sum_{i=1}^{n} N_i u_i \tag{6.38}$$

If u_i is constant, we have

$$\sum_{i=1}^{n} N_i = 1 \tag{6.39}$$

This condition ensures the completeness and is called the partition of unity.

If the interpolation function or the shape function satisfies the conditions of compatibility and completeness, it is known that the approximated solution converges to an accurate solution.

In the following, we will consider the two-dimensional plane strain problem as an example, namely, a triangular element and a four-node quadrilateral element.

6.2.4 Triangular element

Figure 6.4 shows a triangular element with three nodes, i, j, and k. For plane stress and strain problems, the displacement at a node has two components, namely, u_i is the component in the x direction and v_i is the component in the y direction of node i.

The components of the displacement vector at the three nodes constitute a displacement vector $\{\delta\}$ as

$$\{\delta\}^T = [u_i, v_i, u_j, v_j, u_k, v_k] \tag{6.40}$$

When we assume the linear variation in the displacements of the element, the displacements are expressed by the linear polynomials as

$$u = \alpha_1 + \alpha_2 x + \alpha_3 y \tag{6.41}$$

$$v = \beta_1 + \beta_2 x + \beta_3 y \tag{6.42}$$

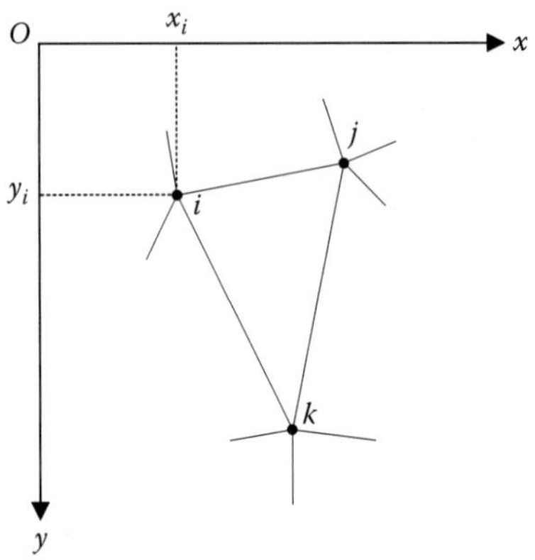

Figure 6.4 Triangular element.

If the displacements at nodes, i,j,k are expressed by (u_i,u_j,u_k), (v_i,v_j,v_k), we have

$$\begin{aligned} u_i &= \alpha_1+\alpha_2 x_i+\alpha_3 y_i, \ v_i = \beta_1+\beta_2 x_i+\beta_3 y_i \\ u_j &= \alpha_1+\alpha_2 x_j+\alpha_3 y_j, \ v_j = \beta_1+\beta_2 x_j+\beta_3 y_j \\ u_k &= \alpha_1+\alpha_2 x_k+\alpha_3 y_k, v_k = \beta_1+\beta_2 x_k+\beta_3 y_k \end{aligned} \tag{6.43}$$

in which x_i are the nodal coordinates.

We can determine six parameters from Equation (6.43) and then we have the following displacement in the x direction by inserting the six parameters:

$$u = \frac{1}{2\Delta}\{(a_i + b_i x + c_i y)u_i + (a_j + b_j x + c_j)u_j + (a_k + b_k x + c_k y)u_k\} \tag{6.44}$$

$$a_i = x_j y_k - x_k y_j, b_i = y_j - y_k, c_i = x_k - x_j \tag{6.45}$$

where

$$2\Delta = \det \begin{vmatrix} 1 & x_i & y_i \\ 1 & x_j & y_j \\ 1 & x_k & y_k \end{vmatrix} \tag{6.46}$$

where Δ is the area of the triangle (i, j, k).

In a similar manner, we have displacement v in the y direction as

$$v = \frac{1}{2\Delta}\{(a_i + b_i x + c_i y)v_i + (a_j + b_j x + c_j y)v_j + (a_k + b_k x + c_k y)v_k\} \qquad (6.47)$$

The strain vector $\{\varepsilon\}$ for the plane problem is given by

$$\{\varepsilon\} = \begin{Bmatrix} \varepsilon_{xx} \\ \varepsilon_{yy} \\ \gamma_{xy} \end{Bmatrix} = \begin{Bmatrix} \dfrac{\partial u}{\partial x} \\ \dfrac{\partial v}{\partial y} \\ \dfrac{\partial u}{\partial y} + \dfrac{\partial v}{\partial x} \end{Bmatrix} \qquad (6.48)$$

where $\gamma_{xy} = 2\varepsilon_{xy}$ and γ_{xy} is the engineering strain.

$\varepsilon_x = \varepsilon_{xx},\ \varepsilon_y = \varepsilon_{yy}.$

$$\{\varepsilon\} = \frac{1}{2\Delta}\begin{bmatrix} b_i & 0 & b_j & 0 & b_k & 0 \\ 0 & c_i & 0 & c_j & 0 & c_k \\ c_i & b_i & c_j & b_j & c_k & b_k \end{bmatrix} \begin{Bmatrix} u_i \\ v_i \\ u_j \\ v_j \\ u_k \\ v_k \end{Bmatrix} = [B]\{\delta\} \qquad (6.49)$$

where $[B]$ is called the $[B]$ matrix.

6.2.5 Isoparametric elements

A shape function of the finite element is defined by the transformation between the master element and the real elements as

$$x_i = N_n(\xi, \eta)x_{in} \qquad (6.50)$$

where x_{in} is the ith coordinate of the node, x_i is the ith coordinate of the point in the element, $N_i(\xi, \eta)$ is a one-to-one mapping of a vector of the nodal coordinates of the parent element onto that of the nodal coordinates of the real element shown, and (ξ, η) are the natural coordinates shown in Figure 6.5.

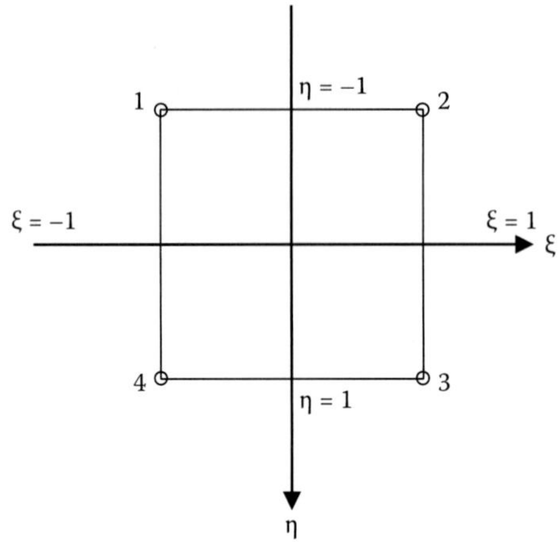

(a)

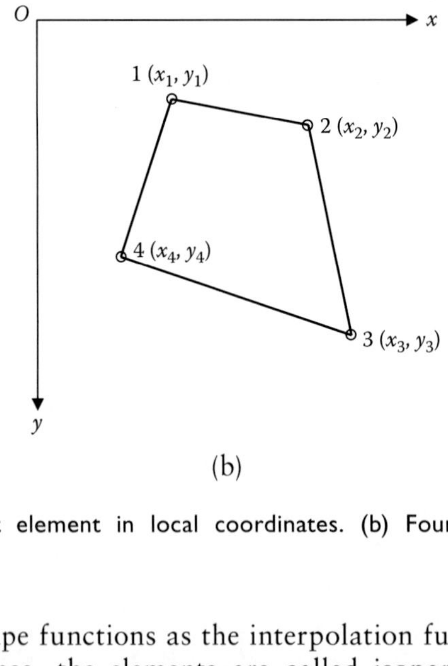

(b)

Figure 6.5 (a) Parent element in local coordinates. (b) Four-noded quadrilateral element.

We can use shape functions as the interpolation functions in the elements. In that case, the elements are called isoparametric elements. Hence, the displacement vector is expressed by the displacement at the nodes.

$$u = \sum_{i=1}^{n} N_i u_i \tag{6.51}$$

Let us consider the parent square element and the real quadrilateral element shown in Equation (6.51), which can be expressed by the components as

$$x = [N_1, N_2, N_3, N_4] \begin{pmatrix} x_1 \\ x_2 \\ x_3 \\ x_4 \end{pmatrix}, \quad y = [N_1, N_2, N_3, N_4] \begin{pmatrix} y_1 \\ y_2 \\ y_3 \\ y_4 \end{pmatrix} \tag{6.52}$$

$$x = \sum_{i=1}^{4} N_i x_i, \quad y = \sum_{i=1}^{4} N_i y_i \tag{6.53}$$

In the finite element method, the variables, such as the displacement, are approximated by the basic interpolation functions as

$$\mathbf{u} = \mathbf{p}^T \mathbf{a}(\mathbf{x}), \; u(\chi) = \{P(\chi)\}\{a\} \tag{6.54}$$

where $\{P(\chi)\}$ is a vector composed of interpolation functions and $\{a\}$ is a coefficient vector.

In general, polynomial functions are adopted for the interpolation functions. For the four-node element, we will use the following interpolation functions:

$$\{P(\chi)\} = \{1, \xi, \eta, \xi\eta\} \tag{6.55}$$

For the parent element shown in Figure 6.5, the vectors are described as

$$P(-1,-1) = P_1 = (1,-1,-1,1) \quad \text{for node No. 1,} \tag{6.56}$$

The displacement at node n is given by

$$\{u_n\} = [P_n]\{a\} \tag{6.57}$$

$$\begin{Bmatrix} u_1 \\ u_2 \\ u_3 \\ u_4 \end{Bmatrix} = \begin{bmatrix} 1 & -1 & -1 & 1 \\ 1 & 1 & -1 & -1 \\ 1 & 1 & 1 & 1 \\ 1 & -1 & 1 & -1 \end{bmatrix} \begin{Bmatrix} a_1 \\ a_2 \\ a_3 \\ a_4 \end{Bmatrix} \tag{6.58}$$

Hence, we have the coefficient vector as

$$\{a\} = [P_n]^{-1}\{u_n\} \tag{6.59}$$

Finally, we have obtained the interpolation functions $\{N(\chi)\}$.

$$\begin{aligned} u(\chi) &= \{P(\chi)\}[P_n]^{-1}\{u_n\} \\ &= \{N(\chi)\}\{u_n\} \end{aligned} \tag{6.60}$$

where $\{N(\chi)\} = \{P(\chi)\}[P_n]^{-1}$.

$$\left[P_n\right]^{-1} = \frac{1}{4}\begin{bmatrix} 1 & 1 & 1 & 1 \\ -1 & 1 & 1 & -1 \\ -1 & -1 & 1 & 1 \\ 1 & -1 & 1 & -1 \end{bmatrix} \tag{6.61}$$

$$\{N(\chi)\} = \{1, \xi, \eta, \xi\eta\}\frac{1}{4}\begin{bmatrix} 1 & 1 & 1 & 1 \\ -1 & 1 & 1 & -1 \\ -1 & -1 & 1 & 1 \\ 1 & -1 & 1 & -1 \end{bmatrix} \tag{6.62}$$

Hence,

$$\begin{aligned} N_1(\chi) &= \frac{1}{4}(1-\xi)(1-\eta), \quad N_2(\chi) = \frac{1}{4}(1+\xi)(1-\eta) \\ N_3(\chi) &= \frac{1}{4}(1+\xi)(1+\eta), \quad N_4(\chi) = \frac{1}{4}(1-\xi)(1+\eta) \end{aligned} \tag{6.63}$$

From the preceding, we have the interpolation functions for the isoparametric elements, also called the shape functions, in Equation (6.51) by

$$N_i = \frac{1}{4}(1+\xi\xi_i)(1+\eta\eta_i) \tag{6.64}$$

$$\left(\begin{matrix} i = 1, 2, 3, 4, & \xi_i = -1, 1, 1, -1 \\ & \eta_i = -1, -1, 1, 1 \end{matrix} \right)$$

Let us consider the quadrilateral isoparametric elements. The displacements in the element are described by the nodal displacements u_i

(displacements in the x direction at node i) and v_i (displacements in the y direction at node i)

$$u = \sum_{i=1}^{4} N_i u_i \ , \quad v = \sum_{i=1}^{4} N_i v_i \tag{6.65}$$

where u and v are the displacements in the x and y directions, respectively.

It is seen that the isoparametric elements satisfy the condition of completeness from Equation (6.39). The isoparametric elements satisfy the aforementioned conditions because the interpolation function is the same as the shape function. If the displacement mode is rigid, displacement u is constant in the element. Equation (6.65) means that the isoparametric elements can describe the rigid body motion.

The other condition for the finite element solutions, which converges to the true solutions, is the condition of compatibility. The compatibility condition is satisfied when the shape function is continuous, the first derivative of that is continuous in the element, and the function is continuous at the boundaries of the elements.

Substituting Equation (6.65) into Equation (6.48), the strain vector is obtained for the plane problem as

$$\{\varepsilon\} = \left\{ \begin{array}{c} \varepsilon_{xx} \\ \varepsilon_{yy} \\ \gamma_{xy} \end{array} \right\} = \begin{bmatrix} \dfrac{\partial [N]}{\partial x} & 0 \\ 0 & \dfrac{\partial [N]}{\partial y} \\ \dfrac{\partial [N]}{\partial y} & \dfrac{\partial [N]}{\partial x} \end{bmatrix} \left\{ \begin{array}{c} \{u\} \\ \{v\} \end{array} \right\} = [B]\{\delta\} \tag{6.66}$$

where $\gamma_{xy} = 2\varepsilon_{xy}$, γ_{xy} is the engineering strain,

$$[N] = [N_1, N_2, N_3, N_4], \ \{u\}^T = \{u_1, u_2, u_3, u_4\}, \ \{v\}^T = \{v_1, v_2, v_3, v_4\} \tag{6.67}$$

We need $\frac{\partial N_i}{\partial x}, \frac{\partial N_i}{\partial y}$ to obtain the $[B]$ matrix.

Since N_i is the function of the natural coordinates (ξ, η) (Figure 6.5), from Equation (6.64), we have

$$\left\{ \begin{array}{c} \dfrac{\partial N_i}{\partial \xi} \\ \dfrac{\partial N_i}{\partial \eta} \end{array} \right\} = \begin{bmatrix} \dfrac{\partial x}{\partial \xi} & \dfrac{\partial y}{\partial \xi} \\ \dfrac{\partial x}{\partial \eta} & \dfrac{\partial y}{\partial \eta} \end{bmatrix} \left\{ \begin{array}{c} \dfrac{\partial N_i}{\partial x} \\ \dfrac{\partial N_i}{\partial y} \end{array} \right\} = [J] \left\{ \begin{array}{c} \dfrac{\partial N_i}{\partial x} \\ \dfrac{\partial N_i}{\partial y} \end{array} \right\} \tag{6.68}$$

From Equations (6.64) and (6.68), Jacobian matrix $[J]$ is given by

$$[J]=\begin{bmatrix} \sum_{j=1}^{4}\frac{\partial N_j}{\partial \xi}x_j & \sum_{j=1}^{4}\frac{\partial N_i}{\partial \xi}y_i \\ \sum_{j=1}^{4}\frac{\partial N_j}{\partial \eta}x_j & \sum_{j=1}^{4}\frac{\partial N_j}{\partial \eta}y_j \end{bmatrix}=\frac{1}{4}\begin{bmatrix} \sum_{j=1}^{4}\xi_j(1+\eta_j\eta)x_j & \sum_{j=1}^{4}\xi_j(1+\eta_j\eta)y_j \\ \sum_{j=1}^{4}\eta_j(1+\xi_j\xi)x_j & \sum_{j=1}^{4}\eta_j(1+\xi_j\xi)y_j \end{bmatrix} \tag{6.69}$$

Then,

$$\begin{Bmatrix} \frac{\partial N_i}{\partial x} \\ \frac{\partial N_i}{\partial y} \end{Bmatrix}=[J]^{-1}\begin{Bmatrix} \frac{\partial N_i}{\partial \xi} \\ \frac{\partial N_i}{\partial \eta} \end{Bmatrix} \tag{6.70}$$

Substituting Equation (6.70) into Equation (6.66), we have

$$\{\varepsilon\}=\begin{bmatrix} \frac{\partial N_1}{\partial x}, & 0, & \frac{\partial N_2}{\partial x}, & 0, & \frac{\partial N_3}{\partial x}, & 0, & \frac{\partial N_4}{\partial x}, & 0 \\ 0, & \frac{\partial N_1}{\partial y}, & 0, & \frac{\partial N_2}{\partial y}, & 0, & \frac{\partial N_3}{\partial y}, & 0, & \frac{\partial N_4}{\partial y} \\ \frac{\partial N_1}{\partial y}, & \frac{\partial N_1}{\partial x}, & \frac{\partial N_2}{\partial y}, & \frac{\partial N_2}{\partial x}, & \frac{\partial N_3}{\partial y}, & \frac{\partial N_3}{\partial x}, & \frac{\partial N_4}{\partial y}, & \frac{\partial N_4}{\partial x} \end{bmatrix}\begin{Bmatrix} u_1 \\ v_1 \\ u_2 \\ v_2 \\ u_3 \\ v_3 \\ u_4 \\ v_4 \end{Bmatrix}=[B]\{\delta\} \tag{6.71}$$

Using the $[B]$ matrix, we have the $[K]$ matrix shown in Equation (6.31) as

$$[K]=\int_V [B]^T[D][B]dV=\int_{-1}^{1}\int_{-1}^{1}[B]^T[D][B]det[J]d\xi d\eta \tag{6.72}$$

$$[B]=[B(\xi,\eta)] \tag{6.73}$$

where (ξ,η) are the natural coordinates used for the shape functions.

If we set $[f(\xi,\eta)]=[B]^T[D][B]det[J]$, we have

$$[K]=\int_{-1}^{1}\int_{-1}^{1}f(\xi,\eta)\,d\xi\,d\eta \tag{6.74}$$

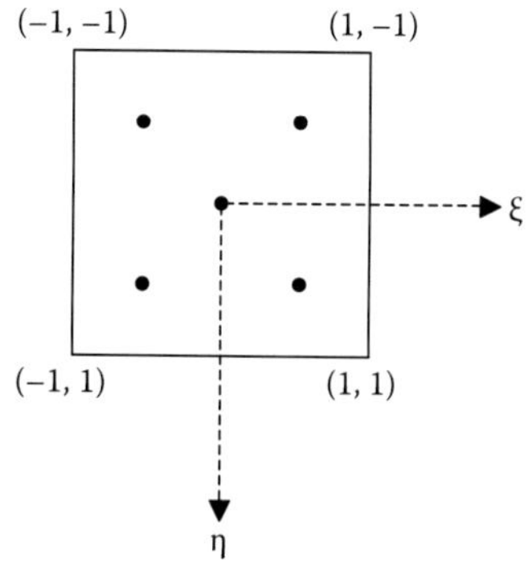

Figure 6.6 Integration points.

In order to perform the integration of Equation (6.74), we can use Gauss–Legendre's numerical as

$$[K]=\int_{-1}^{1}\int_{-1}^{1} f(\xi,\eta)d\xi d\eta=\sum_{p2=1}^{2}\sum_{q2=1}^{2} H_{p2}H_{q2}[f(\xi_{p2},\eta_{p2})] \tag{6.75}$$

For the two-point Gauss quadrature rule for integration, integration points (ξ_{p2},η_{q2}), $p_2,q_2=1,2$ are given in Figure (6.6) as

$$(\xi_{p2},\eta_{q2})=\pm\frac{1}{\sqrt{3}} \tag{6.76}$$

where the weighting coefficients are $(H_{p2},H_{q2})=(1.0,1.0)$.

The volumetric strain is

$$\varepsilon_v=\varepsilon_{xx}+\varepsilon_{yy}=\left[\frac{\partial N_1}{\partial x},\frac{\partial N_1}{\partial y},\frac{\partial N_2}{\partial x},\frac{\partial N_2}{\partial y},\frac{\partial N_3}{\partial x},\frac{\partial N_3}{\partial y},\frac{\partial N_4}{\partial x},\frac{\partial N_4}{\partial y}\right]\{\delta\}=[B_v]^T\{\delta\} \tag{6.77}$$

Then,

$$[K_v]^T=\int_V [B_v]^T dV=\int_{-1}^{1}\int_{-1}^{1}[B_v]^T det[J]d\xi d\eta \tag{6.78}$$

and

$$\int_V [B]^T\{\Delta\sigma^*\}dV=\int_{-1}^{1}\int_{-1}^{1}[B]^T\{\Delta\sigma^*\}det[J]d\xi d\eta \tag{6.79}$$

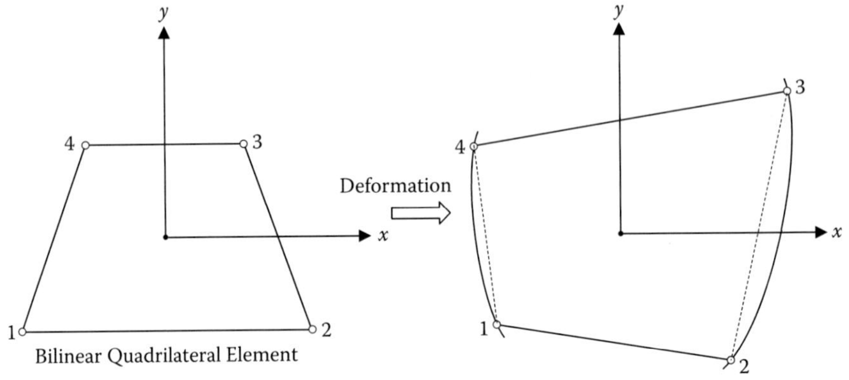

Figure 6.7 Incompatible formation of trapezoidal four-noded quadrilateral element.

If we use bilinear quadrilateral elements as

$$\begin{aligned} u\ (x,y) &= a_1 + a_2 x + a_3 y + a_4 xy \\ v\ (x,y) &= b_1 + b_2 x + b_3 y + b_4 xy \end{aligned} \tag{6.80}$$

we may have incompatible deformation shown in Figure 6.7 (Terada 2008) although the isoparametric elements satisfy the compatibility.

6.3 DYNAMIC PROBLEM

Based on the principle of d'Alembert, the virtual work theorem for a one-phase material becomes

$$\int_V \{\hat{\varepsilon}\}^T \{\sigma\} dV = \int_S \{\hat{u}\}^T \{T_s\} ds + \int_V \{\hat{u}\}^T (\{F_b\} - \rho\{\ddot{u}\}) dV \tag{6.81}$$

where $\{\ddot{u}\}$ is the acceleration vector, $\{F_b\}$ is the body force, $\{\hat{\varepsilon}\}$ is the virtual strain vector, and $\{\hat{u}\}$ is the virtual displacement vector.

Then, from Equations (6.26) and (6.28), we have

$$\begin{aligned} &\int_V [B]^T [D][B]\{\Delta u^*\} dV \\ &= \int [B]^T \{\Delta\sigma^*\} dV + \int_V [N]^T \{F_b\} dV - \int_V [N]^T \rho [N]\{\ddot{u}^*\} dV + \int_S [N]^T \{T_s\} dV \end{aligned} \tag{6.82}$$

and

$$[M]\{\ddot{u}^*\} + [K]\{\Delta u^*\} = \{F_e\} \tag{6.83}$$

$$[M] = \int_V [N]^T \rho [N] dV \tag{6.84}$$

$$[K] = \int_V [B]^T [D][B] dV \tag{6.85}$$

$$\begin{aligned}\{F_e\} = &\int_V [N]^T \{F_b\} dV + \int_S [N]^T \{T_s\} dS \\ &- \int_V [B]^T \{\sigma_{(t)}\} dV + \int_V [B]^T \{\Delta\sigma^*\} dV\end{aligned} \tag{6.86}$$

where $\{\ddot{u}^*\}$ is the acceleration vector at a node and $\{\sigma_{\{t\}}\}$ is a stress at a previous time.

Considering the viscous term, $[C]\{\dot{u}^*\}$, due to the material damping and the energy dispersion, Equation (6.83) becomes

$$[M]\{\ddot{u}^*\} + [C]\{\dot{u}^*\} + [K]\{\Delta u^*\} = \{F_e\} \tag{6.87}$$

6.3.1 Time discretization method

For integrating Equation (6.87), we can use the Newmark β method, linear acceleration method, or Wilson θ method.

6.3.1.1 Linear acceleration method and Wilson θ method

In the Wilson θ method, the velocity and the displacement are approximated by

$$\{\dot{u}_{t+\tau}\} = \{\dot{u}_t\} + \tau[(1-a)\{\ddot{u}_t\} + a\{\ddot{u}_{t+\tau}\}] \tag{6.88}$$

$$\{u_{t+\tau}\} = \{u_t\} + \tau\{\dot{u}_t\} + \frac{\tau^2}{2}((1-b)\{\ddot{u}_t\} + b\{\ddot{u}_{t+\tau}\}) \tag{6.89}$$

where τ, a, b are parameters.

We assume $\tau = \theta\Delta t$, $\theta > 1$ and that the equations of motion hold at time $t + \theta\Delta t$ as

$$[M]\{\ddot{u}^*(t+\theta\Delta t)\} + [C]\{\dot{u}^*(t+\theta\Delta t)\} + [K]\{\Delta u^*(t+\theta\Delta t)\} = \{F_e(t+\theta\Delta t)\} \tag{6.90}$$

From Equations (6.88), (6.89), and (6.90), we have $\ddot{u}_{t+\theta\Delta t}$ and then $\ddot{u}_{t+\Delta t}$ is obtained by the interpolation as

$$\underline{\{\ddot{u}_{t+\Delta t}\}} = \{\ddot{u}\}_t + \frac{1}{\theta}(\{\ddot{u}_{t+\theta\Delta t}\} - \{\ddot{u}_t\}) = \frac{1}{\theta}((\theta-1)\{\ddot{u}_t\} + \{\ddot{u}_{t+\theta\Delta t}\}),\quad 0 \le \theta \le 1 \tag{6.91}$$

The velocity and the displacement are obtained as

$$\underline{\{\dot{u}_{t+\Delta t}\}} = \{\dot{u}_t\} + \Delta t((1-a)\{\ddot{u}_t\} + a\{\ddot{u}_{t+\Delta t}\}) \tag{6.92}$$

$$\underline{\{u_{t+\Delta t}\}} = \{u_t\} + \Delta t\{\dot{u}_t\} + \frac{\Delta t^2}{2}((1-b)\{\ddot{u}_t\} + b\{\ddot{u}_{t+\Delta t}\}) \tag{6.93}$$

$$\theta = 1.4, a = \frac{1}{2}, b = \frac{1}{3}$$

It is known that the Wilson θ method is unconditionally stable when $\theta > 1.37$. However, it may bring about the artificial damping effect. When $\theta = 1.0, a = \frac{1}{2}, b = \frac{1}{3}$, the Wilson method becomes the linear acceleration method.

6.3.1.2 Newmark β method

The extended linear acceleration method is the Newmark β method. Using the constants β and γ, we can calculate the displacement and the velocity as

$$\{u_{t+\Delta t}\} = \{u_t\} + \Delta t\{\dot{u}_t\} + \frac{1}{2}(\Delta t)^2\{\ddot{u}_t\} + \beta(\Delta t)^2(\{\ddot{u}_{t+\Delta t}\} - \{\ddot{u}_t\}) \tag{6.94}$$

$$\{\dot{u}_{t+\Delta t}\} = \{\dot{u}_t\} + \Delta t\{\ddot{u}_t\} + \gamma\Delta t(\{\ddot{u}_{t+\Delta t} - \{\ddot{u}_t\}) \tag{6.95}$$

$$[M]\{\ddot{u}^*(t+\Delta t)\} + [C]\{\dot{u}^*(t+\Delta t)\} + [K]\{\Delta u^*(t+\Delta t)\} = \{F_e(t+\Delta t)\} \tag{6.96}$$

$\gamma = 0.5$ is frequently used and this method is unconditionally stable when $\beta \ge 0.25$. This method is unconditionally stable if $2\beta \ge \gamma \ge 0.5$ for the zero damping system. Even if $\gamma \ge 0.5$ and $\beta \le 1/4$, the system is conditionally stable when $\Delta t \le 1/(\omega_{\max}\sqrt{\gamma/2-\beta})$; $\omega_{\max}$ is the maximum angular frequency.

By simultaneously solving Equations (6.94), (6.95), and (6.96), $\{\ddot{u}_{t+\Delta t}\}$, $\{\dot{u}_{t+\Delta t}\}$, and $\{u_{t+\Delta t}\}$ are obtained.

Rewriting Equation (6.96) by the Newmark β method, we have

$$([M]+\gamma\Delta t[c]+\beta(\Delta t)^2[K])_{t+\Delta t}\{\ddot{u}_{t+\Delta t}\}$$

$$=\{F_e\}_{t+\Delta t}-[C](\{\dot{u}\}+\Delta t\{\ddot{u}\}-\gamma\Delta t\{\ddot{u}\})_t-[K](\{u_t\}+\Delta t\{\dot{u}\}_t+(\frac{1}{2}-\beta)(\Delta t)^2\{\ddot{u}\}_t) \tag{6.97}$$

6.3.1.3 *Central finite difference scheme*

Using the central finite difference scheme, we can get the acceleration and the velocity as

$$\{\ddot{u}_t\}=\frac{1}{(\Delta t)^2}(\{u_{t+\Delta t}\}-2\{u_t\}+\{u_{t-\Delta t}\}) \tag{6.98}$$

$$\{\dot{u}_t\}=\frac{1}{2\Delta t}(\{u_{t+\Delta t}\}-\{u_{t-\Delta t}\}) \tag{6.99}$$

6.3.2 Mass matrix

As has been shown in Equation (6.83), we need mass matrix $[M]$. There are two types of mass matrices, namely, (1) consistent mass matrix method and (2) lumped mass matrix method. The consistent mass matrix, shown in Equation (6.84), is derived rigorously, but the mass matrix includes a nondiagonal component. From a physical point of view, the mass matrix has to be symmetric. For the mass matrix to be symmetric, we can take the average of the masses of the nodes.

The consistent mass is given by Equation (6.84) as $[M]=\int_V [N]^T\rho[N]dV$. The lumped mass matrix can be obtained by equally dividing the mass of the element and distributing it symmetrically. The lumped mass matrix has only diagonal components.

6.4 DYNAMIC ANALYSIS OF WATER-SATURATED SOIL

Let us explain the discretization of the governing equations. In the discretization of the governing equations in space, both the finite element method (FEM) and finite difference method (FDM) are used. For the discretization of the equations of motion of the mixture, FEM is used. On the other hand, for the discretization of the continuity equation of pore fluids, FDM is used in this section for simplified practical use. In the method, the pore water pressure is defined at the center of the element, and the stress and the strain are also calculated at the center of the element with the reduced integration technique. Using this method, it is possible to avoid shear locking under undrained conditions and to reduce the degree of freedom in the computation (Hisada and Noguchi 1995). The Newmark β method is used in the discretization in time. For the accuracy of the numerical analysis, the method has been verified by comparing its results with the analytical solution of BVP for water-saturated elastic media (Oka et al. 1994). Of course we can use FEM for the discretization of the continuity equation. In particular, we need the finite element formulation for the finite deformation analysis (Mirjalili et al. 2011).

6.4.1 Equation of motion

Multiplying Equation (2.32), the arbitrary weighted function δu_i^s (virtual displacement) that satisfies $\delta u_i^s = 0$ on the displacement boundary, we obtain the weak form after integration in domain V:

$$\int_V \left(\rho \ddot{u}_i^s - \frac{\partial \sigma_{ij}}{\partial x_j} - \rho b_i \right) \delta u_i^s dV = 0 \tag{6.100}$$

On the traction boundary S_t, we have

$$\int_{S_t} (\bar{T}_i - \sigma_{ij} n_i) \delta u_i^s \, dS = 0 \tag{6.101}$$

The integration of the second term on the left-hand side of Equation (6.100) leads to the following equation considering the displacement boundary conditions:

$$\int_V \left(\rho \ddot{u}_i^s - \rho b_i \right) \delta u_i^s dV - \int_{S_t} \sigma_{ij} n_j \delta u_i^s dS + \int_V \sigma_{ij} \frac{\partial \delta u_i^s}{\partial x_j} dV = 0 \tag{6.102}$$

where n_j is the outward unit normal vector of surface S_t.

Virtual strain tensor $\delta \varepsilon_{ij}^s$ can be expressed in a similar manner to Equation (2.8):

$$\delta \varepsilon_{ij}^s = \frac{1}{2} \left(\frac{\partial \delta u_i^s}{\partial x_j} + \frac{\partial \delta u_j^s}{\partial x_i} \right) \tag{6.103}$$

Considering the symmetry of stress tensor. $\sigma_{ij} = \sigma_{ji}$, we have

$$\sigma_{ij}\frac{\partial \delta u_i^s}{\partial x_j} = (\sigma'_{ij}\delta\varepsilon_{ij}^s - p\delta\varepsilon_{ii}^s) \tag{6.104}$$

Upon substitution of Equations (6.101) and (6.104) into Equation (6.101), we get the following equation after manipulation:

$$\int_V \rho\ddot{u}_i^s\delta u_i^s dV + \int_V \sigma'_{ij}\delta\varepsilon_{ij}^s dV - \int_V p\delta\varepsilon_{ii}^s dV = \int_V \rho b_i\delta u_i^s dV + \int_{S_t} \bar{T}_i\delta u_i^s dS \tag{6.105}$$

This equation is a weak form of the equation of motion in domain V.

Here, the vector expression of Equation (6.105) for the formulation by FEM becomes

$$\int_V \rho\{\delta u^s\}^T\{\ddot{u}^s\}dV + \int_V \{\delta\varepsilon^s\}^T\{\sigma'\}dV - \int_V p\delta\varepsilon_v^s dV$$
$$= \int_V \rho\{\delta u^s\}^T\{b\}dV + \int_{S_t} \{\delta u^s\}^T\{\bar{T}\}dS \tag{6.106}$$

where variables without subscripts correspond to the vectors in Equation (6.105) and the pore water pressure $p(=-u_w)$ is positive in compression. In order to use the incremental nonlinear constitutive law shown in Equation (2.11), the effective stress vector, $\{\sigma'\}$, is expressed as

$$\{\sigma'\} = \{\sigma'\}_{|t} + \{\Delta\sigma'\} \tag{6.107}$$

where $\{\sigma'\}_{|t}$ is an effective stress vector at a previous time t when the current time is $t + \Delta t$.

The substitution of Equation (6.107) into Equation (6.106) gives

$$\int_V \rho\{\delta u^s\}^T\{\ddot{u}^s\}dV + \int_V \{\delta\varepsilon^s\}^T(\{\sigma'\}_{|t} + \{\Delta\sigma'\})dV - \int_V p\delta\varepsilon_v^s dV$$
$$= \int_V \rho\{\delta u^s\}^T\{b\}dV + \int_{S_t} \{\delta u^s\}^T\{\bar{T}\}dS \tag{6.108}$$

In the following, the displacement vector, $\{u_N\}$, at the nodes and pore water pressure, p_E, at the gravitational center of the element (pore pressure increment during earthquake in the final setting p_{dE}) are taken as unknown variables in the discretized equation of motion.

The displacement vector of the solid phase, $\{u^s\}$, is given by the displacement vector at the nodes $\{u_N\}$ in terms of the shape function matrix $[N]$ as

$$\{u^s\} = [N]\{u_N\} \tag{6.109}$$

The pore water pressure is assumed to be constant in the element (namely, the 0th order shape function is used) based on Christian's method (1970).

Using the pore water pressure at the gravitational center of the element, the pore water pressure is given by

$$p = N_p p_E = p_E \tag{6.110}$$

where N_p is the zeroth-order shape function, that is, equal to 1.

From Equation (2.9), the strain vector of the element $\{\varepsilon^s\}$ is given as

$$\{\varepsilon^s\} = [L]\{u^s\} \tag{6.111}$$

where $[L]$ is a matrix that transforms displacements into strains.

Upon substitution of Equation (6.109) into Equation (6.111), we have the following equation:

$$\{\varepsilon^s\} = [L][N]\{u_N\} = [B]\{u_N\} \tag{6.112}$$

The virtual volumetric strain, $\delta\varepsilon^s_{ii}$, of the element is expressed using the virtual displacement vector, $\{\delta u_N\}$, as

$$\delta\varepsilon^s_{ii} = \delta\varepsilon^s_v = [B_v]^T\{\delta u_N\} = \{\delta u_N\}^T[B_v] \tag{6.113}$$

where $\{B_v\}$ is a vector that transforms the nodal displacements of the element into the volumetric strain of the element.

Finally, Equation (2.11) can be written in matrix form as

$$\{\Delta\sigma'\} = [D]\{\Delta\varepsilon^s\} = [D][B]\{\Delta u_N\} \tag{6.114}$$

where $\{\Delta u_N\}$ is the displacement increment vector.

Substituting Equations (6.109), (6.110), (6.112), (6.113), and (6.114) into Equation (6.108), we obtain the following equation:

$$\{\delta u_N\}^T \int_V \rho[N]^T[N]dV\{\ddot{u}_N\} + \{\delta u_N\}^T \int_V [B]^T[D][B]dV\{\Delta u_N\} - \{\delta u_N\}^T \int_V [B_v]dVp_E$$

$$= \{\delta u_N\}^T \int_V \rho[N]^T\{b\}dV + \{\delta u_N\}^T \int_{S_t} [N]^T\{\bar{T}\}dS - \{\delta u_N\}^T \int_V [B]^T\{\sigma'\}_{|t}d^{\text{`}} \quad (6.115)$$

Rearranging the terms with respect to the virtual displacement vector, $\{\delta u_N\}$, we have

$$\{\delta u_N\}^T \left(\int_V \rho[N]^T[N]dV\{\ddot{u}_N\} + \int_V [B]^T[D][B]dV\{\Delta u_N\} - \int_V [B_v]dVp_E \right)$$

$$= \{\delta u_N\}^T \left(\int_V \rho[N]^T\{b\}dV + \int_{S_t} [N]^T\{\bar{T}\}dS - \int_V [B]^T\{\sigma'\}_{|t}dV \right. \quad (6.116)$$

Since this equation holds for the arbitrary unrestricted virtual displacement vector, $\{\delta u_N\}$, the stiffness equation of the element is given by

$$\int_V \rho[N]^T[N]dV\{\ddot{u}_N\} + \int_V [B]^T[D][B]dV\{\Delta u_N\} - \int_V [B_v]dVp_E$$

$$= \int_V \rho[N]^T\{b\}dV + \int_{S_t} [N]^T\{\bar{T}\}dS - \int_V [B]^T\{\sigma'\}_{|t}d^{\text{`}} \quad (6.117)$$

Herein, we set the following terms as

$$[M] = \int_V \rho[N]^T[N]dV \quad (6.118)$$

$$[K] = \int_V [B]^T[D][B]dV \quad (6.119)$$

$$\{K_v\} = \int_V [B_v]dV \quad (6.120)$$

$$\{F\} = \int_V \rho[N]^T\{b\}dV + \int_{S_t} [N]^T\{\bar{T}\}dS \tag{6.121}$$

$$\{R\}_{|t} = \int_V [B]^T\{\sigma'\}_{|t}dV \tag{6.122}$$

These matrices are calculated for the element. Then, assembling the stiffness matrices of all the elements, the total stiffness matrix is derived.

In the calculation of the integration included in the stiffness equation of the element, the Gauss numerical integration method is used.

For Equations (6.119) and (6.122), the stresses are calculated at the Gauss integration points. Although four integration points are, in general, used for the square element, there is a tendency for the shear stiffness to be overestimated under the undrained conditions and the deformation is underestimated (shear locking).

To avoid shear locking under undrained conditions, the reduced integration technique is adopted, namely, the stress levels are calculated at the gravitational center of the element. Due to this method, the pore water pressure defined at the center of gravitational center corresponds one-to-one with the stress.

The mass matrix $[M]$ of Equation (6.118) is a consistent mass matrix. In the program LIQCA (Oka et al. 1994, 2004; LIQCA Research and Development Group 2005), a lumped mass matrix is used as

$$[M] = \frac{\rho V_e}{n_e}[I_e] \tag{6.123}$$

where V_e is the volume of the element, n_e is the number of nodal points, and $[I_e]$ is a unit matrix that corresponds to the degree of freedom of the nodal points of the element $(2n_e \times 2n_e)$.

In Equation (6.117), the damping of the system is given by the hysteretic damping of the constitutive model. However, the material damping of the system is difficult to reproduce only by the hysteretic damping of the model in both the small vibration and in the high frequency region.

Hence, we use the Rayleigh damping matrix [C] that is proportional to the velocity vector of the nodal points. Rayleigh damping is described by the linear combination of the mass matrix $[M]$ and the stiffness matrix $[K]$ as

$$[C] = \alpha_0[M] + \alpha_1[K] \tag{6.124}$$

where α_0 and α_1 are constants.

These two constants can be determined by solving the following simultaneous equation derived by specifying damping for two particular frequencies (the ith and the jth order characteristic frequencies)

$$\begin{cases} \alpha_0/\omega_i + \alpha_1\omega_i & = & 2h_i \\ \alpha_0/\omega_j + \alpha_1\omega_j & = & 2h_j \end{cases} \tag{6.125}$$

where ω_i and ω_j are the ith and the jth natural circular frequencies, and h_i and h_j are damping for the ith and the jth modes. For the joint element, Rayleigh damping is not considered, which is defined in the entire domain.

In addition, the damping matrix [C] is used for setting the viscous boundary. At the viscous boundary, the following terms are given at the nodal points:

- For the degree of freedom vertical to the boundary, $\rho_b V_p \ell / 2$
- For the degree of freedom horizontal to the boundary, $\rho_b V_s \ell / 2$

where ρ_b is the density of the semi-infinite ground, V_p and V_s are the velocity of the P wave and the velocity of the S wave of the semi-infinite ground, and ℓ is the length of the side of the element at the boundary.

From the above, the substitution of Equations (6.118) to (6.124) into Equation (6.117) leads to

$$[M]\{\ddot{u}_N\} + [C]\{\dot{u}_N\} + [K]\{\Delta u_N\} + \{K_v\}p_E = \{F\} - \{R\}_{|t} \tag{6.126}$$

Next, we will consider the external force vector. The external forces applied to the system can be derived into two parts: the force that has been applied before earthquakes and the force that is applied during earthquakes. The former is a gravitational force and the latter is the inertia force due to earthquakes.

The external force expressed by Equation (6.121) can be divided into the force having been applied before earthquakes and the incremental force during earthquakes, as follows:

$$\begin{aligned} &\{F\} = \{F_s\} + \{F_d\} \\ &= \left(\int_V \rho[N]^T \{b_s\} dV + \int_{S_t} [N]^T \{\bar{T}_s\} dS \right) + \left(\int_V \rho[N]^T \{b_d\} dV + \int_{S_t} [N]^T \{\bar{T}_d\} dS \right) \end{aligned} \tag{6.127}$$

where $\{F_s\}$ is the self-weight force vector due to gravity and $\{F_d\}$ is the loading vector due to earthquakes.

Similarly, the body force vector, $\{b\}$, can be divided into $\{b_s\}$ and $\{b_d\}$, and the surface traction vector, $\{\bar{T}\}$, can be divided into $\{\bar{T}_s\}$ and $\{\bar{T}_d\}$.

Now, we consider the static analysis (i.e., initial stress analysis) due to the self-weight force.

The equilibrium equation is obtained by subtracting the dynamic term from Equation (6.126) as

$$[K]\{\Delta u_N\} - \{K_v\} p_{E|t=0} = \{F_s\} - \{R\}_{|t} \tag{6.128}$$

where $p_{E|t=0}$ is the initial pore water pressure before an earthquake. In this equation, the nonlinearity of the stiffness matrix is taken into consideration.

Using the initial effective stress vector before earthquakes, $\{\sigma'\}_{|t=0}$, the equilibrium equation becomes

$$\int_V [B]^T \{\sigma'\}_{|t=0} dV - \{K_v\} p_{E|t=0} = \{F_s\} \tag{6.129}$$

Upon substitution of Equations (6.127) and (6.129) into Equation (6.126),

$$[M]\{\ddot{u}_N\} + [C]\{\dot{u}_N\} + [K]\{\Delta u_N\} - \{K_v\} p_E = \int_{V_e} [B]^T \{\sigma'\}_{|t=0} dV - \{K_v\} p_{E|t=0} + \{F_d\} - \{R_d\}_{|t} \tag{6.130}$$

The residual vector, $\{R_d\}_{|t}$, can be rewritten considering Equation (6.122) as

$$\{\bar{R}_d\}_{|t} = \int_V [B]^T (\{\sigma'\}_{|t} - \{\sigma'\}_{|t=0}) dV \tag{6.131}$$

Substituting Equations (6.131) into Equation (6.130), after manipulation we obtain

$$[M]\{\ddot{u}_N\} + [C]\{\dot{u}_N\} + [K]\{\Delta u_N\} - \{K_v\}(p_E - p_{E|t=0}) = \{F_d\} - \{\bar{R}_d\}_{|t} \tag{6.132}$$

As noted earlier, we will use Equation (6.132) considering the initial stress due to the self-weight force before the earthquake.

6.4.2 Continuity equation

In the general formulation of the finite element method, the general weighted function has been used to obtain the weak form of the governing equations. For the selected point collocation method, the unit value function at several selected points is used; for the subdomain collocation method, weighted function, w, is used in which $W = 1$ in the domain and $W = 0$ outside of the domain; for the Bubnov Galerkin method, the shape function is used as a weighted function. The weak form of the governing equation can be discretized based on the appropriated weighting function, w, which is locally defined.

6.4.2.1 Galerkin method

We can employ the shape function, which is used as the interpolation function for the isoparametric elements as $W = [N_h]$. Using the Galerkin method, we obtain the weak form of the continuity equation (Equation 2.45) as

$$\int_V [N_h]^T \left(-\frac{k}{\gamma_w}\rho^f \ddot{\varepsilon}^s_{ii} - \frac{k}{\gamma_w}\frac{\partial^2 p}{\partial x_i^2} + \dot{\varepsilon}^s_{ii} + \frac{n}{K^f}\dot{p} \right) dV = 0 \tag{6.133}$$

Using the Gauss theorem, we have

$$\int_V -[N_h]^T \frac{k}{\gamma_w}\rho^f \ddot{\varepsilon}^s_{ii} dv + \int_V \frac{k}{\gamma_w}[\nabla N_h]^T \{\nabla p\} dv + \int_V [N_h]^T \dot{\varepsilon}^s_{ii} dv + \int_V \frac{n}{K^f}[N_h]^T \dot{p} dv$$
$$= \int_S \frac{k}{\gamma_w}[N_h]^T \{\nabla p\}\{n\} ds \tag{6.134}$$

When permeability, porosity, and the volumetric modulus of water are constant in the element, we can discretize the equation as

$$-\frac{k}{\gamma_w}\rho^f [K_v]^T \{a_N\} + [K_v]^T \{v_N\} + \frac{k}{\gamma_w}[K_h]\{p_N\} + \frac{n}{K^f}[K_p]\{\dot{p}_N\}$$
$$= \frac{k}{\gamma_w}\int_S [N_h]\{\nabla p\}\{n\}\, ds \tag{6.135}$$

$$p = [N_h]\{p_N\}, \qquad \{\nabla p\} = [\nabla N_h]\{p_N\} = [B_h]\{p_N\} \tag{6.136}$$

$$\dot{\varepsilon}_{ii} = [B_v]^T \{v_N\},\ \ddot{\varepsilon}_{ii} = [B_v]^T \{a_N\} \tag{6.137}$$

$$[K_v]^T = \int_V [N_h]^T [B_v]^T dv \tag{6.138}$$

$$[K_h] = \int_V [B_h]^T [B_h] dv \tag{6.139}$$

$$[K_p] = \int_V [N_h]^T [N_h] dv \tag{6.140}$$

where $\{v_N\}$ is the velocity vector, $\{a_N\}$ is the acceleration vector, and $\{p_N\}$ is the pore pressure vector.

6.4.2.2 Finite volume method

When the pore water pressure is constant in the element, it is useful to apply a variant of the finite volume method, which is similar to the subdomain collaboration method. In the finite volume method, the node is defined at the center of the element.

The continuity equation is discretized by a weight function as

$$\begin{aligned} W_n &= 1 \quad (\textit{in the element}) \\ W_n &= 0 \quad (\textit{outside the element}) \end{aligned} \tag{6.141}$$

When we consider the water pressure boundary, we will have the same results even if disregarding the initial hydrostatic pressure, that is, using the excess pressure that is obtained by subtracting the initial hydrostatic pressure from the total water pressure.

Multiplying the continuity equation, Equation (2.45), by the weighted function, W_n, and integrating it over domain V, we have the following weak form as

$$\int_V \left(-\frac{k}{\gamma_w} \rho^f \ddot{\varepsilon}^s_{ii} - \frac{k}{\gamma_w} \frac{\partial^2 p_d}{\partial x_i^2} + \dot{\varepsilon}^s_{ii} + \frac{n}{K^f} \dot{p}_d \right) W_n dV = 0 \tag{6.142}$$

where p_d is the pore pressure.

Using the Gauss theorem, we have

$$\int_V \left(-\frac{k}{\gamma_w} \rho^f \ddot{\varepsilon}^s_{ii} + \dot{\varepsilon}^s_{ii} + \frac{n}{K^f} \dot{p}_d \right) W_n dV - \int_S \frac{k}{\gamma_w} \frac{\partial p_d}{\partial x_i} n_i W_n dS = 0 \tag{6.143}$$

In the following, we will develop the equations with the unknowns of the nodal displacement vector of the element, $\{u_N\}$, and pore water pressure increment, p_{dE}, during earthquakes.

As is the case with Equation (6.113), the term including the volumetric strain of solid phase, ε^s_{ii}, in Equation (6.143) can be written using the nodal displacement vector as Equation (6.23): $\varepsilon^s_{ii} = [B_v]^T\{u_N\}$.

Expressing the first term on the left-hand side of Equation (6.143) in a vector form and substituting Equation (6.137) into it, we have the following equation:

$$-\int_V \frac{k}{\gamma_w}\rho^f[B_v]^T dV\{\ddot{u}_N\} + \int_V [B_v]^T dV\{\dot{u}_N\} + \int_V \frac{n}{K^f} dV\dot{p}_{dE} - \int_S \frac{k}{\gamma_w}\frac{\partial p_{dE}}{\partial x_i} n_i dS = 0 \tag{6.144}$$

Herein, using Equation (6.120), Equation (6.144) becomes

$$-\frac{\rho^f k}{\gamma_w}\{K_v\}^T\{\ddot{u}_N\} + \{K_v\}^T\{\dot{u}_N\} + A\dot{p}_{dE} - \int_S \frac{k}{\gamma_w}\frac{\partial p_{dE}}{\partial x_i} n_i dS = 0 \tag{6.145}$$

where coefficient A is given by

$$A = \int_V \frac{n}{K^f} dV \tag{6.146}$$

It is worth noting that Equation (6.145) can be applicable to any shape of quadrilateral element.

For rectangular elements, the second gradients of the pore water pressure can be approximated by the finite difference scheme as

$$\frac{\partial^2 p_d}{\partial x^2} = \frac{p_d(x+\Delta x) - 2p_d(x) + p_d(x-\Delta x)}{\Delta x^2} \tag{6.147}$$

However, we can use the finite volume method for the continuity equations. For the finite volume method, we can use a polygon, such as a quadrilateral element, and the nodes are located in the center of the volume, that is, elements. First, we consider the two adjacent square elements shown in Figure 6.8. The pore water pressure at the gravitational center of the element concerned is p_{dE}, and the pore water pressure at the gravitational center of the adjacent element i is p_{dEi}. The permeability coefficient of the element concerned is k, and the permeability of the adjacent element i is k_i. The length of the boundary adjacent to the neighboring element is b_i, and its components in the x and y directions are b_{xi} and b_{yi}, respectively. The distance between

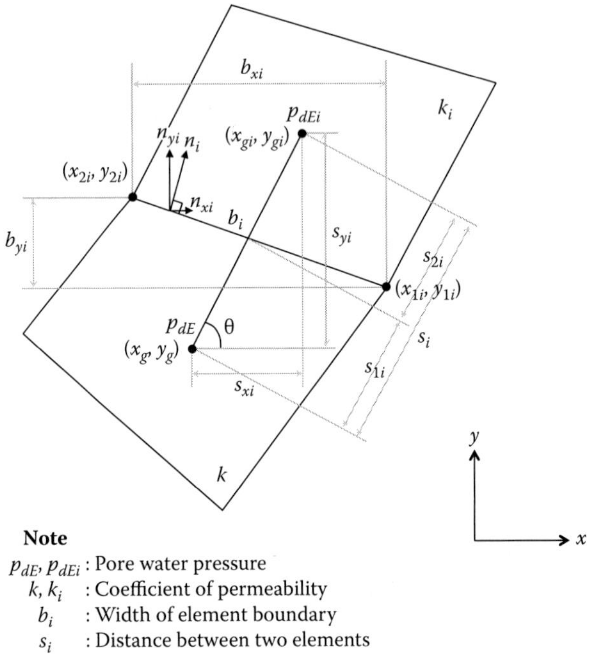

Figure 6.8 Discretization of pore water flow equation.

the gravitational center of the element and adjacent element i is s_i, and its components in the x and y directions are s_{xi} and s_{yi}, respectively. s_i is given by adding distance s_{1i} to distance s_{1i}, as shown in Figure 6.8.

Since the permeability of the element may be different from that of the adjacent element, we use the average permeability coefficient, $\bar{k}_i$. The average permeability coefficient between elements, $\bar{k}_i$, is given by Darcy's law as

$$\frac{s_i}{\bar{k}_i} = \frac{s_{1i}}{k} + \frac{s_{2i}}{k_i} \tag{6.148}$$

Using the unit normal vector to the side of element n_i in the coordinate system shown in Figure 6.8, the third term on the left-hand side of Equation (6.145) is expressed as

$$\int_S \frac{k}{\gamma_w} \frac{\partial p_{dE}}{\partial x_i} n_i dS = \sum_{i=1}^{4} \int_{b_i} \left(v_{xi} n_{xi} + v_{yi} n_{yi} \right) db \tag{6.149}$$

where n_{xi} is the component of the unit normal n_i in the x direction and n_{yi} is the component of the unit normal n_i in the y direction.

The velocity of the in-flow of pore water into the center element from element i is given by

$$v_i = \frac{k}{\gamma_w}\frac{p_{dEi} - p_{dE}}{s_i} \tag{6.150}$$

The components in the x and y directions are

$$v_{xi} = \frac{k}{\gamma_w}\frac{p_{dEi} - p_{dE}}{s_i}\cos\theta = \frac{k}{\gamma_w}\frac{p_{dEi} - p_{dE}}{s_i}\frac{S_{xi}}{s_i} \tag{6.151}$$

$$v_{yi} = \frac{k}{\gamma_w}\frac{p_{dEi} - p_{dE}}{s_i}\sin\theta = \frac{k}{\gamma_w}\frac{p_{dEi} - p_{dE}}{s_i}\frac{S_{yi}}{s_i} \tag{6.152}$$

The components of the unit normal n_i to the boundary are expressed by

$$n_{xi} = \frac{b_{yi}}{b_i}, \quad n_{yi} = \frac{b_{xi}}{b_i} \tag{6.153}$$

The in-flow per unit time into the element from the adjacent elements is calculated as

$$\begin{aligned}\int_{b_i}(v_{xi}n_{xi} + v_{yi}n_{yi})db &= \frac{k}{\gamma_w}(p_{dEi} - p_{dE})\frac{s_{xi}}{s_i^2}b_{yi} + \frac{k}{\gamma_w}(p_{dEi} - p_{dE})\frac{s_{yi}}{s_i^2}b_{xi} \\ &= \frac{k}{\gamma_w}(p_{dEi} - p_{dE})\frac{s_{xi}b_{yi} + s_{yi}b_{xi}}{s_i^2}\end{aligned} \tag{6.154}$$

Finally, the total in-flow of the pore water into the elements becomes

$$\sum_{i=1}^{4}\frac{k}{\gamma_w}(p_{dEi} - p_{dE})\frac{s_{xi}b_{yi} + s_{yi}b_{xi}}{s_i^2} = -\alpha p_{dE} + \sum_{i=1}^{4}\alpha_i p_{dEi} \tag{6.155}$$

where

$$\alpha_i = \frac{\bar{k}}{\gamma_w}\left(\frac{b_{xi}s_{yi} - b_{yi}s_{xi}}{s_i^2}\right), \quad \alpha = \sum_{i=1}^{4}\alpha_i \tag{6.156}$$

$$-\frac{\rho^f k}{\gamma_w}\{K_v\}^T\{\ddot{u}_N\} + \{K_v\}^T\{\dot{u}_N\} + A\dot{p}_{dE} - \left(\alpha P_{de} - \sum_{i=1}^{4}\alpha_i P_{dEi}\right) = 0 \tag{6.157}$$

6.4.3 Time discretization

The discretized equation of motion, Equation (6.132), and the continuity equation, Equation (6.157), at current time $t + \Delta t$ are given by

$$[M]\{\ddot{u}_N\}_{|t+\Delta t} + [C]_{|t+\Delta t}\{\dot{u}_N\}_{|t+\Delta t} + [K]_{|t+\Delta t}\{\Delta u_N\}_{|t+\Delta t} - \{K_v\}p_{dE|t+\Delta t} = \{F_d\}_{|t+\Delta t} - \{R_d\}_{|t} \tag{6.158}$$

$$-\frac{\rho^f k}{\gamma_w}\{K_v\}^T\{\ddot{u}_N\}_{|t+\Delta t} + \{K_v\}^T\{\dot{u}_N\}_{|t+\Delta t} + A\dot{p}_{dE|t+\Delta t} - \left(\alpha p_{dE|t+\Delta t} - \sum_{i=1}^{4}\alpha_i p_{dEi|t+\Delta t}\right) = 0 \tag{6.159}$$

For the time discretization, the Newmark β method is used. The nodal displacement vector and the velocity vector at the current time are described as

$$\{u_N\}_{|t+\Delta t} = \{u_N\}_{|t} + \Delta t\{\dot{u}_N\}_{|t} + \frac{(\Delta t)^2}{2}\{\ddot{u}_N\}_{|t} + \beta(\Delta t)^2(\{\ddot{u}_N\}_{|t+\Delta t} - \{\ddot{u}_N\}_{|t}) \tag{6.160}$$

$$\{\dot{u}_N\}_{|t+\Delta t} = \{\dot{u}_N\}_{|t} + \Delta t\{\ddot{u}_N\}_{|t} + \gamma\Delta t(\{\ddot{u}_N\}_{|t+\Delta t} - \{\ddot{u}_N\}_{|t}) \tag{6.161}$$

where Δt is the time increment, and β and γ are parameters. In many cases, 0.3025 and 0.6 are taken for β and γ, respectively. With these values and a time increment of 0.001, the present method gives a stable solution for a sandy ground with a smaller permeability than 0.1 cm/s since γ has to satisfy the following relations based on the stability of the u-p formulation (Kato 1995; LIQCA Research and Development Group 2005).

$$\gamma \geq \frac{k}{g\Delta t} + \frac{1}{2} \tag{6.162}$$

where Δt is the time increment, k is the permeability coefficient and g is the gravitational acceleration.

Using these relations, we have the equations with unknown variables of nodal acceleration vector $\{\ddot{u}_N\}_{|t+\Delta t}$ and the excess pore water pressure of element $p_{dE|t+\Delta t}$.

6.4.3.1 Equation of motion

Upon substitution of Equations (6.160) and (6.161) into Equation (6.158), we obtain

$$[M]\{\ddot{u}_N\}_{|t+\Delta t} + [C]_{|t+\Delta t}\left(\{\dot{u}_N\}_{|t} + \Delta t\{\ddot{u}_N\}_{|t} + \gamma\Delta t(\{\ddot{u}_N\}_{|t+\Delta t} - \{\ddot{u}_N\}_{|t})\right)$$

$$+[K]_{|t+\Delta t}\left(\Delta t\{\dot{u}_N\}_{|t} + \frac{(\Delta t)^2}{2}\{\ddot{u}_N\}_{|t} + \beta(\Delta t)^2(\{\ddot{u}_N\}_{|t+\Delta t} - \{\ddot{u}_N\}_{|t})\right)$$

$$-\{K_v\}p_{dE|t+\Delta t} = \{F_d\}_{|t+\Delta t} - \{R_d\}_{|t} \tag{6.163}$$

After the transposition of unknown nodal acceleration vector $\{\ddot{u}_N\}_{|t+\Delta t}$ and the excess pore water pressure of element $p_{dE|t+\Delta t}$ to the left-hand side and the manipulation, we have the following equation:

$$\left([M] + \gamma\Delta t[C]_{|t+\Delta t} + \beta(\Delta t)^2[K]_{|t+\Delta t}\right)\{\ddot{u}_N\}_{|t+\Delta t} - \{K_v\}p_{dE|t+\Delta t}$$

$$= \{F_d\}_{|t+\Delta t} - \{R_d\}_{|t} - [C]_{|t+\Delta t}\left(\{\dot{u}_N\}_{|t} + (1-\gamma)\Delta t\{\ddot{u}_N\}_{|t}\right)$$

$$-[K]_{|t+\Delta t}\left(\Delta t\{\dot{u}_N\}_{|t} + \left(\frac{1}{2} - \beta\right)(\Delta t)^2\{\ddot{u}_N\}_{|t}\right) \tag{6.164}$$

This is the discretized equation of motion.

6.4.3.2 Continuity equation

The substitution of Equation (6.161) into Equation (6.159) gives

$$-\frac{\rho^f k}{\gamma_w}\{K_v\}^T\{\ddot{u}_N\}_{|t+\Delta t} + \{K_v\}^T\left(\{\dot{u}_N\}_{|t} + \Delta t\{\ddot{u}_N\}_{|t} + \gamma\Delta t(\{\ddot{u}_N\}_{|t+\Delta t} - \{\ddot{u}_N\}_{|t})\right)$$

$$+A\dot{p}_{dE|t+\Delta t} - \left(\alpha p_{dE|t+\Delta t} - \sum_{i=1}^{4}\alpha_i p_{dEi|t+\Delta t}\right) = 0 \tag{6.165}$$

Here, we apply the backward Euler finite difference scheme to the third term on the left-hand side of the equation:

$$\dot{p}_{dE|t+\Delta t} = \frac{p_{dE|t+\Delta t} - p_{dE|t}}{\Delta t} \tag{6.166}$$

Substituting $\gamma_w = \rho^f g$ and Equation (6.166) into Equation (6.165) yields

$$-\frac{k}{g}\{K_v\}^T\{\ddot{u}_N\}_{|t+\Delta t} + \{K_v\}^T\left(\{\dot{u}_N\}_{|t} + \Delta t\{\ddot{u}_N\}_{|t} + \gamma\Delta t(\{\ddot{u}_N\}_{|t+\Delta t} - \{\ddot{u}_N\}_{|t})\right)$$
$$+\frac{A}{\Delta t}(p_{dE|t+\Delta t} - p_{dE|t}) - \left(\alpha p_{dE|t+\Delta t} - \sum_{i=1}^{4}\alpha_i p_{dEi|t+\Delta t}\right) = 0 \tag{6.167}$$

Rearranging the unknown nodal acceleration vector $\{\ddot{u}_N\}_{|t+\Delta t}$ and the excess pore water pressure of element $p_{dE|t+\Delta t}$ in Equation (6.167), we have

$$-\left(\frac{k}{g} - \gamma\Delta t\right)\{K_v\}^T\{\ddot{u}_N\}_{|t+\Delta t} + \left(\frac{A}{\Delta t} - \alpha\right)p_{dE|t+\Delta t} + \sum_{i=1}^{4}\alpha_i p_{dEi|t+\Delta t}$$
$$= -\{K_v\}^T\left(\{\dot{u}_N\}_{|t} + (1-\gamma)\Delta t\{\ddot{u}_N\}_{|t}\right) + \frac{A}{\Delta t}p_{dE|t} \tag{6.168}$$

As noted earlier, the continuity equation can be discretized. Combining Equations (6.164) and (6.168), we will have an antisymmetric total matrix because the coefficients of the pore water pressure in Equations (6.164) and (6.168) at time $t+\Delta t$ are different.

Hence, we will rearrange Equation (6.168) to be symmetric although stiffness matrix K is not symmetric, because the elastoplastic matrix is derived based on the nonassociated flow rule.

In Equation (6.168), the coefficients are replaced as

$$\alpha' = \frac{1}{\left(\frac{k}{g} - \gamma\Delta t\right)}\alpha \tag{6.169}$$

$$\alpha'_i = \frac{1}{\left(\frac{k}{g} - \gamma\Delta t\right)}\alpha_i \tag{6.170}$$

$$A' = \frac{1}{\Delta t\left(\frac{k}{g} - \gamma\Delta t\right)}A \tag{6.171}$$

The substitution of Equations (6.169), (6.170), and (6.171) into Equation (6.168) gives

$$-\{K_v\}^T\{\ddot{u}_N\}_{|t+\Delta t}+(A'-\alpha')p_{dE|t+\Delta t}+\sum_{i=1}^{4}\alpha'_i p_{dEi|t+\Delta t}$$

$$=\frac{-1}{\left(\frac{k}{g}-\gamma\Delta t\right)}\{K_v\}^T\left(\{\dot{u}_N\}_{|t}+(1-\gamma)\Delta t\{\ddot{u}_N\}_{|t}\right)+A'p_{dE|t} \tag{6.172}$$

Finally, combining the equation of motion, Equation (6.164), and the continuity equation, Equation (6.172), we have

$$\begin{bmatrix}[M]+\gamma\Delta t[C]_{|t+\Delta t}+\beta(\Delta t)^2[K]_{|t+\Delta t} & -\{K_v\}\\ -\{K_v\}^T & A'-\alpha'\end{bmatrix}\begin{Bmatrix}\{\ddot{u}_N\}_{|t+\Delta t}\\ p_{dE|t+\Delta t}\end{Bmatrix}+\begin{Bmatrix}0\\ \sum_{i=1}^{4}\alpha'_i p_{dEi|t+\Delta t}\end{Bmatrix}$$

$$=\begin{Bmatrix}\{F_d\}_{|t+\Delta t}-\{R_d\}_{|t}-[C]_{|t+\Delta t}\left\{\{\dot{u}_N\}_{|t}+(1-\gamma)\Delta t\{\ddot{u}_N\}_{|t}\right\}\\ -[K]_{|t+\Delta t}\left(\Delta t\{\dot{u}_N\}_{|t}+\left(\frac{1}{2}-\beta\right)(\Delta t)^2\{\ddot{u}_N\}_{|t}\right)\\ \frac{-1}{\left(\frac{k}{g}-\gamma\Delta t\right)}\{K_v\}^T\left(\{\dot{u}_N\}_{|t}+(1-\gamma)\Delta t\{\ddot{u}_N\}_{|t}\right)+A'p_{dE|t}\end{Bmatrix} \tag{6.173}$$

Since the pore water pressure of the two adjacent elements is unknown, after the rearrangements we have

$$\begin{bmatrix}[M]+\gamma\Delta t[C]_{|t+\Delta t}+\beta(\Delta t)^2[K]_{|t+\Delta t} & \{K_v\} & 0\\ -\{K_v\}^T & A'-\alpha' & \{\alpha'_i\}^T\end{bmatrix}\begin{Bmatrix}\{\ddot{u}_N\}_{|t+\Delta t}\\ p_{dE|t+\Delta t}\\ \{p_{dEi|t+\Delta t}\}\end{Bmatrix}$$

$$=\begin{Bmatrix}\{F_d\}_{|t+\Delta t}-\{R_d\}_{|t}-[C]_{|t+\Delta t}\left\{\{\dot{u}_N\}_{|t}+(1-\gamma)\Delta t\{\ddot{u}_N\}_{|t}\right\}\\ -[K]_{|t+\Delta t}\left(\Delta t\{\dot{u}_N\}_{|t}+\left(\frac{1}{2}-\beta\right)(\Delta t)^2\{\ddot{u}_N\}_{|t}\right)\\ \frac{-1}{\left(\frac{k}{g}-\gamma\Delta t\right)}\{K_v\}^T\left(\{\dot{u}_N\}_{|t}+(1-\gamma)\Delta t\{\ddot{u}_N\}_{|t}\right)+A'p_{dE|t}\end{Bmatrix} \tag{6.174}$$

Although this equation seems to be nonsymmetric, the total matrix becomes globally symmetric, because the adjacent elements have the same coefficient vector α_i when we construct the total matrix. It should be noted, however, that the stiffness matrix is not symmetric because of the nonassociated flow rule.

6.5 FINITE DEFORMATION ANALYSIS FOR FLUID–SOLID TWO-PHASE MIXTURES

In this chapter, we formulate a three-dimensional finite element method based on Biot's two-phase mixture theory (1956) in the framework of the finite deformation theory. A finite deformation analysis is necessary to simulate large deformation problems, such as strain localization problems. The strain localization phenomenon is a geometrically nonlinear problem since the deformation in shear bands is large. The constitutive equation for clay used in this study is a nonlinear elastoviscoplastic model and it is formulated in an incremental form. For simplicity, both the grain particles and the fluid are assumed to be incompressible.

In order to deal with such a large nonlinear deformation problem, applying an incremental constitutive model, an updated Lagrangian method is used. There are several methods for the finite deformation theory, namely, the total Lagrangian method is suitable for solids and the Eulerian method has been used for fluids. Granular materials, such as soil, are characterized as materials between solids and fluids. Hence, the updated Lagrangian method is appropriate for describing soils. Since the deformation is large and the reference configuration is updated at each step, it is necessary to use an objective stress rate. Hence, in this section, the objective Jaumann rate of Cauchy stress is adopted. Of course, we can use another type of objective rate, such as Nagdhi's rate. The Jaumann stress rate is used since the strain level is moderate and less than 100% (Johnson and Bammann 1984). In addition, the Jaumann rate of Cauchy stress has been derived based on the double slip theory for granular materials (Anand 1983).

6.5.1 Effective stress and fluid–solid mixture theory

Using Terzaghi's concept of effective stress, the total stress tensor and the time rate of stress are given as

$$T_{ij} = T'_{ij} + u_w \delta_{ij} \tag{6.175}$$

$$\dot{T}_{ij} = \dot{T}'_{ij} + \dot{u}_w \delta_{ij} \tag{6.176}$$

in which T_{ij} denotes the total Cauchy stress tensor, $\dot{T}_{ij}$ denotes the effective Cauchy stress tensor, u_w denotes the pore water pressure, tension is positive ($= u_w = -p$; p is pressure and compression is positive for p), δ_{ij} is the second-order identity tensor, and the superimposed dots indicate the time differentiation.

6.5.2 Equilibrium equation

When we consider an arbitrary domain V with a boundary S, the conservation of linear momentum for the whole fluid–solid mixture in the current configuration is given by the following equation, as presented in Chapter 1:

$$\frac{D}{Dt}\int_V \rho v_i \, dv = \int_S t_i ds + \int_V \rho b_i dv \tag{6.177}$$

in which D/Dt is the material time derivative, ρ is the mass density, v_i is the velocity vector, t_i is the surface traction vector, and b_i is the body force vector.

When we deal with quasi-static problems, the acceleration can be assumed to be zero. Consequently, this assumption provides the equilibrium equation resulting from Equation (6.177) as

$$\int_S t_i ds + \int_V \rho b_i dv = 0 \tag{6.178}$$

When the body force is constant, the rate type of equilibrium equation is expressed as follows:

$$\frac{D}{Dt}\int_S t_i ds = 0 \tag{6.179}$$

Now let us discuss the equilibrium equation with respect to the reference configuration. The balance of linear momentum with respect to the reference configuration,

$$\frac{D}{Dt}\int_V \rho_0 v_i \, dV = \int_S \Pi_{ji} N_j \, dS + \int_V \rho_0 b_i \, dV \tag{6.180}$$

where Π_{ij} is the nominal stress tensor, N_{ij} is the outward normal with respect to the reference configuration, ρ_0 is the mass density at the reference configuration, and b_i the body force vector.

For the quasi-static problems, taking a time derivative of Equation (6.180) gives

$$\int_S \dot{\Pi}_{ji} N_j \, dS + \int_V \rho_0 \dot{b}_i \, dV = 0 \tag{6.181}$$

When we take the current configuration as the reference configuration and use the Gauss theorem, Equation (6.181) becomes

$$\int_V \dot{S}_{ji,j} \, dv + \int_V \rho \dot{b}_i \, dv = 0 \tag{6.182}$$

where the following relations presented in Chapter 1 are used:

$$\dot{\Pi}_{ji} = J \frac{\partial X_j}{\partial x_k} \dot{S}_{ki} \tag{6.183}$$

$$\dot{S}_{ki} \equiv \dot{T}_{ki} - T_{kp} L_{ip} + L_{pp} T_{ki} \tag{6.184}$$

in which $\dot{S}_{ki}$ is the nominal stress rate tensor with respect to the current configuration.

Without any body force increment, the material time derivative of the equilibrium equations becomes

$$\int_V \dot{S}_{ji,j} dv = 0 \tag{6.185}$$

The effective nominal stress rate tensor, $\dot{S}'_{ij}$, is given by the following equation:

$$\dot{S}'_{ij} = \dot{T}'_{ij} - T'_{ip} L_{jp} + L_{pp} T'_{ij} \tag{6.186}$$

To obtain the relation between $\dot{S}_{ki}$ and $\dot{S}'_{ij}$, we substitute the definition for the effective Cauchy stress and the effective Cauchy stress rate, Equations (6.175) and (6.176), respectively, into Equation (6.184), namely,

$$\dot{S}_{ij} = \dot{S}'_{ij} + \dot{u}_w \delta_{ij} + L_{pp} u_w \delta_{ij} - u_w L_{ji} \tag{6.187}$$

By letting

$$U_{ij} = L_{pp}u_w\delta_{ij} - u_w L_{ji} \tag{6.188}$$

Equation (6.187) becomes

$$\dot{S}_{ij} = \dot{S}'_{ij} + \dot{u}_w\delta_{ij} + U_{ij} \tag{6.189}$$

When we consider closed domain V, the weak form of the rate type of equilibrium equation is given as follows:

$$\int_V \dot{S}_{ji,j}\delta v_i dv = 0 \tag{6.190}$$

in which δv_i is the virtual velocity vector.

The boundary of domain V is composed of a velocity boundary and a traction boundary. The velocity boundary is denoted by S_u if the velocity is prescribed; S_t denotes the traction boundary if traction is prescribed.

$$v_i = \bar{v}_i \quad \text{on} \quad S_u \tag{6.191}$$

$$\dot{s}_i = \dot{S}_{ji}n_j = \bar{\dot{s}}_i \quad \text{on} \quad S_t \tag{6.192}$$

in which v_i is the velocity vector, n_i indicates the unit normal to the body, $\dot{s}_i$ is the nominal traction rate vector, and the specified values are designated by a superposed bar.

By taking the Gauss theorem and the compatibility condition, Equation (6.190) can be written as

$$\int_S \dot{S}_{ji}\delta v_i n_j ds - \int_V \dot{S}_{ji}\,\delta L_{ij}\,dv = 0 \tag{6.193}$$

Considering the weak form of the boundary conditions on S_t and substituting Equations (6.186) and (6.189) into the second term of Equation (6.193), and transforming the first term by Equation (6.192) yields

$$\begin{aligned}&\int_V \dot{T}'_{ij}\delta D_{ij}dv + \int_V D_{kk}T'_{ij}\delta L_{ij}dv - \int_V T'_{jp}L_{ip}\delta L_{ij}dv + \int_V \dot{u}_w\delta D_{kk}dv\\ &+ \int_V U_{ji}\delta L_{ij}dv - \int_S \dot{s}_i\delta v_i ds = 0\end{aligned} \tag{6.194}$$

in which D_{ij} is the stretching tensor and the symmetry of the effective Cauchy stress is used.

An updated Lagrangian method is used in this formulation based on the finite deformation theory. Thus, the Jaumann rate of effective Cauchy stress tensor, $\hat{T}'_{ij}$, is employed as an objective stress rate, which leads to

$$\hat{T}'_{ij} = \dot{T}'_{ij} - W_{ik}T'_{kj} + T'_{ik}W_{kj} \tag{6.195}$$

where W_{ij} is the spin tensor.

The elastoviscoplastic constitutive model can be written as

$$\hat{T}'_{ij} = C_{ijkl}D_{kl} - C_{ijkl}D_{kl}^{vp} \tag{6.196}$$

where D_{ij} is the total stretching tensor, D_{ij}^{vp} is the viscoplastic stretching tensor, and C_{ijkl} is the elastic stiffness tensor of the fourth order.

6.5.3 Continuity equation

For describing the motion of pore water, Biot's two-phase mixture theory (1956) is used in the analysis with a v_i^f (velocity) $-u_w$ (pore pressure; tension is positive) formulation. When the soil particles and the pore water are incompressible, Darcy's law and the conservation of mass for the mixture yield the continuity equation as

$$\frac{k}{\gamma_w}\nabla^2 u_w + D_{kk}=0 \tag{6.197}$$

where k is the coefficient of permeability and spatially constant, and γ_w is the unit weight of the pore water.

The boundary for the pore water pressure is decomposed into two parts, S_p and S_q. S_p is the boundary at which the pore pressure is specified and S_q is the boundary at which the flow of pore water is specified as

$$u_w = \bar{u}_w \quad \text{on} \quad S_p \tag{6.198}$$

$$q_i = \frac{k}{\gamma_w}\nabla u_w = \bar{q}_i^f \quad \text{on } S_q \tag{6.199}$$

in which the specified values are designated by a superposed bar and $\bar{q}_i^f$ is the flow velocity of the pore water through the boundary surface.

Weak forms of the continuity equation and boundary conditions are given by

$$\frac{k}{\gamma_w}\int_V \nabla^2 u_w \hat{u}_w \, dv + \int_V D_{kk} \hat{u}_w \, dv = 0 \tag{6.200}$$

$$\int_{S_p} \hat{u}_w (u_w - \bar{u}_w) \, ds = 0 \tag{6.201}$$

$$\int_{S_q} \hat{u}_w (q_i^f - \bar{q}_i^f) \, ds = 0 \tag{6.202}$$

In order to derive a weak form of the continuity equation, we will use a test function that is zero "$\hat{u}_w = 0$" at the boundaries, S_p where the pore pressure is prescribed because the positive definiteness of the resultant coefficient matrix (Atluri 2004) can be assured and the solution exists, and in practice we will consider the prescribed value at the boundaries during the construction of the final matrix. Of course we can use trial functions, that is, approximation functions that satisfy the prescribed boundary conditions such as S_p (Zienkiewicz and Taylor 1997; Lewis and Schrefler 1998).

Considering the aforementioned test function, $\hat{u}_w$, we can obtain the weak form of the continuity equation with the boundary conditions as

$$\frac{k}{\gamma_w}\int_V \nabla^2 u_w \hat{u}_w \, dv + \int_V D_{kk} \hat{u}_w \, dv - \int_{S_q} \hat{u}_w (q_i^f - \bar{q}_i^f) n_i ds = 0 \tag{6.203}$$

By applying the Gauss theorem, the following equation can be obtained:

$$-\frac{k}{\gamma_w}\int_V \nabla \hat{u}_w \cdot \nabla u_w \, dv + \int_V \hat{u}_w D_{kk} \, dv + \int_{S_q} \hat{u}_w \bar{q}_i^f n_i ds = 0 \tag{6.204}$$

It is worth noting that if test function $\hat{u}_w = 0$ on S_p or the trial function satisfies the boundary condition on S_p and zero on other boundary, the positive definiteness of the resultant matrix is assured, that is, the solution exists.

6.5.4 Discretization of the weak forms for the equilibrium equation and the continuity equation

6.5.4.1 *Discretization of the weak forms for the equilibrium equation*

In the following, we will use the Galerkin method in which the interpolation function such as the shape function is used as a test function. For the discretization of the weak form of the equilibrium equation, the following relations are used:

$$\{v\} = [N]\{v^*\}, \tag{6.205}$$

in which $\{v\}$ is the velocity vector in an element, $\{v^*\}$ is the nodal velocity vector, and $[N]$ is the shape function of the element. It should be noted that the square brackets [ ] denote the matrix and the braces { } denote the vector.

$$\{D\} = [B]\{v^*\}, \tag{6.206}$$

in which $[B]$ is the matrix which transforms the nodal velocity vector to the vector form of the stretching tensor $\{D\}$.

$$\{L\} = [B_M]\{v^*\}, \quad \{\delta L\} = [B_M]\{\delta v^*\} \tag{6.207}$$

where $[B_M]$ is the matrix that transforms the nodal velocity vector into the vector form of the velocity gradient vector $\{L\}$.

$$D_{kk} = [B_v]^T\{v^*\} \tag{6.208}$$

where $[B_v]$ is the vector that transforms the nodal velocity into the trace of D_{ij}.

$$\dot{u}_w = [N_h]\{\dot{u}_w^*\} \tag{6.209}$$

in which $\{\dot{u}_w\}$ represents the pore pressure rate, $\{\dot{u}_w^*\}$ represents the nodal pore pressure rate vector, and $[N_h]$ represents the shape function.

In the analysis, the tangent modulus method for the rate-dependent material (Peirce et al. 1984) is used in order to evaluate the viscoplastic stretching tensor, D^{vp}. This method can be classified as a forward gradient method and has been well adopted in viscoplastic analyses (see Oka et al. 1995; Kimoto et al. 2004; Higo et al. 2006). It is known that the method is stable and yields accurate results for time steps larger than those for the Euler explicit method (Peirce et al. 1984). This method is identical to the tangent modulus of the semi-implicit backward Euler scheme on the first iteration (Belytschko et al. 2000). In the formulation, the trapezoidal rule can be used to evaluate the value of the viscoplastic strain, in which the value of trapezoidal parameter θ (Peirce et al. 1984) is taken as 0.5 in the present analysis. $\theta = 1$ corresponds to the usual tangent modulus.

The relation between the rate of effective stress and the stretching tensor for the elastoviscoplastic model presented in Chapter 5 can be written in matrix form, as shown in the following equation:

$$\{\hat{T}'\} = [C]\{D\} - \{Q\} \tag{6.210}$$

where [C] is the elastoviscoplastic tangential stiffness matrix and {Q} can be called the relaxation stress vector.

Substituting Equation (6.195) into Equation (6.210) yields

$$\{\dot{T}'\} = [C]\{D\} - \{Q\} + \{W^*\} \tag{6.211}$$

where $\{W^*\}$ is the column vector related to the spin tensor and the components are computed via $W^*_{ij} = W_{ik}T'_{kj} - T'_{ik}W_{kj}$.

By all the matrix and vector relations obtained previously, and based on the theory of virtual displacement, we have obtained the following relation considering the arbitrariness of the unconstrained virtual nodal velocities:

$$[K]\{v^*\} - \int_V [B]^{\mathrm{T}}\{Q\}dv + \int_V [B]^{\mathrm{T}}\{W^*\}dv + [K_L]\{v^*\} + [K_v]\{\dot{u}^*_w\} = \{\dot{F}\} \tag{6.212}$$

in which

$$[K] = \int_V [B]^{\mathrm{T}}[C][B]dv \tag{6.213}$$

$$[K_L] = \int_V [B_M]^{\mathrm{T}}[D'_s][B_M]dv + \int_V [B_M]^{\mathrm{T}}[U][B_M]dv + \int_V [B_M]^{\mathrm{T}}\{T'\}\{B_v\}^{\mathrm{T}}dv \tag{6.214}$$

$$[K_v] = \int_V [B_v][N_h]dv \tag{6.215}$$

$$\{\dot{F}\} = \int_{S_t} [N]^{\mathrm{T}}\{\dot{s}\}ds \tag{6.216}$$

where $\{\dot{s}\}$ is the traction vector at the boundary.

In the preceding equations, $[D'_s]\{L\} = -T'_{ik}L_{jk}$, $[U]\{L\} = L_{kk}u_w\delta_{ij} - u_w\delta_{ik}L_{jk}$.
And, these matrix forms are given by

$$-T'_{ik}L_{jk} = [D'_s][B_M]\{v^*\} \tag{6.217}$$

$$L_{kk}u_w\delta_{ij} - u_w\delta_{ik}L_{jk} = [U][B_M]\{v^*\} \tag{6.218}$$

where $[B_M]$ is the nodal velocity- L^T matrix and $[D'_s]$ is the matrix that expresses the matrix form of $-T_{ik}L_{jk}$ by L^T.

The relation between nodal velocity vector $\{v^*\}$ and nodal displacement increment vector $\{\Delta u^*\}$ can be obtained by using Euler's approximation as

$$\{v^*\} \approx \frac{\{\Delta u^*\}}{\Delta t} \tag{6.219}$$

Similarly, the pore water pressure can be obtained as

$$\left\{\dot{u}_w^*\right\} \approx \frac{\left\{u_w^*\right\}_{t+\Delta t} - \left\{u_w^*\right\}_t}{\Delta t} \tag{6.220}$$

Substituting Equations (6.219) and (6.220) into Equation (6.212), the weak form of the equilibrium equations is obtained, that is,

$$[[K]+[K_L]]\left\{\Delta u^*\right\} + [K_v]\left\{u_w^*\right\}_{t+\Delta t} = \Delta t\{\dot{F}\} + [K_v]\left\{u_w^*\right\}_t$$

$$+\Delta t\int_V [B]^{\mathrm{T}}\{Q\}dv - \Delta t\int_V [B]^{\mathrm{T}}\left\{W^*\right\}dv \tag{6.221}$$

For the discretization of the continuity equation, Equation (6.204), the following vectors and matrices are used:

$$u_w = [N_h]\left\{u_w^*\right\}, \quad \hat{u}_w = [N_h]\left\{\hat{u}_w^*\right\} \tag{6.222}$$

where $\left\{u_w^*\right\}$ is the nodal pore pressure vector and $\{N_h\}$ is the shape function of the element.

The spatial gradient of the pore water pressure, u_w, is discretized by

$$\nabla u_w = \nabla[N_h]\left\{u_w^*\right\} = [B_h]\left\{u_w^*\right\} \tag{6.223}$$

in which $[B_h]$ is the matrix that transforms the nodal pore pressure into the spatial derivative of the pore pressure.

6.5.4.2 Discretization of the weak form for the continuity equation

Substituting Equations (6.222), (6.223), and (6.199) into Equation (6.204), the following relation is given considering the arbitrariness of the unconstrained virtual nodal pore water pressures:

$$-\frac{k}{\gamma_w}\int_V [B_h]^{\mathrm{T}}[B_h]dv\left\{u_w^*\right\}+\int_V [N_h]^{\mathrm{T}}[B_v]^{\mathrm{T}}dv\left\{v^*\right\}+\int_{\Gamma_p}[N_h]^{\mathrm{T}}\left\{\bar{q}_i^f\right\}^{\mathrm{T}}\left\{n\right\}ds=0 \tag{6.224}$$

Using Equation (6.219), the discretization of the continuity equation is obtained as follows:

$$[K_v]^{\mathrm{T}}\left\{\Delta u^*\right\}-\Delta t[K_h]\left\{u_w^*\right\}_{t+\Delta t}=\Delta t[V] \tag{6.225}$$

where

$$[K_h]=\frac{k}{\gamma_w}\int_V [B_h]^{\mathrm{T}}[B_h]dv \tag{6.226}$$

$$[K_v]^{\mathrm{T}}=\int_V [N_h]^{\mathrm{T}}[B_v]^{\mathrm{T}}dv \tag{6.227}$$

$$[V]=-\int_{\Gamma_q}[N_h]^{\mathrm{T}}\left\{\bar{q}_i^f\right\}^{\mathrm{T}}\left\{n\right\}ds \tag{6.228}$$

Finally, the weak form of the rate type of equilibrium equation, Equation (6.221), and the weak form of the continuity equation, Equation (6.225), are given in the following matrix form:

$$\begin{bmatrix} [K]+[K_L] & [K_v] \\ [K_v]^T & -\Delta t[K_h] \end{bmatrix}\begin{bmatrix} \{\Delta u^*\} \\ \{u_w^*\}_{t+\Delta t} \end{bmatrix}$$

$$=\begin{bmatrix} \Delta t\{\dot{F}\}+\left[K_v\right]^{\mathrm{T}}\left\{u_w^*\right\}_t+\Delta t\int_V\left[B\right]^{\mathrm{T}}\left\{Q\right\}dv-\Delta t\int_V\left[B\right]^{\mathrm{T}}\left\{W^*\right\}dv \\ \Delta t\left\{V\right\} \end{bmatrix} \tag{6.229}$$

In the formulation of Equation (6.229), the incremental governing equations are evaluated at $t+\Delta t$. As shown in Equation (6.225), the pore water pressure is implicitly evaluated at $t+\Delta t$.

For the element type, for example, in the three-dimensional analysis of Chapter 8, a twenty-node quadrilateral isoparametric element with a reduced Gaussian eight-point integration is used to eliminate shear locking as well as to reduce the appearance of a spurious hourglass mode. The pore water pressure is defined by an eight-node quadrilateral isoparametric element.

REFERENCES

Anand, L. 1983. Plane deformations of ideal granular materials, *J. Mech. Phys. Solids*, 31(2):105–122.

Atluri, S.N. 2004. *The Meshless Method for Domain and BIE Discretizations*, Techno Science Press, Chapter 1.

Bathe, K.J. 1996. *The Finite Element Procedure*, Prentice Hall.

Belytschko, T., Liu, W.K., and Moran, B. 2000. *Nonlinear Finite Elements for continua and Structures*, John Wiley & Sons, New York 290–291.

Biot, M.A. 1956. Theory of propagation of elastic waves in a fluid-saturated porous solid, *J. Acoust. Soc. Am.*, 28(2):168–178.

Christian, J.T., and Boehmer, J.W. 1970. Plane strain consolidation by finite elements, Proc. ASCE, 96(SM4):1435–1457.

Dhatt, G., and Touzot, G. 1984. *The Finite Element Method Displayed*, New York: John Wiley & Sons.

Higo, Y., Oka, F., Jiang, M., and Fujita, Y. 2005. Effect of transport of water and material heterogeneity on strain localization analysis of fluid-saturated gradient-dependent viscoplastic geomaterial, *Int. J. Numer. Anal. Meth. Geomech.* 29:495–523.

Higo, Y., Oka, F., Kodaka, T., and Kimoto, S. 2006. Three-dimensional strain localization of water-saturated clay and numerical simulation using an elasto-viscoplastic model, *Philos. Mag.*, 86(21–22):3205–3240.

Hisada, T., and Noguchi, H. 2002. *Fundamentals and Application of Non-Linear Finite Element Method*, Maruzen, Tokyo (In Japanese).

Johnson, G.C., and Bammann, D. 1984. A discussion of stress rates in finite deformation problems, *Int. J. Solids Struct.*, 20(8):725–737.

Kato, M. 1995. Study on multi-dimensional liquefaction analysis method and its application, Doctoral thesis, Gifu University (In Japanese).

Kimoto, S., Oka, F., and Higo, Y. 2004. Strain localization analysis of elasto-viscoplastic soil considering structural degradation, *Int. J. Comput. Meth. Appl. Mech. Eng.*, 193(27–29):2845–2866.

Lewis, R.W., and Schrefler, B.A. 1998. *The Finite Element Method in the Static and Dynamic Deformation and Consolidation of Porous Media*, Chichester, UK: Wiley.

LIQCA Research and Development Group (Representative Fusao Oka, Kyoto University). 2005. User's manual for LIQCA2D04 (2005 released print), http// nakisuna2.kuciv.kyoto-u.ac.jp/liqca.htm

Mirjalili, M., Kimoto, S., Oka, F., and Higo, Y. 2011. Elasto-visoplastic modeling of Osaka soft clay considering destructuration and its effect on the consolidation analysis of an embankment, *Geomech. Geoeng.: Int. J.*, 6(2):69–89.

Oka, F. 2000. *Elasto-Viscoplastic Constitutive Model for Geomaterials*, Morikita Syuppan (In Japanese).

Oka, F., Adachi, T., and Okano, Y. 1986. Two-dimensional consolidation analysis using an elasto-viscoplastic constitutive equation, *Int. J. Numer. Anal. Meth. Geomech.*, 10:1–16.

Oka, F., Adachi, T., and Yashima, A. 1995. A strain localization analysis using a viscoplastic softening model for clay, *Int. J. Plasticity*, 11(5):523–545.

Oka, F., Higo, Y., and Kimoto, S. 2002. Effect of dilatancy on the strain localization of water-saturated elasto-viscoplastic soil, *Int. J. Solids Struct.*, 39(13–14):3625–3647.

Oka, F., Kodaka, T., and Kim, Y.-S. 2004. A cyclic viscoelastic-viscoplastic constitutive model for clay and liquefaction analysis of multi-layered ground, *Int. J. Numerical Anal. Methods Geomech.*, 28(2):131–179.

Oka, F., Yashima, A., and Kohara, I. 1992. A finite element analysis of clay foundation based on finite elasto-viscoplasticity, Proceedings of 4th International Symposium on Numerical Models in Geomechanics, Swansea, G.N. Pande, and S. Pietruszczak, eds., Balkema, 2:915–922.

Oka, F., Yashima, A., Shibata, T., Kato, M., and Uzuoka, R. 1994. FEM–FDM coupled liquefaction analysis of a porous soil using an elasto-plastic model, *Appl. Sci. Res.*, 52:209–245.

Peirce, D., Shih, C., and Needleman, A. 1984. A tangent modulus method for rate dependent solids, *Comput. Struct.*, 18(5):845–887.

Schrefler, B.A., Majorana, C.E., and Sanavia, L. 1995. Shear band localization in saturated porous media, *Arch. Mech.*, 3:577–599.

Terada, K. 2008. Common knowledge of computational mechanics, *JSCE*, Maruzen, 44–45 (In Japanese).

Togawa, H. 1975. *Vibration Analysis by Finite Element Method*, Saiennsu Sya. (In Japanese)

Vardoulakis, I., and Sulem, J. 1995. *Bifurcation Analysis in Geomechanics*, Blackie Academic & Professional.

Washizu, H. 1982. *Variational Methods in Elasticity and Plasticity,* 3rd ed., Pergamon Press, Oxford.

Zienkiewicz, O.C. 1977. *The Finite Element Method,* 3rd ed., McGraw-Hill, London.

Zienkiewicz, O.C., and Taylor, R.L. 1997. *The Finite Element Method,* 4th ed., McGraw-Hill, London.

Chapter 7

Consolidation analysis

It is well known that there are two types of time-dependent behavior for soil. One is consolidation and the other is brought about by the inherent viscous nature of the soil skeleton. The interaction between the pore water and the soil skeleton results in consolidation. The viscous properties of the soil skeleton are related to the microstructure of the soil particles. Although many problems due to the consolidation of various types of soil have been solved, some problems still exist. One of them is the interaction between the viscosity and the changes in the soil structure. In the following, two problems will be discussed. One is the influence of the soil specimen thickness on consolidation and the other is the interaction between the viscoplastic properties and the strain softening due to structural changes.

7.1 CONSOLIDATION BEHAVIOR OF CLAYS

It has been reported that the influence of specimen thickness on consolidation plays an important role in the prediction of the actual settlements due to the consolidation (Aboshi 1973; Ladd et al. 1977; Aboshi and Matsuda 1981; Oka et al. 1986; Leroueil 1995; Mesri et al. 1995; Tang and Imai 1995; Oka 2005). As is well known, in the general report for the 9th ICSMFE, Ladd et al. (1977) showed two hypotheses for consolidation behavior by compiling the previous results (Figure 7.1). Curve A is supported by Ladd et al. (1977) and Mesri and Rokhsar (1974). Curve B is based on the hypothesis that there is a unique stress–strain–time relationship with respect to time-dependent characteristics and that creep deformation occurs from the beginning of the consolidation. Curve C is between curves A and B, and appears to correspond to the experimental results of Aboshi (1973, 1995, 2004), shown in Figure 7.2.

Many researchers have reported that the anomalous pore pressure behavior, pore pressure stagnation, or a continuous increase after all the fill placement, which appears to be associated with the collapse of the soil structure, can be recognized during the consolidation process. Bishop and

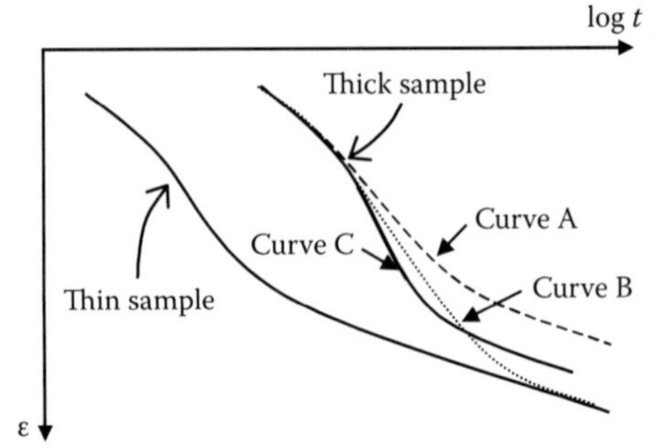

Figure 7.1 Schematic diagram of the average strain to time.

Lovenbury (1969) conducted constant stress creep tests under drained conditions on undisturbed clay and observed a sudden increase in the strain rate. Concerning field cases, the anomalous pore pressure behavior during the consolidation of soft clay has been reported by many researchers. Mesri and Choi (1979) first pointed out that this kind of problem was based on the in situ measured results of the pore water pressure. Mitchell (1986) reported this behavior as a surprising behavior in soil mechanics due to a structural breakdown during compression. Furthermore, Leroueil (1988) observed increases in the pore water after the completion of the construction of test embankments, reflecting the fact that the effective stress temporarily diminished in the stress–strain curve. The prediction of these phenomena by

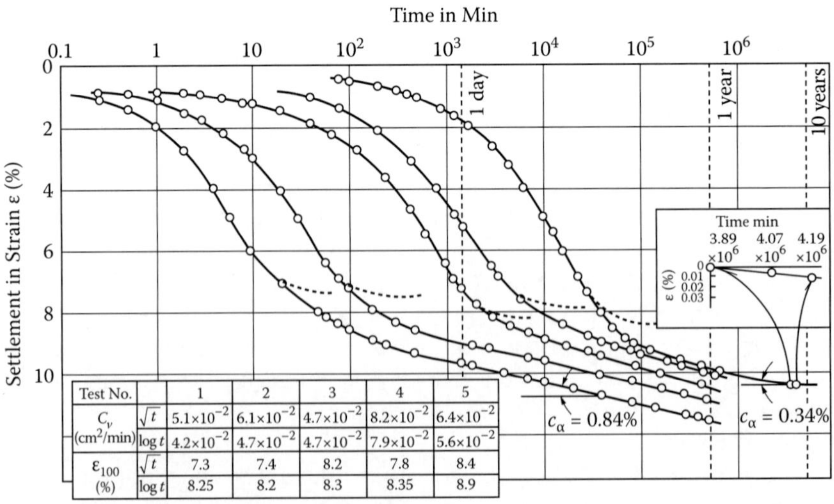

Test No.		1	2	3	4	5
C_v (cm²/min)	$\sqrt{t}$	5.1×10^{-2}	6.1×10^{-2}	4.7×10^{-2}	8.2×10^{-2}	6.4×10^{-2}
	$\log t$	4.2×10^{-2}	4.7×10^{-2}	4.7×10^{-2}	7.9×10^{-2}	5.6×10^{-2}
ε_{100} (%)	$\sqrt{t}$	7.3	7.4	8.2	7.8	8.4
	$\log t$	8.25	8.2	8.3	8.35	8.9

Figure 7.2 Comparison of settlement curves. (From Aboshi, H., 1973, Proc. 8th ICSMFE, Moscow, 4(3):88.)

numerical methods has been studied since the 1980s. Kabbaj et al. (1986) analyzed one-dimensional creep tests by the finite difference method using an elastoviscoplastic constitutive model (Oka 1981). They showed that the strain rate remained momentarily constant during creep simulations around the preconsolidation pressure.

7.2 CONSOLIDATION ANALYSIS: SMALL STRAIN ANALYSIS

7.2.1 One-dimensional consolidation problem

To examine the effect of a sample thickness of clay on consolidation phenomena, we used the constitutive equations introduced in the previous section, which were developed by Adachi and Oka (1982), for the consolidation analysis of clay strata of different heights. The finite element mesh is shown in Figure 7.3. Four-node isoparametric elements are used. The material

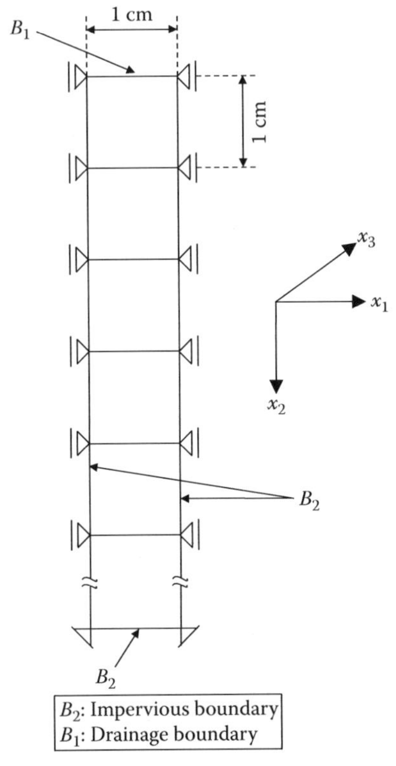

Figure 7.3 Finite element mesh.

Table 7.1 Material parameters

λ	κ	M_m^*	m'	C_k^*	k_0 *(m/sec)*
0.231	0.05	0.865	25.0	0.1	1.16×10^{-9}
G_0 (kPa)	e_0	K_0	$\sigma'_{22(0)}$ (kPa)	$u_{w(0)}$ (kPa)	
3675.0	1.5	0.5	196.0	98.0	

parameters are listed in Table 7.1. The coefficient of permeability is assumed to be determined by

$$k = k_0 \exp[(e - e_0)/C_k^*] \tag{7.1}$$

where k_0 is the permeability coefficient, C_k^* is a material constant, and the subscript 0 indicates the initial value.

For a one-dimensional consolidation problem, the fully implicit scheme is used to calculate the viscoplastic strain increment.

Vertical settlement–time curves for different heights are shown in Figure 7.4. The effect of the length of the drainage path has been well investigated. Various hypotheses, which can explain the previously mentioned scale effect, are illustrated schematically in Figure 7.1. Curve A is supported by Ladd (1973) and by Mesri and Rokhsar (1974); curve B is supported by Barden (1969); curve C appears to correspond to the experimental results of Aboshi. The calculated results denoted by the solid lines in Figure 7.4 are nearly equal to those of curve B. In this case, the initial viscoplastic strain

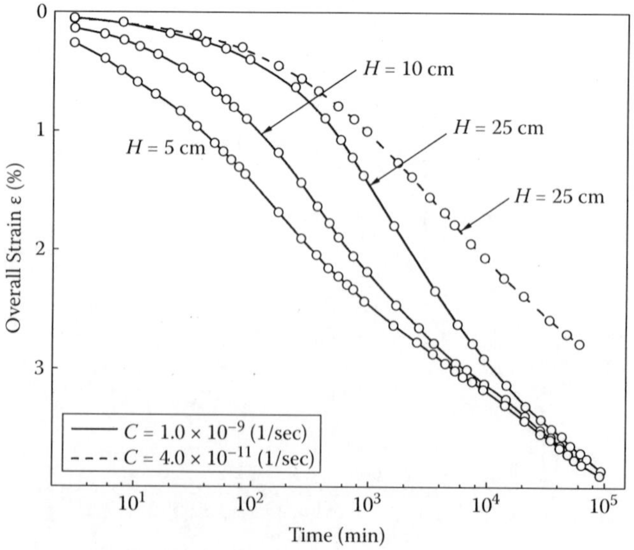

Figure 7.4 Settlement–time curve.

rates are assumed to be equal. In the experimental study by Aboshi (1973), howeve, the initial strain on the four clay specimens of different heights is almost equal, but the initial volumetric strain rate for the thick sample is lower than that for the thin sample.

As has been pointed out by Sekiguchi and Toriihara (1976), the value of the initial volumetric strain rate, $\dot{v}^p_{(0)}(=CM^*_m)$, plays an important role in the consolidation analysis. Predictably, the values of $\dot{v}^p_{(0)}$ for the thick clay deposits in the field may differ from those for the thin samples used in laboratory tests. If we hypothesize that $\dot{v}^p_{(0)}$ is proportional to H^{-2} (H is the height of the specimen), the broken line in Figure 7.4 is obtained. In this case, the tendency is for the settlement–time curve to approach the value of curve A in Figure 7.1. The calculated results are very sensitive to the value of $C=\dot{v}^p_{(0)}/M^*_m$, as seen in Figure 7.4. Clearly, it is then necessary to accurately determine the value of C, in other words, the initial viscoplastic strain rate, when we want to obtain good predictions of the settlement–time curve. If a unique relation among stress, strain, and the strain rate exists, that is, if the concept of isotaches (Suklje 1969) holds true, theoretically we can predict that the effect of sample thickness is described by curve B in Figure 7.1.

Parameter C also reflects the effect of the duration of the preceding secondary compression, that is, quasi-overconsolidation. A heavily aged clay stratum has a large value for the initial volumetric strain in comparison to the value for normally consolidated clay, that is, young clay. Conversely, if we assume that the value of the initial volumetric strain is zero for both types of clay at the beginning of consolidation, then the value of C for aged clay is smaller than that for young clay.

The theoretical effect of the value of C on the excess pore water pressure–time profile is shown in Figure 7.5. The calculated results are supported by the experimental results of Barden (1969) and of Mesri and Choi (1979). The curve that corresponds to the smaller value of C is S-shaped and has an inflection point. Leroueil, Lebihan, and Tavenas (1980) reported similar results in one-dimensional tests on undisturbed samples of Champlain clay in relation to the determination of the preconsolidation pressure. They took into account the effective stress at the inflection point of the u_W-t curve of the preconsolidation pressure. The $e-\ln\sigma'_m$ relation, in which the value of σ'_m at point A′, corresponding to point A in Figure 7.5, is given in Figure 7.6. From the above numerical study, we conclude that the proposed theory can also predict the one-dimensional consolidation behavior of undisturbed aged clay.

Next, we discuss the stress path during the one-dimensional consolidation of a clay stratum. Akai and Adachi (1965) and Moore and Spencer (1972) have shown that the stress ratio of vertical to horizontal effective stress is nearly constant during the one-dimensional consolidation of normally consolidated clay. The solid line in Figure 7.7 shows the stress path when H is 5 cm. In the early stages of consolidation, the stress ratio decreases, after which it approaches the line defined by the initial stress

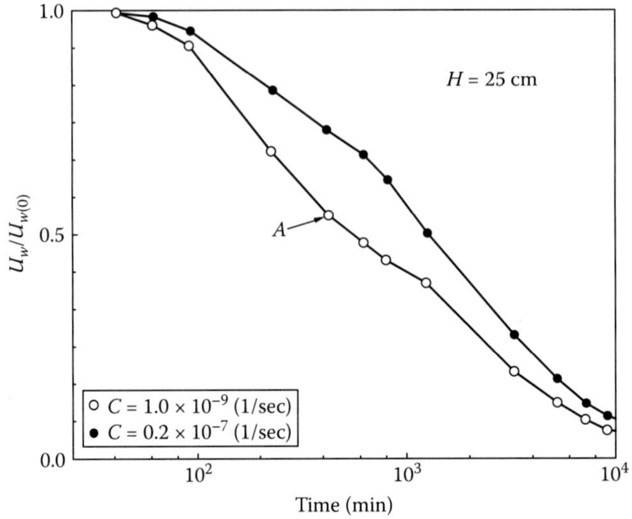

Figure 7.5 Pore water pressure–time relations.

ratio, the so-called K_0-line. Comparing the two curves, we conclude that the separation of the stress path from the K_0-line is caused by the elasticity of the soil and that the stress path approaches the K_0-line because of the viscoplasticity of the soil. Similar experimental results have been obtained by Akai and Adachi (1965). A detailed interpretation of these results, based on Equation (5.15) follows.

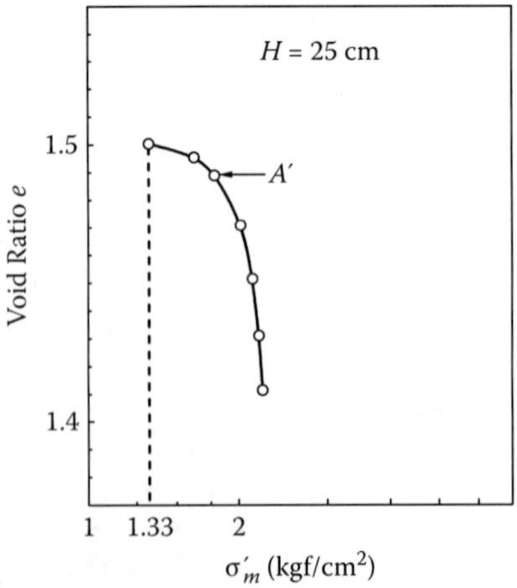

Figure 7.6 Void ratio–mean effective stress relation.

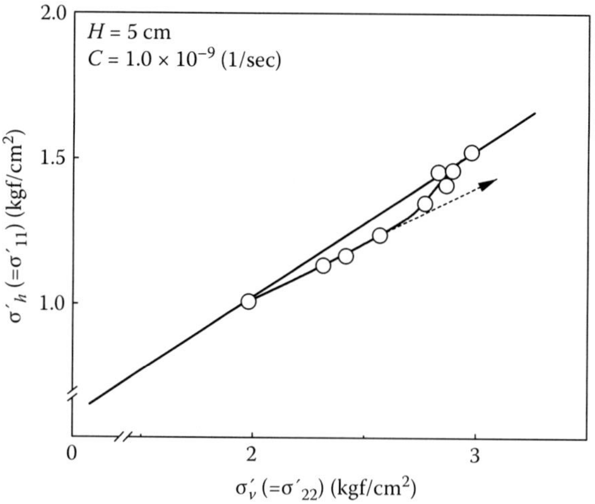

Figure 7.7 Effective stress path.

When the lateral deformation is constrained, that is, $\Delta\varepsilon_{11} = 0$ under plane strain conditions, the stress increments are given by

$$\Delta\sigma'_{11} = \left(K - \frac{2}{3}G\right)\Delta\varepsilon_{22} - \frac{2}{3}GCC_3\left[-\frac{1}{\bar{\eta}^*}\{q/\sigma'_m - (q/\sigma'_m)_{(0)}\}\right]\Delta t$$

$$- KCC_3\left[M^*_m - \frac{s_{kl}(\eta^*_{kl} - \eta^*_{kl(0)})}{\bar{\eta}^*\sigma'_m}\right]\Delta t \tag{7.2}$$

$$\Delta\sigma'_{22} = \left(K + \frac{4}{3}G\right)\Delta\varepsilon_{22} - \frac{2}{3}GCC_3\left[\frac{2}{\bar{\eta}^*}\{q/\sigma'_m - (q/\sigma'_m)_{(0)}\}\right]\Delta t$$

$$- KCC_3\left[M^*_m - \frac{s_{kl}(\eta^*_{kl} - \eta^*_{kl(0)})}{\bar{\eta}^*\sigma'_m}\right]\Delta t \tag{7.3}$$

$$\Delta\sigma'_{33} = \Delta\sigma'_{11} \tag{7.4}$$

in which $q = \sigma_{22} - \sigma_{11}$ and subscript (0) denotes the values at the end of the anisotropic consolidation,namely,

$$K = \frac{(1+e)\sigma'_m}{\kappa} \quad \text{and} \quad C_3 = \exp[m'\{\bar{\eta}^*_{(0)}/M^*_m + \ln(\sigma'_m/\sigma'_{me}) - (1+e)v^p/(\lambda-\kappa)\}].$$

In this case, the shear stress is negligible. When $[q/\sigma'_m - (q/\sigma'_m)_{(0)}]$ is positive, the second term on the right-hand side of Equation (7.2) is positive and that of Equation (7.3) is negative. Therefore, the difference between σ'_{22} and σ'_{11} decreases and the stress path asymptotically approaches the state defined by the K_0-line. The stress state of clay element returns to its original stress ratio in all cases. If K_0 is 0.5, the assumption that the effective stress ratio is constant during the one-dimensional consolidation is true only when the elastic Poisson's ratio is 1/3. This means that the memory of the initial stress ratio does not fade in soil defined by the proposed constitutive equations for clay.

7.2.2 Two-dimensional consolidation problem

The response of clay foundations during the construction of embankments has been explained in detail on the basis of field data collected by Tavenas and Leroueil (1980). They reported that none of the existing analytical methods can universally describe field data. In this section, we have tried to account theoretically for the real behavior of clay foundations during the construction of embankments.

The behavior of clay foundations under plane strain conditions during the construction of embankments can be analyzed numerically using elastoviscoplastic constitutive equations and Biot's consolidation theory. The finite element mesh and the boundary conditions are shown in Figure 7.8. The material parameters and the initial conditions are listed in Table 7.2. The time increment in the calculation is 8 h. Coefficient of permeability, k, and elastic shear modulus, G, are determined as in the one-dimensional

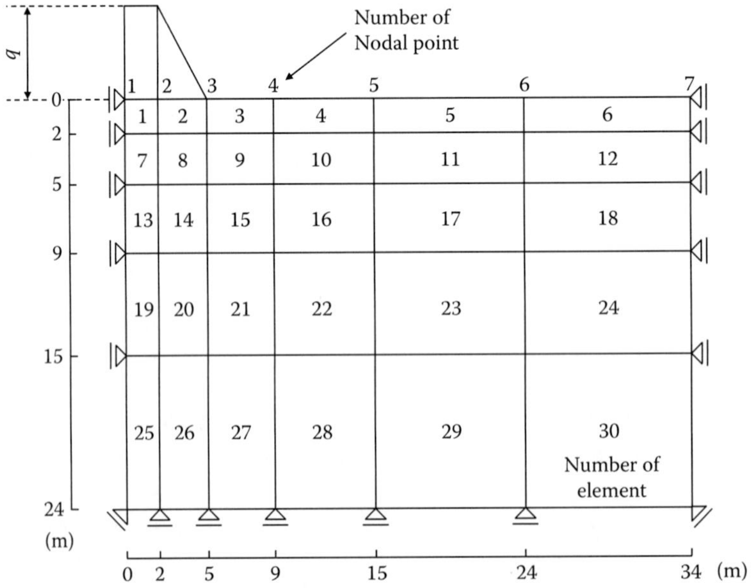

Figure 7.8 Finite element mesh.

Table 7.2 Material parameters

λ	κ	M_m^*	m'	C_k^*	k_0 *(m/sec)*
0.231	0.05	0.865	25.0	0.1	1.16×1.0^{-10}
G_0 (kPa)	e_0	K_0	$\sigma'_{22(0)}$ (kPa)	C (1/s)	
1960.0	1.5	0.5	98.0	1.0×1.0^{-12}	

problem. Euler's time integration method and the fully implicit scheme are used to calculate the strain increment.

The following type of loading rate, dq/dt, is also used:

$dq/dt = 0.2$ ton/m/day (calculation being stopped at 100 days)

The relation between the excess pore water pressure, u_w, and the total vertical stress increment, $\Delta\sigma'_v$, of element 1 in case A is shown in Figure 7.9. The ratio of excess pore water pressure to the vertical stress increment, $\Delta u_w / \Delta\sigma_v$, is almost 0.75 during the first 33 days after construction of the embankment has begun; thereafter, $\Delta u_w / \Delta\sigma_v$ increases and becomes equal to 1.0.

After compiling many field observations, Tavenas and Leroueil (1980) found the typical pattern for the ratio of the excess pore water pressure increment under the center of fill to the applied vertical stress on the clay foundation during the construction of an embankment. It is similar to the calculated result given earlier.

They explained this phenomenon as follows: as natural clay is initially characterized by the typical properties of overconsolidated clay (high

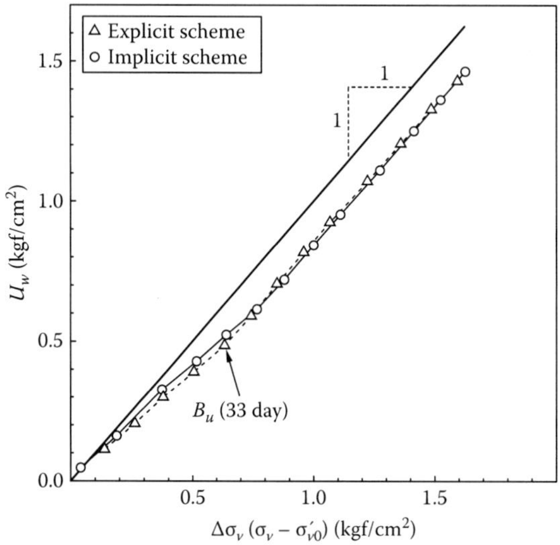

Figure 7.9 Excess pore water pressure–total vertical stress relations.

stiffness), the rate of pore water pressure dissipation is high at the beginning of construction. At that time, $\Delta u_w / \Delta \sigma_v$, is smaller than unity. As the construction advances, however, the pore water pressure increment becomes equal to the embankment load increment, $\Delta u_w / \Delta \sigma_v$, being almost equal to unity due to the passage of the clay to the state of normally consolidated clay. One possible reason why the proposed theory can explain the phenomenon described is that the proposed constitutive equations can describe the quasi-overconsolidated effect of natural clays due to aging, as discussed in the previous section on one-dimensional consolidation.

Tavenas and Leroueil (1980) suggested that it is appropriate to describe the lateral displacement behavior by the relation of maximum lateral displacement, y_m, below the toe of the embankment, to the settlement s of the center line. It can been seen in Figure 7.11 that dy_m/ds is small in the early stage of loading, but it increases after 33 days. Tavenas and Leroueil (1980) reported that inflection point B_u in Figure 7.9 corresponds to point B_s in Figure 7.11. In this case, the two points do agree.

The difference between the settlement–lateral displacement curves obtained by Euler's integration method and by the fully implicit scheme increases with the increase in settlement. The settlement–time profile and the excess pore water pressure–time profile of element l are given in Figure 7.10. Point C corresponds to the inflection points B_u and B_s in Figures 7.9 and 7.11. The distributions with depth of the lateral displacement and the excess pore water pressure build-up during construction are shown in Figure 7.12.

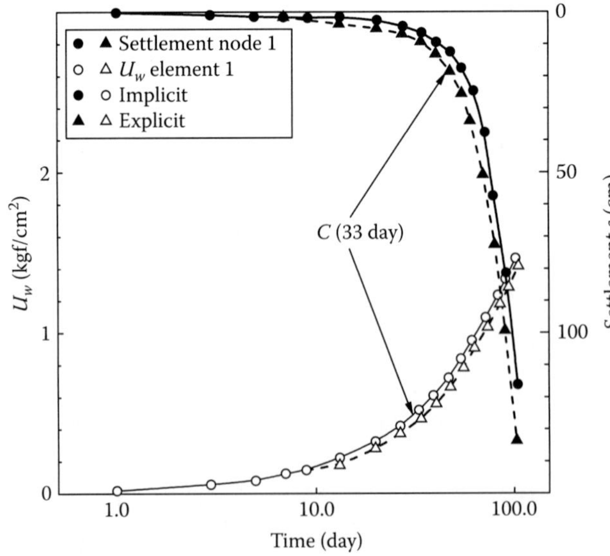

Figure 7.10 Excess pore water pressure–time profile and settlement–time profile.

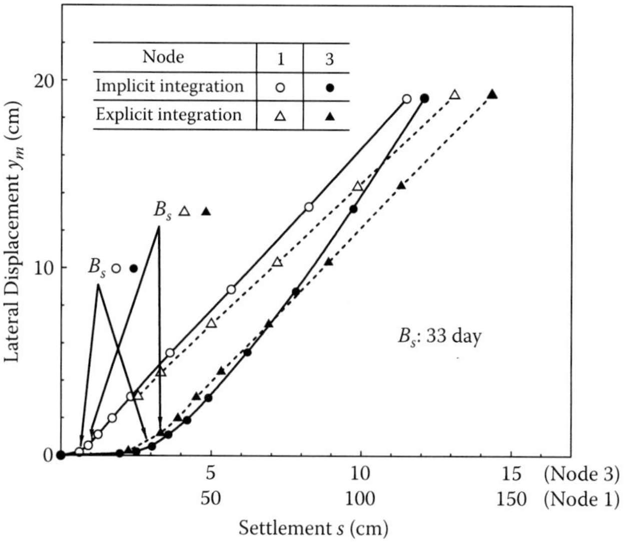

Figure 7.11 Lateral displacement–settlement relations.

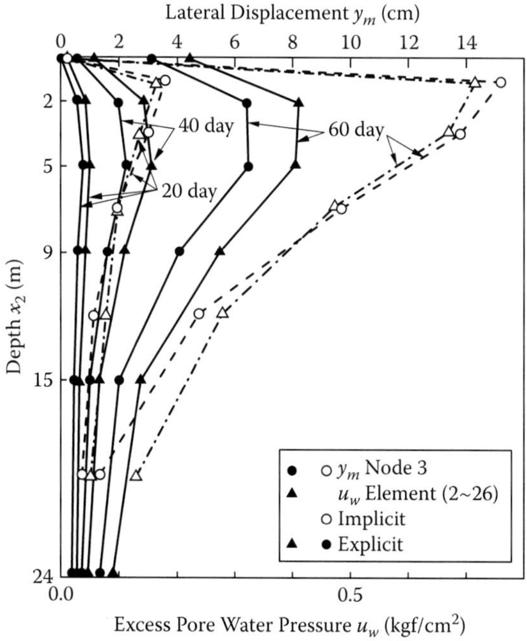

Figure 7.12 Distributions of lateral displacement and excess pore water pressure with depth.

Both lateral displacement and excess pore water pressure are at maxima between 2 and 5 m.

7.2.3 Summary

One-dimensional consolidation

1. The theory can describe the effect of the length of the drainage path on the settlement–time curve. We also confirmed that the initial volumetric inelastic strain rate has an important function in the prediction of the settlement–time curve.
2. Parameter C in Equation (5.19) reflects how long the clay stratum has been aged in situ. Heavily aged clay has a smaller C value than young clay.
3. The excess pore water pressure–time profile, which is peculiar to aged clay, can be simulated by the proposed constitutive equations.
4. According to our proposed theory, the assumption that the ratio of horizontal to vertical effective stress is nearly constant is true only when the elastic material constants take special values.

Two-dimensional consolidation

1. The trend in the calculated relations between the lateral displacement at the toe of the embankment and the settlement under the center of the embankment during construction is in good agreement with the empirical relations given by Tavenas and Leroueil (1980) based on their field observations.
2. The typical empirical relation between the applied vertical stress increment and the excess pore water pressure increment in a clay foundation during embankment construction, proposed by Tavenas and Leroueil (1980), is very similar to the calculated relations between the total vertical stress increment and the excess pore water pressure increment.
3. With the theory, it is possible to simulate the behavior of a clay foundation during the construction of an embankment.

7.3 CONSOLIDATION ANALYSIS WITH A MODEL CONSIDERING STRUCTURAL DEGRADATION

In this section of the present study, one-dimensional consolidation behavior is simulated using an elastoviscoplastic constitutive model taking into account the effect of structural degradation. We will discuss the influence of sample thickness with respect to the soil structure and the initial strain rate.

One-dimensional consolidation has been numerically examined by a finite element analysis. For boundary value problems related to the soil–water coupled consolidation problem, Biot's two-phase mixture theory is adopted.

The infinitesimal strain is valid for these problems since large deformations are not expected. A four-node quadrilateral element with a reduced Gaussian integration is used for the displacement, and the pore water pressure is defined at the center of each element. The top of the specimen is set to be permeable while the bottom and the sides are set to be impermeable. The size of each element is 0.4 cm × 0.4 cm for all the calculations.

7.3.1 Effect of sample thickness

Simulations have been performed for normally consolidated clay. The initial stress conditions of the calculations and the material parameters are shown in Table 7.3. The elastic modulus, G_0, was set at 36100 kPa in the previous calculations (Kimoto 2002), and the settlement during the primary consolidation was much smaller than that obtained during the secondary consolidation. For comparison, G_0 is supposed to 360 kPa in the present study. Viscoplastic parameter C, which describes the initial viscoplastic strain rate, is set to be 1.0×10^{-13} (1/s) (=C_0) at first. An excess porepressure level of 1160 kPa, which is twice as large as the compression yield stress, is applied as the initial loading for all the analyses.

In the first calculation, viscoplastic parameter C is assumed to be C_0 for samples with different heights (H = 2, 20 cm). Vertical strain reaches almost the same value at 6×10^6 seconds (= 70 days). The results correspond to curve B as shown in Figure 7.1.

In the next calculation, an initial viscoplastic strain rate C is assumed to be inversely proportional to H^2, that is, $C = C_0 (H_0/H)^2$ (H is the height of the specimen and H_0 is set at 2 cm). The results are shown in Figure 7.13.

Table 7.3 Initial conditions and material parameters

Initial mean effective stress	$\sigma'_{m(0)}$ = 580 kPa
Coefficient of earth pressure	K_0 = 0.5
Coefficient of permeability	$k_0 = 0.8 \times 10^{-9}$ m/s
Permeability change index	C^*_k = 0.1
Elastic shear modulus	G_0 = 360 kPa
Compression index	λ = 0.508
Swelling index	κ = 0.0261
Initial void ratio	e_0 = 1.70
Compression yield stress	σ'_{mai} (=σ'_{mai}) = 580 kPa
Stress ratio at maximum compression	M^*_m = 1.09
Viscoplastic parameter	m' = 18.5
Viscoplastic parameter	$C = 1.3 \times 10^{-13}$ 1/s (=C_0)
Structural parameter	σ'_{maf} = 290 kPa
Structural parameter	β = 0, 5, 20, 40

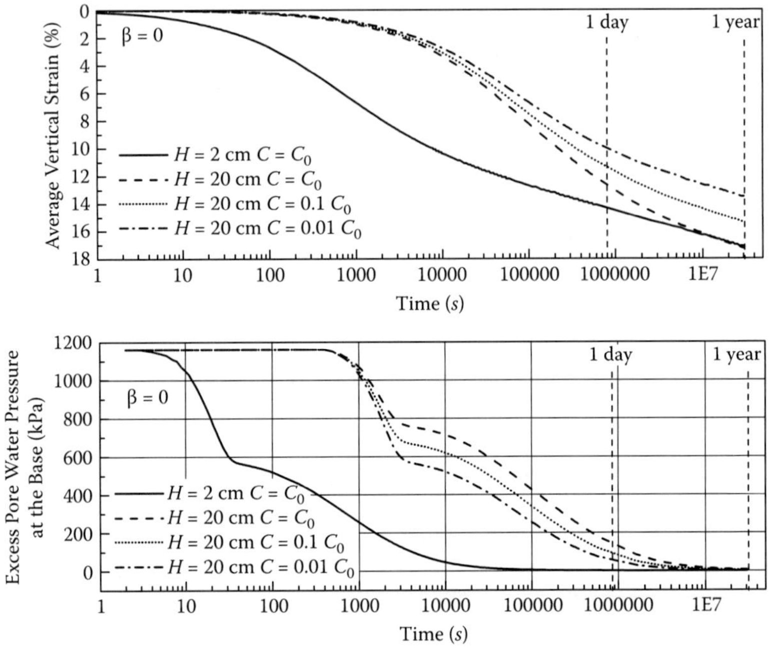

Figure 7.13 Effects of parameter *C* (without structure $\beta = 0$, $C_0 = 1.3 \times 10^{-13}$).

The two lines ($C = C_0$ for $H_0 = 2$ and $C = 0.01C_0$ for $H = 20$) are parallel to each other; this corresponds to curve A.

Aboshi (1973) experimentally observed that the initial strain rate for the thick sample is lower than that for the thin one. The pre-compression as a preparatory consolidation was performed with the stress change between 0 and 19.6 kPa. The data of the compression by Aboshi are shown as follows (from private communication with Aboshi in 1983, presented in Oka 2000).

The initial strains, $\varepsilon_{v(0)}$, are just after the preparatory consolidation. The initial strain rates, $\dot{\varepsilon}_{v(0)}$, are average values between consolidation stresses of 9.8–19.6 kPa, except No. 5. For No. 5, the strain rate is calculated between consolidation pressures of 0–19.6 kPa. The average initial water content is 80%:

Specimen No. 1 ($H = 2$ cm)
- Pressure 0–9.8 kPa : $\varepsilon_v = 1.3\%$ (duration time: 1440 min)
- Pressure 9.8–19.6 kPa: $\varepsilon_v = 3.4\%$ (1440 min)
- Average strain rate $\dot{\varepsilon}_{v(0)} = 2.36 \times 10^{-3}\%/\text{min}$ (9.8–19.6 kPa)
- Total strain after preparatory consolidation $\varepsilon_{v(0)} = 4.7\%$

Specimen No. 2 ($H = 4.8$ cm)
- Pressure 0–9.8 kPa: $\varepsilon_v = 2.8\%$ (1440 min)

Pressure 9.8–19.6 kPa: $\varepsilon_v = 2.5\%$ (1440 min)
Average strain rate $\dot{\varepsilon}_{v(0)} = 1.736\times 10^{-3}\%/\text{min}$ (9.8–19.6 kPa)
Total strain after preparatory consolidation $\varepsilon_{v(0)} = 5.3\%$

Specimen No. 3 ($H = 20$ cm)
Pressure 0–9.8 kPa: $\varepsilon_v = 1.9\%$ (11000 min)
Pressure 9.8–19.6 kPa: $\varepsilon_v = 2.2\%$ (21000 min)
Average strain rate $\dot{\varepsilon}_{v(0)} = 1.05\times10^{-4}\ \%/\text{min}$ (9.8–19.6 kPa)
Total strain after preparatory consolidation $\varepsilon_{v(0)} = 4.1\%$

Specimen No. 4 ($H = 40$ cm)
Pressure 0–9.8 kPa: $\varepsilon_v = 2.3\%$ (20000 min)
Pressure 9.8–19.6 kPa: $\varepsilon_v = 2.7\%$ (42000 min)
Average strain rate $\dot{\varepsilon}_{v(0)} = 6.43\times 10^{-5}\ \%/\text{min}$ (9.8–19.6 kPa)
Total strain after preparatory consolidation $\varepsilon_{v(0)} = 5.0\%$

Specimen No. 5 ($H = 100$ cm)
Pressure 0–19.6 kPa: $\varepsilon_v = 4.7\%$ (62000 min)
$\dot{\varepsilon}_{v(0)} = 7.58\times 10^{-5}\%/\text{min}$ (0–19.6 kPa)

Total strain after preparatory consolidation $\varepsilon_{v(0)} = 4.7\%$

As shown, the initial strain rate of the thick sample just before consolidation is smaller than that of the thin sample, although the strain just after the preparatory consolidation is almost equal. The reason for the difference in strain rates is that the periods of preconsolidation are different, namely, from one day for the thinner sample to four months for the thicker one.

In the calculation in Figure 7.13, we assumed that $C_0(H_0/H)^2 < C = C_0(H_0/H) < C_0$, namely, $C = 0.1\,C_0$ for $H = 20$ cm. Considering Equations (5.27) and (5.67), the assumption leads to viscoplastic parameter C depending on the initial viscoplastic strain rates and $C_0\left(\dot{\varepsilon}_{i0}/\dot{\varepsilon}_i\right)^2 < C = C_0\left(\dot{\varepsilon}_{i0}/\dot{\varepsilon}_i\right) < C_0$, where $\dot{\varepsilon}_i$ is the initial strain rate before consolidation and $\dot{\varepsilon}_{i0}$ is the initial strain rate before consolidation of the sample with a height of H_0, although the strain rate in Equations (5.27) and (5.67) is the viscoplastic volumetric strain rate.

In this case, curve C is obtained when the value of C is larger than $C_0\left(\dot{\varepsilon}_{i0}/\dot{\varepsilon}_i\right)^2$ and smaller than C_0, which is consistent with Aboshi's experiments. Consequently, it was found that the effect of the sample thickness depends on the value of C, that is,

$$C = C_0\left(\dot{\varepsilon}_{i0}/\dot{\varepsilon}_i\right)^2 \quad \text{(for curve } A\text{)}$$

$$C_0\left(\dot{\varepsilon}_{i0}/\dot{\varepsilon}_i\right)^2 < C = C_0\left(\dot{\varepsilon}_{i0}/\dot{\varepsilon}_i\right) < C_0 \quad \text{(for curve } C\text{)}$$

$$C = C_0 \quad \text{(for curve } B\text{)}$$

The effect of sample thickness on the time–settlement curve is mainly due to the difference in strain rates before consolidation. This explanation was first pointed by Oka, Adachi, and Okano (1986), and was confirmed by Tang and Imai (1995). Of course there might be other reasons such as sample disturbance and natural variability (Mesri and Choi 1985; Leroueil 1995).

7.3.2 Simulation of Aboshi's experimental results

In this section, Aboshi's consolidation test results with different thicknesses have been numerically simulated.

7.3.2.1 Determination of material parameters

Aboshi (1973) performed consolidation tests on clays with different thicknesses. The specimens were obtained from Hiroshima Bay and reconstituted in a trench, 10 m × 15 m with a depth of 1.5 m, for more than 6 years with a thin sand layer on it. The material contains 27% clay, 68% silt, and 5% sand, and the plasticity index, I_p, is 42% (Aboshi 1973). There are two elastic parameters, six viscoplastic parameters, and two parameters for hydraulic conductivity. Some assumptions are needed since there are not enough data, such as triaxial test results. The determination of the parameters is shown in the following.

7.3.2.2 Elastic parameters

The swelling index, κ, is assumed as 0.2 of compression index λ. λ is obtained as 0.304 from the *e-logp* curve in Matsuda and Aboshi (1983). Initial shear modulus G_0 is determined by assuming that Poisson's ratio is equal to 0.3 at the initial state of the main loading.

7.3.2.3 Viscoplastic parameters

The consolidation yield stress, σ'_{mbi}, is equal to the initial mean effective stress in the normally consolidation region. The stress ratio at maximum compression, M^*_m, which is equal to the failure stress ratio in the model, can be assumed to be around 0.73 to 0.98 from the relationship between the plasticity index and the friction angle for the clays shown in Leroueil and Hight (2003, p. 196). We take the mean value 0.85 for M^*_m. Viscoplastic parameter *m'* is calculated from Equation (5.66) related to the secondary compression rate $C_\alpha = de/d\log t$. Aboshi and Matsuda (1981) mentioned that the average value of the secondary compression coefficient $C_{\varepsilon\alpha} = d\varepsilon/d\log t$ is about 0.95%. However, $C_{\varepsilon\alpha}$ differs for the different thicknesses of the samples and also gradually becomes smaller during the secondary consolidation process. We take the mean value of $C_{\varepsilon\alpha}$ for $H = 1$ and

50 cm from the experimental results by Aboshi and Matsuda (1981). We performed simulations for the value of m' from 22.0 to 33.0, which correspond to $C_{\varepsilon\alpha}$ from 0.98% and 0.63%. Finally, we used 31.0 of m'. The viscoplastic parameter, C_0, is determined by fitting the settlement curve for the specimen since we do not have the triaxial test data. For the void ratio dependency of the permeability coefficient, k, following Tang and Imai (1995), C_k^* is assumed to be 0.478.

7.3.2.4 Consolidation analysis

Simulations of one-dimensional consolidation have been conducted by the finite element analysis. The simulation method, that is, the soil–water coupled consolidation analysis method based on Biot's two-phase mixture theory presented in Chapter 6, is used. A four-node quadrilateral element with a reduced Gaussian one-point integration is used for the displacement, and the pore water pressure is defined at the center of each element. The infinitesimal strain is assumed. The finite element meshes and boundary conditions are shown in Figure 7.14. The top of the specimen is set to be permeable, while the bottom and the side are impermeable. The sample thickness is assumed to be 1.0, 2.4, 10, 20, and 50 cm. First, the loading is simulated from 9.8 kPa to 19.6 kPa as a preparatory consolidation. The material parameters are listed in Table 7.4. From the results of the preparatory consolidation, the strain rates at the end of the primary consolidation have been obtained. Figure 7.15 indicates the relationship between the strain rate at the end of the primary consolidation and the height of the

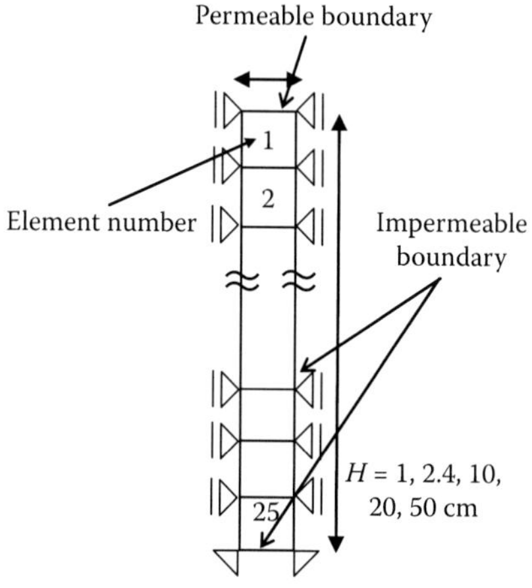

Figure 7.14 Finite element meshes and boundary conditions.

Table 7.4 Material parameters for the preparatory consolidation (9.8–19.6 kPa)

Coefficient of static earth pressure, K_0	0.5
Initial mean effective stress, σ'_{mo} (kPa)	6.53
Initial void ratio	2.22
Initial permeability, k (m/s)	1.9×10^{-09}
Permeability changing index, C^*_k	0.478
Compression index, λ	0.304
Swelling index, κ	0.0608
Initial shear modulus, G_0 (kPa)	221
Compression yield stress, $\sigma'_{mbi} / \sigma'_{m0}$	1.0
Stress ratio at maximum compression, M^*_m	0.85
Viscoplastic parameter, m'	31.0
Viscoplastic parameter, C (1/s)	4.00×10^{-13}

specimen for the preparatory consolidation in which the stress increment of 9.8 kPa has been applied to the specimen with an initial stress of 9.8 kPa, that is, the final stress level is 19.6 kPa.

From Figure 7.15, the average strain rate at the end of the primary consolidation is proportional to $H^{-1.1}$. Hence, in the following analysis of the main consolidation, we assumed that the initial strain rate is proportional to $\dot{\varepsilon}^{-1.1}$, that is, $H^{-1.1}$. This is consistent with the fact that Aboshi (1973) observed that it takes a much longer time for the preparatory

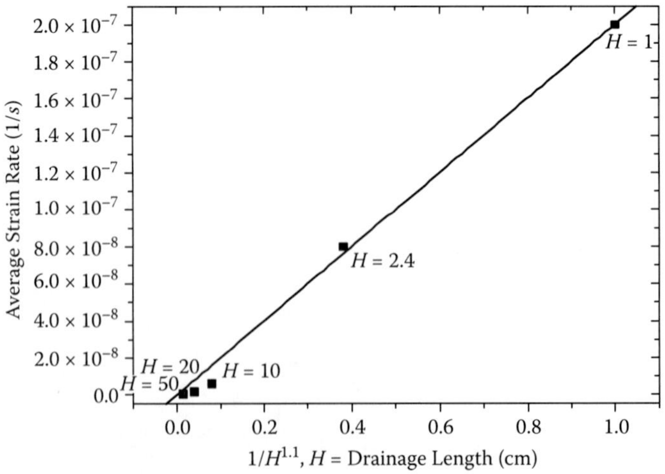

Figure 7.15 Relationship between strain rates at the end of primary consolidation and $H^{-1.1}$ during the preparatory consolidation.

Table 7.5 Material parameters for the main consolidation

Coefficient of static earth pressure, K_0	0.5
Initial mean effective stress, σ'_{m0} (kPa)	13.1
Initial void ratio	2.15
Initial permeability, k (m/s)	1.64×10^{-09}
Permeability changing index, C_k^*	0.478
Compression index, λ	0.304
Swelling index, κ	0.0608
Initial shear modulus, G_0 (kPa)	313
Compression yield stress, $\sigma'_{mbi}/\sigma'_{m0}$	1.0
Stress ratio at maximum compression, M_m^*	0.85
Viscoplastic parameter, m'	31.0
Viscoplastic parameter C (1/s) (for H = 1 cm)	1.00×10^{-14}
Height dependency of C (for H = 2.4, 10, 20, 50)	0, 1.1, 2.0

consolidation of a thicker sample, although the strain levels are almost equal. The material parameters used in the analysis of the main consolidation are listed in Table 7.5. The applied stress increment is 58.8 kPa, and the final stress level is 78.4 kPa. Figure 7.16 shows the simulation of the consolidation with different thicknesses. This figure indicates the assumption that the viscoplastic parameter C is proportional to $\dot{\varepsilon}^{-1.1}$, that is, $H^{-1.1}$, which is taken above is effective. The simulation results obtained with the above assumption reproduce Aboshi's experiments very well, whereas the simulation results based on the other assumptions do not.

7.3.3 Effect of degradation

In the calculations in Figure 7.16, β is set at zero. Next, let us discuss the effect of the parameter on structural degradation β. Calculations are performed for different values of parameter β (= 0, 5, 20, 40). The case of $\beta = 0$ corresponds to the original model without considering structural changes. The vertical settlement with time and the pore water generation with time are shown in Figure 7.17. The excess pore pressure shows a temporary increase when the rate of structural changes takes a high value, in other words, β = 40 (Figure 7.17b). The average void ratio e-log σ'_m relations for β = 0, 20, 40 are shown in Figure 7.18. It shows a temporary decrease in the mean effective stress for the e-log σ'_m relations for β = 20 and 40. A similar tendency is obtained from the test embankment (Leroueil 1988). The inflexion points in the curve correspond to points where the pore water pressure begins to be stagnant or to increase in Figure 7.17b, at about 40 s for β = 0 and 66 s for β = 40.

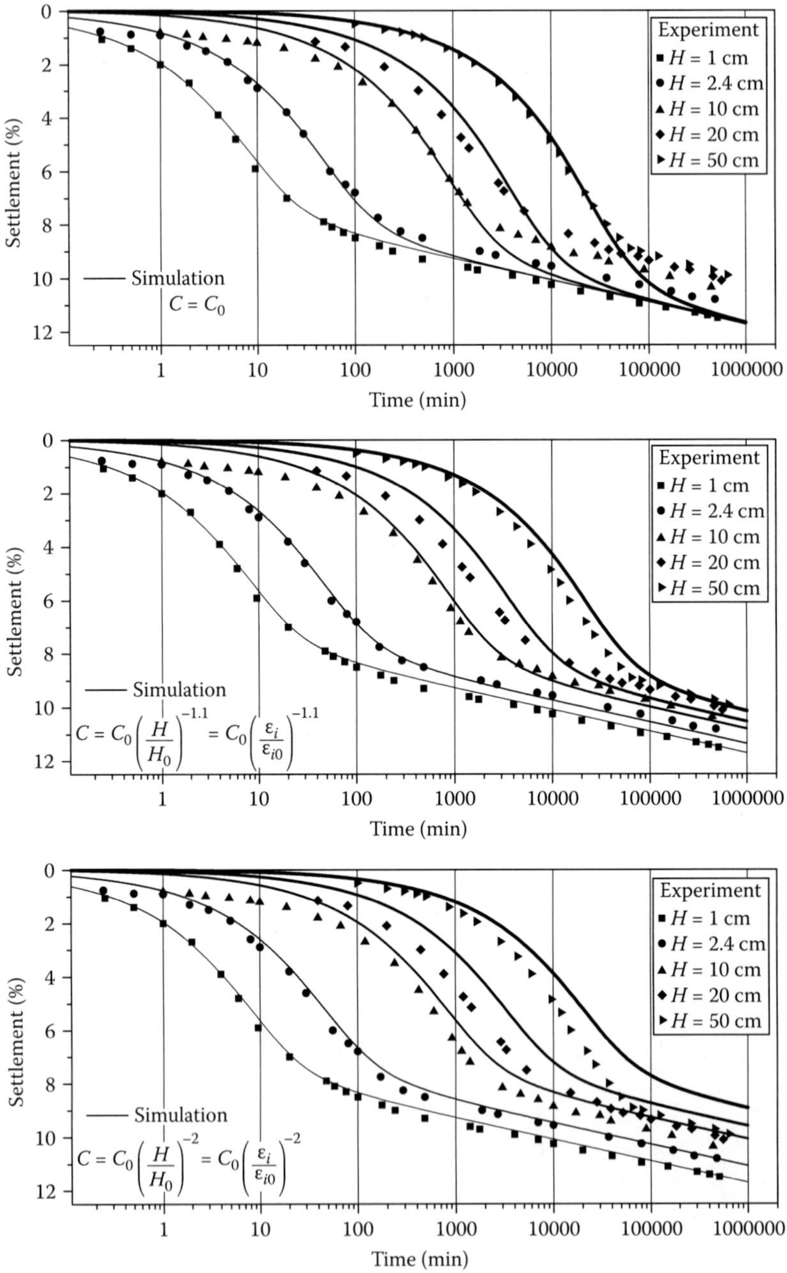

Figure 7.16 Settlement–time relations during the consolidation with different viscoplastic parameters.

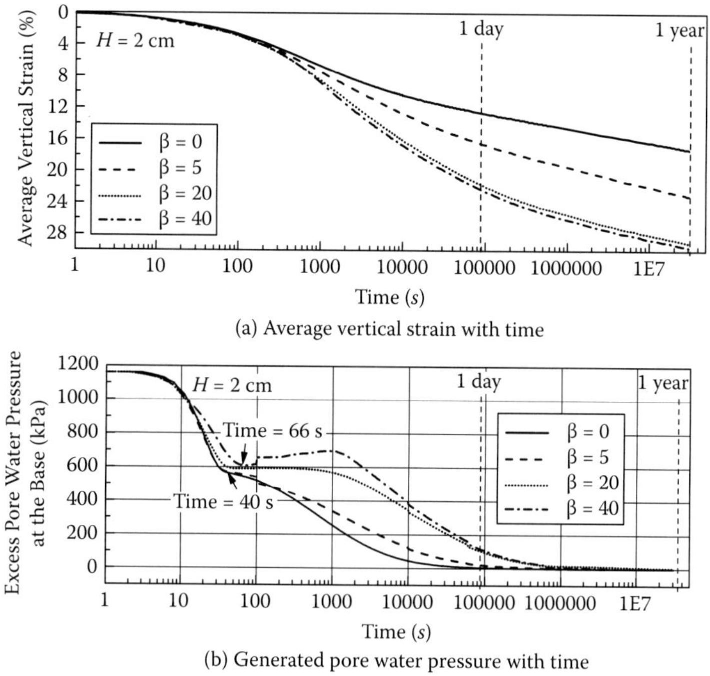

Figure 7.17 Effects of structural parameter β. (a) Average vertical strain with time. (b) Generated pore water pressure with time.

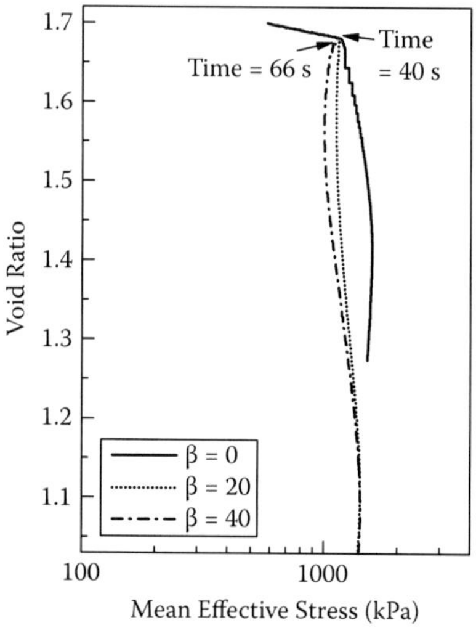

Figure 7.18 Average e-log σ'_m curve.

7.4 CONSOLIDATION ANALYSIS OF CLAY FOUNDATION

7.4.1 Introduction

Stagnation or an increase in the pore water pressure after loading and during the consolidation of soft clay is called anomalous pore pressure. It has been observed after loading and during consolidation by J.K. Mitchell in his 20th Terzaghi Lecture (1986). This problem has hitherto been studied, but it has not yet been fully solved. The main reason is the rate-dependent structural degradation of soft clay, as discussed in the previous section. Oka et al. (1991) numerically analyzed such a phenomenon observed in the clay foundation at St. Alban's test embankment D (Leroueil et al. 1978; Tavenas et al. 1974) using an elastoviscoplastic model. They used a model with volumetric strain softening and a comparatively better solution than the conventional model. However, the stagnation or the temporary increase in pore water pressure after the construction of the embankment could not be reproduced. In this section, we have analyzed the same problem using an elastoviscoplastic model with the aforementioned structural degradation.

7.4.2 Consolidation analysis of soft clay beneath the embankment

A comprehensive numerical investigation has been carried out for the consolidation behavior of soft sensitive clay in order to reproduce the field response. The soil parameters and the experimental results are related to the clay foundation beneath the test embankments, constructed in 1972 at Saint Alban, west of Quebec City, Canada (Tavenas et al. 1974; Leroueil et al. 1978; Oka et al. 1994). The current simulation method is based on the elastoviscoplastic constitutive model with structural changes (Kimoto and Oka 2005), as described in Chapter 5. Based on the finite deformation theory presented in Chapter 6, a two-dimensional finite element code (COMVI3D) has been used. The results of the analysis are compared with the experimental ones. It is noted here that the embankment stiffness and foundation roughness are not considered in the present study for simplicity of the numerical implementation.

7.4.2.1 Soil parameters

The soil parameters and the model parameters for the Champlain clay beneath the test embankments at St. Alban are listed in Table 7.6.

Table 7.6 Soil and material parameters

Layer	*Depth (m)*	λ	e_0	$\sigma'_{22(0)}$ *(kPa)*	σ'_p *(kPa)*	G_0 *(kPa)*	*Others*
1	0~0.66	0.100	1.1	981.6	72.5	880.0	$\kappa = 0.0045(1+e)$
2	0.66~1.5	0.363	1.7	1280.0	57.1	1500.9	$C_k^* = 0.5e$
3	1.5~3.0	0.495	2.3	1770.7	46.0	2079.7	$m'=27.24$
4	3.0~4.8	0.411	1.8	2550.3	70.6	3011.7	$M_m^* = 0.98$
5	4.8~6.7	0.282	1.8	3436.6	88.3	4041.2	$K_0 = 0.8$
6	6.7~9.6	0.175	1.4	4417.2	13.7	5601.5	$k_0 = 1.05 \times 10^{-8}$ m/s
7	9.6~13.5	0.100	1.4	6871.3	17.7	10987.2	$\gamma_w = 9.81\ kN/m^3$

Viscoplastic parameter C is chosen based on the viscoplastic strain, as shown in Table 7.7.

Instead of Equation (5.49), the degradation rule for the soil structure is governed by $\sigma'_{ma} = \sigma'_{maf} + (\sigma'_{mai} - \sigma'_{maf})\exp(-\beta z^a)$ and $a = 0.5$ is assumed. The value of β is taken as 5.856, and the other structural degradation parameters are chosen based on the following relations: $\sigma'_{mai} = \sigma'_{m(0)} OCR_{mean}$ and $\sigma'_{maf} = \sigma'_{m(0)}$. We assumed that the top layer is modeled as being elastic with a Poisson ratio $\nu = 0.25$ and $\kappa = 0.01125(1+e)$ (2.5 times higher than the other layers). The basis for the selection of the values for the parameters is to reproduce the field responses as closely as possible.

7.4.2.2 Soil response beneath embankment

Figure 7.19a shows the optimal finite element mesh configuration for the finite deformation analysis for the soil response beneath embankment *D*. The construction history of embankment *D* is shown in Figure 7.19b. As for the boundary conditions, the bottom boundary of the clay foundation is

Table 7.7 Viscoplastic parameter *C*

Volumetric viscoplastic strain ε^{vp}*(%)*	*C (1/sec)*
$\varepsilon^{vp} < 0.01$	1.2×10^{-12}
$0.01 \le \varepsilon^{vp} \le 22.2$	5.9×10^{-11}
$\varepsilon^{vp} > 22.2$	5.9×10^{-11}

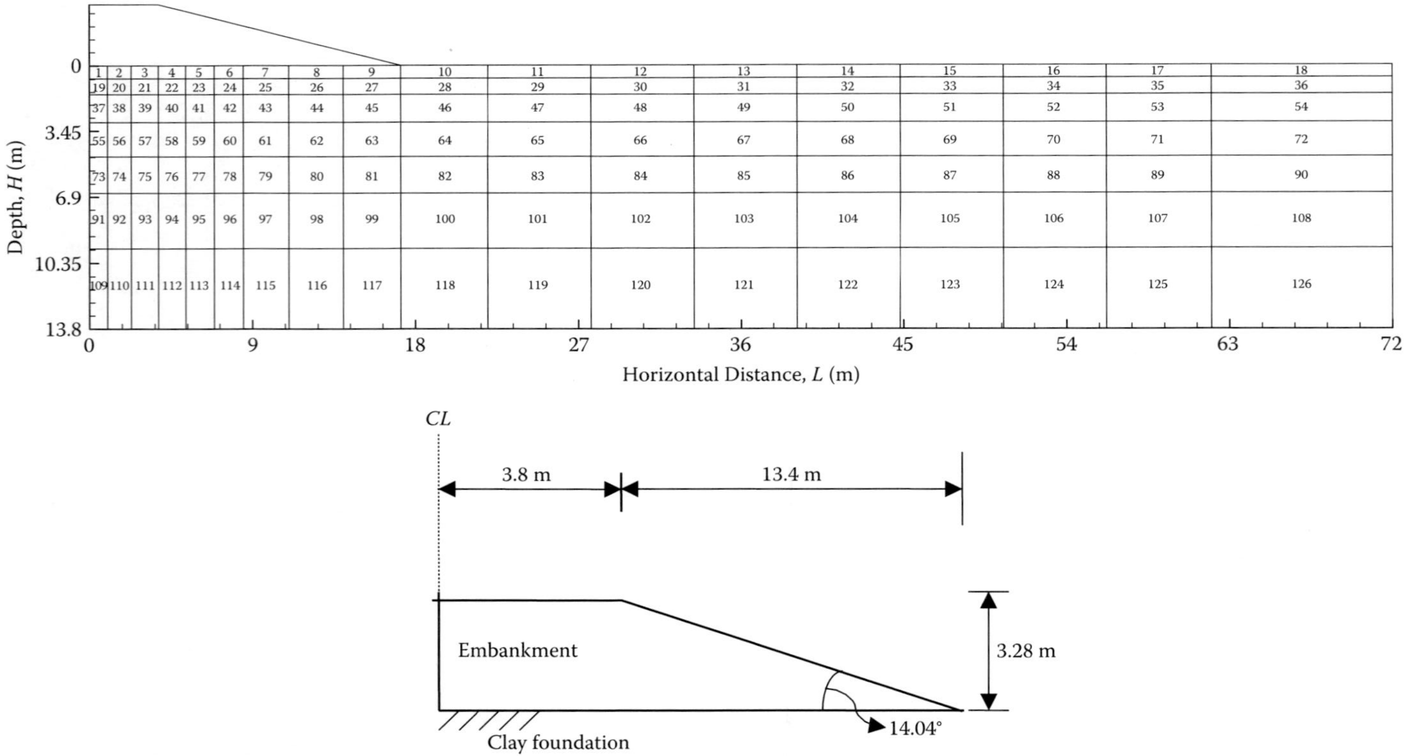

Figure 7.19 (a) Shape of the embankment D at St. Alban and FEM mesh. (b) Construction process for embankment D. (Adapted from Leroueil, S., Tavenas, F., Trak, B., Rochelle, P. L., and Roy, M., 1978, *Can. Geotech. J.*, 15:54–65.)

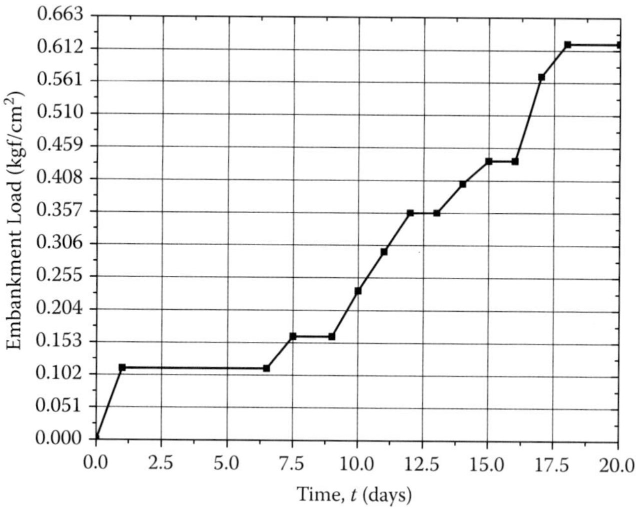

Figure 7.19 (Continued)

assumed to be completely rigid in both directions, while the vertical boundaries are only horizontally restrained. The top surface is free and permeable, while the other boundaries are impermeable because the hydraulic conditions are not clear although it has been reported that the sandy soil layer exists below the depth of 13.8 m (Leroueil et al., 1978). The effect of the boundary conditions have to be examined more in detail in the future. The automatic time increment selection scheme by Karim and Oka (2010) was used throughout the analysis.

The elastoviscoplastic finite element method (FEM) model provides reasonable predictions for the soil responses during and after the construction of the embankment. Figure 7.20 shows the time history of the vertical settlements at node 1 (at the center line of embankment D and on the surface), which depicts an excellent agreement between the FEM model and the test results. However, during the initial period of time, the predicted values by the FEM model are slightly smaller than the field responses. Lateral displacements at the toe of the embankment are shown in Figure 7.21. Although just before the end of the construction, the predicted lateral displacements are higher than the field ones, the overall agreement is quite acceptable. Figure 7.22 shows the distribution of pore water pressure with time at a depth of 2.25 m. Some time after the end of the construction, the pore water pressure becomes stagnant and then slowly increases with time. The distribution of pore water pressure also complies reasonably well with the field response.

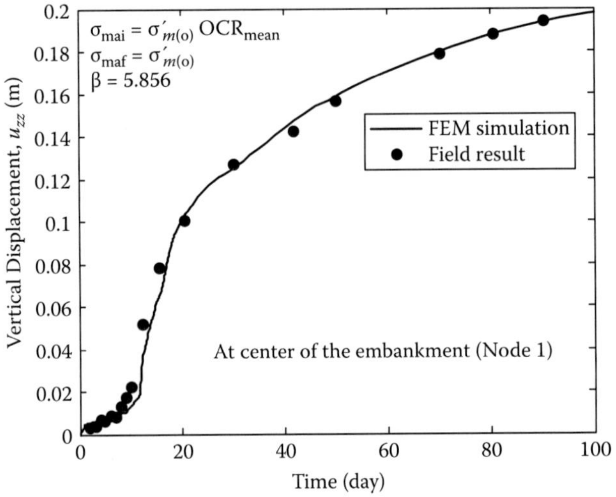

Figure 7.20 Observed and calculated displacement below the centerline of the embankment (node 1).

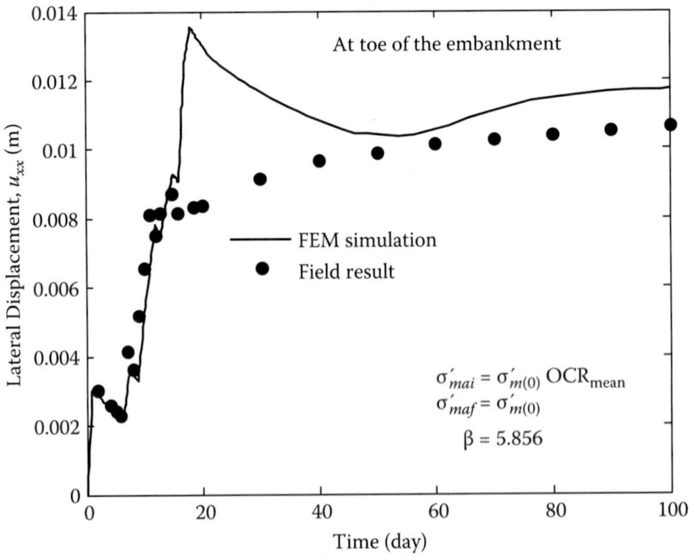

Figure 7.21 Observed and calculated lateral displacement at the toe of the slope.

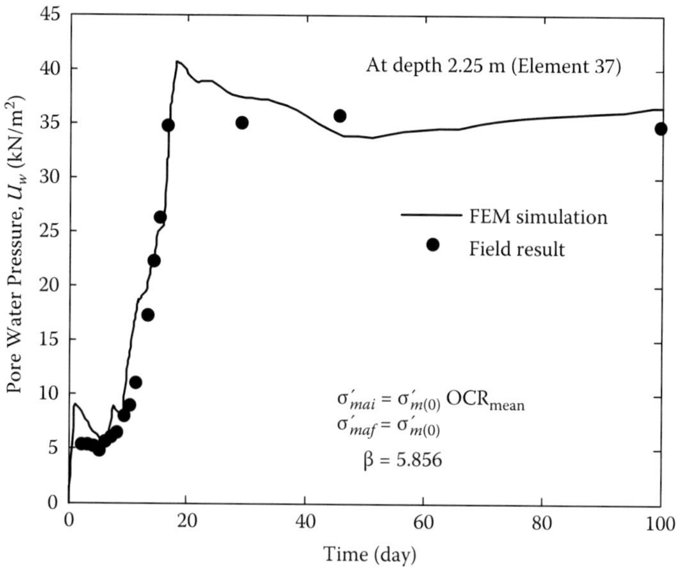

Figure 7.22 Observed and calculated pore–water pressure levels of element 37 beneath the center of the embankment at a depth of 2.25 m.

7.5 CONSOLIDATION ANALYSIS CONSIDERING CONSTRUCTION OF THE EMBANKMENT

The simplest approximation for the simulation of the new material placing in a finite element analysis is by applying the increment of the weight of the fill as the external load on the nodes of the embankment foundation interface. However, for a more accurate analysis, the embankment construction procedure should be implemented, in particular for the large deformation of soft clay deposits, by which the stiffness and the consolidation of the embankment can be considered in addition to the embankment loading.

For the construction of the embankment, the following procedure can be used and is recommended (Potts and Zdravkovic 1999):

1. Divide the analysis into a set of increments. For a particular increment, the element to be constructed is inserted and given a constitutive model appropriate to the material behavior during placing. This often means that the material has low stiffness. In the current study, the elastic behavior with low stiffness equal to 75% of the original stiffness of the material is assumed for the material during placing.

2. Nodal forces due to the self-body forces of the constructed materials are calculated and applied on the corresponding nodes.
3. The global stiffness matrix and all the other boundary conditions are assembled for the increment. The analysis is performed to obtain the incremental changes in displacements, strains, and stresses.
4. Before applying the next increment, the constitutive model for the elements just constructed is changed to represent the behavior of the fill material once placed. The incremental displacements of any nodes that are only connected to the constructed elements (i.e., not connected to elements that were active at the previous increment) are zeroed.
5. Apply the next increment of the analysis.

The incremental procedure is schematically depicted in Figure 7.23.

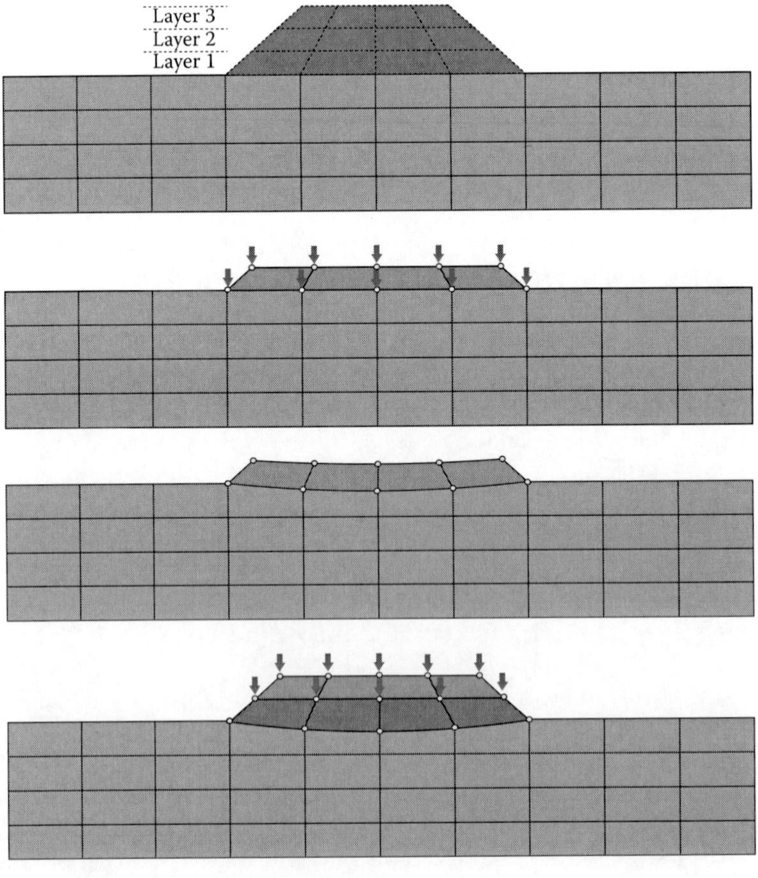

Figure 7.23 The procedure for implementing the embankment construction.

7.5.1 Numerical example

The linearized construction sequence of the embankment is schematically shown in Figure 7.24. As shown, it was assumed that the embankment would be constructed in four layers within 40 consecutive days. A consolidation analysis was performed until 1000 days after the end of the construction.

The geometry and finite element mesh of the problem are presented in Figure 7.25. Table 7.8 gives the material parameters that are determined considering the data of Osaka soft clay (Mirjalili et al. 2011) used in the finite element analysis. In order to discuss the effects on two factors, structural degradation and strain dependent shear modulus, analyses for three cases have been performed (see Table 7.9). In addition, M_m^* is a function of Lode's angle in the analysis. The calculated results of vertical settlements–time profile at the ground level are presented in Figure 7.26a. The results clearly indicate that the effect of both the structural degradation and the strain dependency of the elastic shear modulus. The settlement profiles at the ground level at several times for case 3 are illustrated in Figure 7.26b. Despite the increase in ground settlement beneath the embankment, during and after construction, the surface heave around the toe decreases during consolidation. Figure 7.26c indicates the overall deformation pattern that shows the distorted mesh beneath the embankment. In Figure 7.27 we can see the pore water pressure–time profile at several points. The secondary generated pore water pressure around 40 day at points C and D, located in the area with strain localization are higher than those in points A and B.

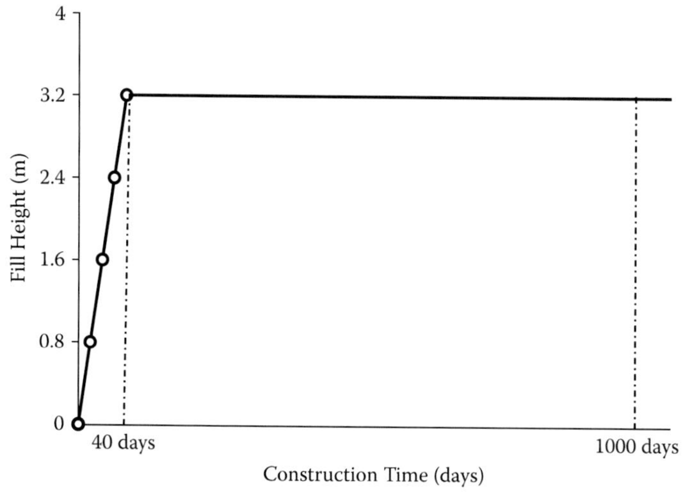

Figure 7.24 Loading profile based on the construction stages.

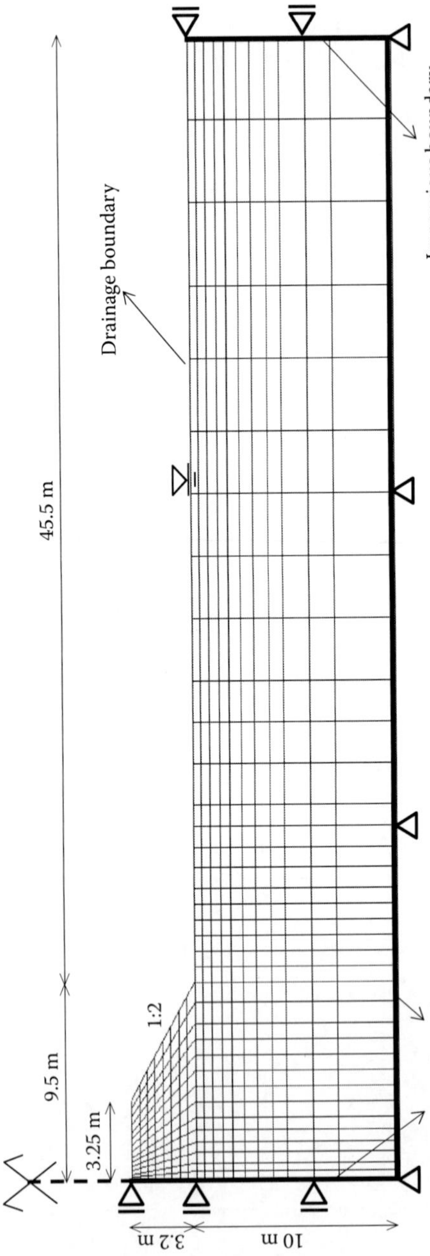

Figure 7.25 Finite element mesh and boundary conditions for the embankment construction.

Table 7.8 Material parameters for the embankment and the ground layer

Parameters	*k (m/s)*	γ_t (kN/m³)	e_0	G_0 (kPa)	*OCR*	λ	κ	M^*_{mc}	m'	C_1, C_2 (1/s)	*n*	β	$\propto$
Embankment	1.00×10^{-5}	19.8	0.8	4300									
Ground	3.85×10^{-10}	16.0	1.65	3930	1.10	0.593	0.027	1.18	28.2	1.85×10^{-11}	0.67	10,15[(1)]	0,10

* Embankment is modeled as an elastic material.

Table 7.9 Specifications for each case in the 2D numerical analysis

	Structural degradation	*Strain dependency of shear modulus*
Case 1 (No structural degradation: $\beta = 0, \propto = 0$)	×	×
Case 2 (Structural degradation: $\beta = 10, a = 0$)	✓	×
Case 3 (Structural degradation + Strain dependency of G: $\beta = 15, a = 10$)	✓	✓

Note: ✓, Considered, ×, Not considered.

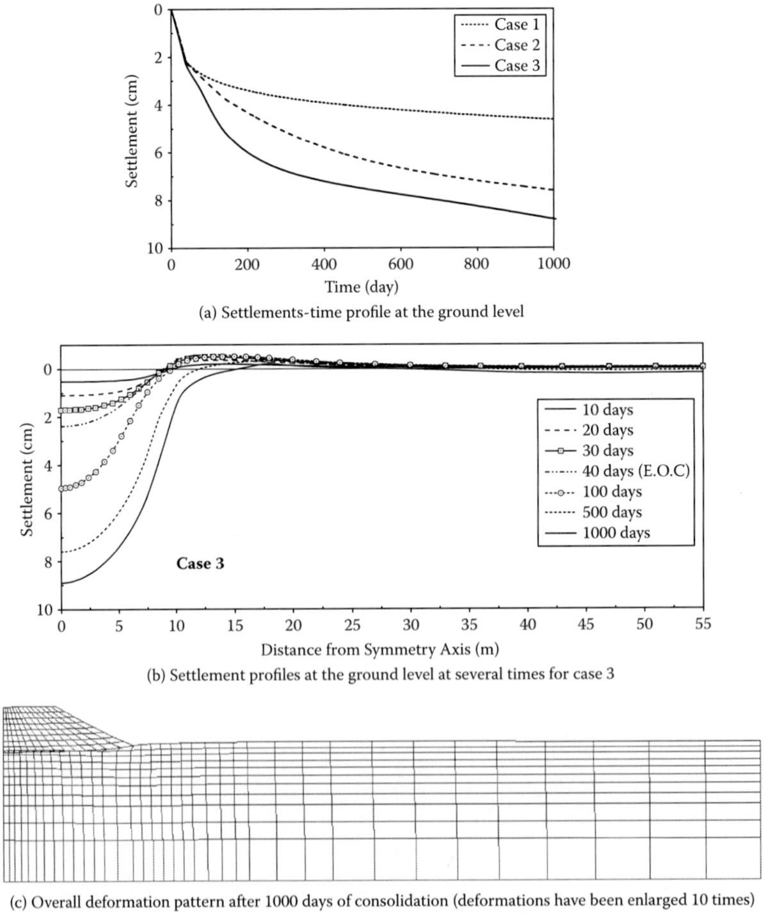

Figure 7.26 Deformation of the embankment. (a) Settlements–time profile at the ground level. (b) Settlement profiles at the ground level at several times for Case 3. (c) Overall deformation pattern after 1000 days of consolidation (deformations have been enlarged 10 times).

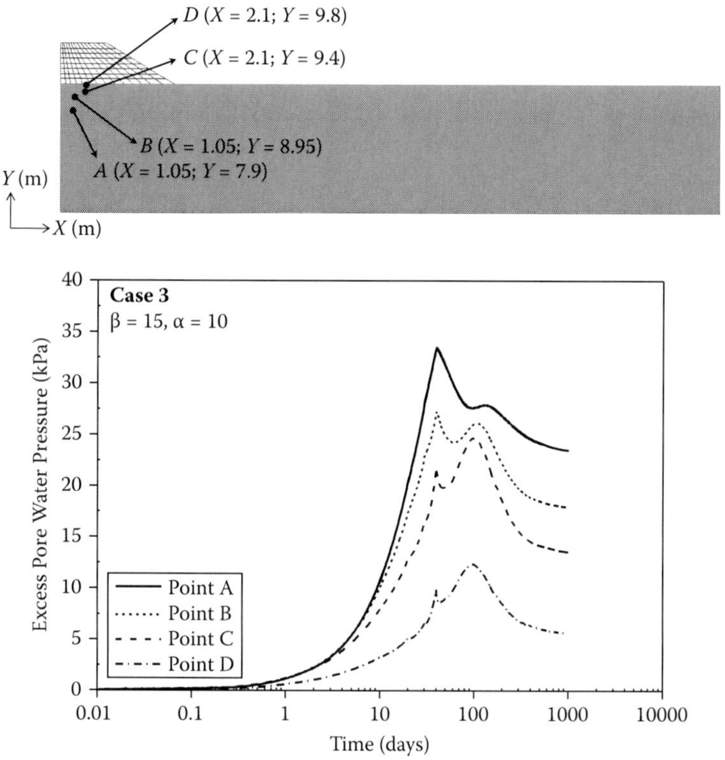

Figure 7.27 Excess pore water pressure–time profile.

REFERENCES

Aboshi, H. 1973. An experimental investigation on the similitude of consolidation of clay including secondary creep settlement, Proc. 8th ICSMFE, Moscow, 4(3):88.

Aboshi, H., and Matsuda, H. 1981.Secondary compression of clays and its effect on settlement analysis, Tsuchi-to-Kiso *JSSMFE*, 29(3):19–24 (In Japanese).

Aboshi, H. 1995. Creep records of long-term measurement of consolidation settlement and their predictions, Compression and consolidation of clayey soils, H. Yoshikuni and O. Kusakabe eds., Rotterdam, Balkema, 2:847–872.

Aboshi, H. 2004. Long-term effect of secondary consolidation on consolidation settlement of marine clays, Advances in geotechnical engineering, The skempton conference, ICE, Thomas Telford, 1:345–355.

Adachi, T., and Oka, F. 1982. Constitutive equations for normally consolidated clay based on elasto-viscoplasticity, *Soils and Foundations*, 22(4):57–70.

Adachi, T., Oka, F., and Tange, Y. 1982. Finite element analysis of two-dimensional consolidation using an elasto-viscoplastic constitutive equation,

Proceedings of 4th International Conference on Numerical Methods in Geomechanics, Z. Zisenstein, ed., Canada Edmonton, Balkema Rotterdam, 2:287–296.

Akai, K., and Adachi, T. 1965. Study on the one-dimensional consolidation and the shear strength characteristics of fully saturated clay, in terms of effective stress, Proc. 6th ICSMFE, Montreal, 1:146–150.

Akai, K., and Tamura, T. 1979. Study of two-dimensional consolidation accompanied by an elasto-plastic constitutive equation, Proc. JSCE, 269:95–104 (In Japanese).

Barden, L. 1969. Time dependent deformation of normally consolidated clays and peats, Proc. ASCE, 95(SMI):1–31.

Berre, T., and Iversen, K. 1972. Oedometer tests with different specimen heights on a clay exhibiting large secondary compression, *Géotechnique*, 22(1):53–70.

Bishop, A.W., and Lovenbury, H.T. 1969. Creep characteristics of two undisturbed clays, Proc. 7th Int. Conf. Soil Mech. Found. Eng., Mexico City, 1:29–37.

Christian, J.T., and Boehmer, J.W. 1970. Plane strain consolidation by finite elements, Proc. ASCE, 96(SM4):1435–1457.

Garlanger, J.E. 1972. The consolidation of soils exhibiting creep under constant effective stress, *Géotechnique*, 22(1):71–78.

Kabbaj, M., Oka, F., Leroueil, S., and Tavenas, F. 1986. "Consolidation of natural clays and laboratory testing," Consolidation of soils: testing and evaluation, ASTM STP 892, R.N. Yong and F.C. Townsend, eds., 378–404.

Kabbaj, M., Tavenas, F., and Leroueil, S. 1988. In situ and laboratory stress-strain relationships, *Géotechnique*, 38(1):83–100.

Karim, Md Rezaul. 2006. Simulation of long term consolidation behavior of soft sensitive clay using an elasto-viscoplastic constitutive model, PhD thesis, Kyoto University.

Karim, M. R., and Oka, F. 2010. An automatic time increment selection scheme for simulation of elasto-viscoplastic consolidation of clayey soils, *Geomech. Geoeng.*, 5(3):153–177.

Kimoto, S. 2002. Constitutive models for geomaterials considering structural changes and anisotropy, PhD thesis, Kyoto University, Japan.

Kimoto, S., Oka, F., and Higo, Y. 2004. Strain localization analysis of elasto-viscoplastic soil considering structural degradation, *Comput. Meth. Appl. Mech. Eng.*, 193:2845–2866.

Kimoto, S. and Oka, F. 2005. An elasto-viscoplastic model for clay considering destructuralization and consolidation analysis of unstable behavior, *Soils and Foundations*, 45(2):29–42.

Ladd, C.C. 1973. Settlement analysis for cohesive soils, Research Report R71-2(272), Department of Civil Engineering, Massachusetts Institute of Technology, 115.

Leroueil, S., and Hight, D.W. 2003. Behavior and properties of natural soils and soft rocks, In Characterisation and Engineering Properties of Natural Soils, T.S. Tan, K.K. Phoon, D.W. Hight, and S. Leroueil, eds., Balkema, 1:29–254.

Leroueil, S. 1988. Recent developments in consolidation of natural clays, *Canadian Geotechnical Journal*, 25(1):85–107.

Leroueil, S. 1995. Could it be that clays have no unique way of behaving during consolidation? Compression and consolidation of clayey soils, H. Yoshikuni and O. Kusakabe eds., Rotterdam, Balkema, 2:1049–1048.

Leroueil, S., Lebihan, J.P., and Tavenas, F. 1980. An approach for determination of the preconsolidation pressure in sensitive clays, *Can. Geotech. J.*, 17(3):446–453.

Leroueil, S., Tavenas, F., Trak, B., Rochelle, P. L., and Roy, M. 1978. Construction pore water pressures in clay foundations under embankments. Part I: the Saint-Alban test fills, *Can. Geotech. J.*, 15:54–65.

Matsuda, H. and Aboshi, H. 1983. A study on preconsolidation by separate-type consolidometer, Research report of the Department of Engineering, Yamaguchi University, 33(2):39–48.

Mesri, G., and Choi, Y.K. 1979. Excess pore water pressure during consolidation, Proc. 6th Asian Reg. Conf. on Soils Mech. Found. Eng., 1:151–154.

Mesri, G., and Choi, Y.K. 1980. (Discussion) Excess pore water pressure and pre-consolidation effect developed in normally consolidated clay of same age, *Soils and Foundations*, 20(4):143–148.

Mesri, G., and Rokhsar, A. 1974. Theory of consolidation for days, J. Geotech. Eng. Div., ASCE, 100(GT8):889–904.

Mesri, G. and Choi, Y.K. 1985. The uniqueness of the end-of-primary (EOP) void ratio-effective stress relationship, Proc. 11th ICSMFE, San Francisco, ISSMFE, 2:587–590.

Mesri, G., Shahien, M. and Feng, T.W. 1995. Compressibility parameters during primary consolidation, Compression and consolidation of clayey soils, H. Yoshikuni and O. Kusakabe eds., Rotterdam, Balkema, 2:1021–1037.

Mirjalili, M. 2010. Numerical analysis of a large scale levee on soft soil deposits using two-phase finite deformation theory, PhD thesis, Kyoto University.

Mirjalili, M., Kimoto, S., Oka, F., and Higo, Y. 2011. Elasto-viscoplastic modeling of Osaka soft clay considering destructuration and its effect on the consolidation analysis of an embankment, *Geomech. Geoeng., Int. J.*, 6(2), 69–89.

Mirjalili, M., Kimoto, S., Oka, F., and Hattori, T. In press. Long-term consolidation analysis of a large-scale embankment construction soft clay deposits using an elasto-viscoplastic model, *Soils and Foundations*.

Mitchell, J.K. 1986. Practical problems from surprising soil behavior, 20th Terzaghi Lecture, ASCE, *J. Geotech. Engrg.*, 112(3):259–289.

Moore, P.J., and Spencer, G.K. 1972. Lateral pressures from soft clay, Proc. ASCE, 98 (SM11):1225–1244.

Oka, F. 1981. Prediction of time dependent behaviour of clay, Proc. 10th ICSMFE, Stockholm, 1:215–218.

Oka, F. 2000. *Elasto-viscoplastic constitutive models of geomaterials*, Morikita Shuppan, (in Japanese).

Oka, F. 2005.Computational modeling of large deformations and the failure of geomaterials, Proc. 16th ICSMGE, Osaka, 1:95–122.

Oka, F., Adachi, T., and Okano, Y. 1986. Two-dimensional consolidation analysis using an elasto-viscoplastic constitutive equation, *Int. J. Numer. Anal. Meth. Geomech*, 10:1–16.

Oka, F., Tavenas, F., and Leroueil, S. 1991. An elasto-viscoplastic FEM analysis of sensitive clay foundation beneath embankment, Proceedings of 7th International Conference on Computer Method and Advances in Geomechanics, G. Beer, J.R. Booker, and J.P. Carter, eds., Cairns, Balkema, 2:1023–1028.

Perzyna, P. 1963. The constitutive equations for work-hardening and rate sensitive plastic materials, Proc. Vibrational Probl., Warsaw, 4(3):281–290.

Potts, D.M., and Zdravkovic, L. 1999. *Finite Element Analysis in Geotechnical Engineering Theory*, Thomas Telford.

Roscoe, K.H., Schofield, A.N., and Thurairlljah, A. 1963. Yielding of clays in states wetter than critical, *Géotechnique*, 13(3):211–240.
Sekiguchi, H., and Ohta, H. 1977. Induced anisotropy and time dependency in clays, Proc. Speciality Session 9, 9th ICSMFE, Tokyo, 229–2381.
Sekiguchi, H., and Shibata, T. 1979. Undrained behavior of soft clay under embankment loading, Proc. 3rd International Conference on Numerical Methods in Geomechanics, Achen, W. Wittke, ed., Balkema, 2:717–724.
Sekiguchi, H., and Toriihara, M. 1976.Theory of one-dimensional consolidation of clay with consideration of their rheological properties, *Soils and Foundations*, 16(1):27–44.
Suklje, L. 1969. *Rheological Aspects of Soil Mechanics*, New York: Wiley-Interscience.
Tang, Y.X., and Imai, G. 1995. A constitutive relation with creep and its application to numerical analysis of one-dimensional consolidation, In Compression and Consolidation of Clayey Soils, H. Yoshikuni and O. Kusakabe, eds., Balkema, Rotterdam, 465–472.
Tavenas, F., Chapeau, C., Rochelle, P. L., and Roy, M. 1974. Immediate settlements of three test embankments on Champlain clay, *Can. Geotech. J.*, 11(4):109–141.
Tavenas, F., and Leroueil, S. 1980. The behavior of embankments on clay foundations, *Can. Geotech. J.*, 17(2):236–260.
Zienkiewicz, O.C. 1977. *The Finite Element Method*, 3rd ed., New York: McGraw-Hill, 450–499.

Chapter 8

Strain localization

It is well known that the strain localization of geomaterials causes such important problems as slope failure. In slope failure, deformation occurs in a narrow zone. In other words, strain localization is closely related to the onset of failure (Scott 1987). In the present chapter, strain localization problems are reviewed, and the instability and the numerical simulation of strain localization will be presented.

8.1 STRAIN LOCALIZATION PROBLEMS IN GEOMECHANICS

In order to analyze failure, we have to deal with strain localization near and/or after failure. The development of a failure surface is a classical issue in soil mechanics. Coulomb (1773) considered a failure surface in order to determine the collapse load in his famous work. Sokolovsky (1942) analyzed a slip plane as a stress characteristic at the limit equilibrium. As has been pointed out by Taylor (1948), the failure phenomenon of geomaterials is progressive. Hence, strain localization is a precursor to the development of a failure surface and is a very important subject to investigate. Over the last three decades, the problems of strain localization in geomaterials, that is, the formation of shear band of soil and rock, have been extensively studied within the context of experimental, theoretical, and numerical approaches. Many researchers (Hill 1962; Rice 1975, 1976; Rudnicki and Rice 1975; Vardoulakis 1980; Vermeer 1982; Mühlhaus and Vardoulakis 1987; Wang et al. 1990; Oka et al. 1994; Muir Wood 2002; Borja 2004; Gudehus and Nübel 2004) have been working in this area from both experimental and analytical points of view. Rice (1976) and Rudnicki and Rice (1975) pointed out that the nature of this problem can be solved under the general framework of bifurcation problems and that localization problems should be studied within a wider framework of mechanics, including the rapid degradation of the material strength.

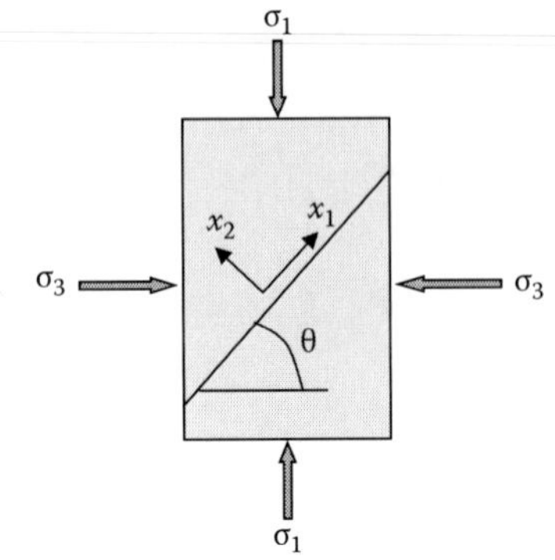

Figure 8.1 Angle of shear band ($\sigma_1 > \sigma_3$).

8.1.1 Angle of shear band

The classical Mohr–Coulomb law gives the angle of the shear band (failure surface) as

$$\theta_M = \frac{\pi}{4} + \frac{\phi}{2} \tag{8.1}$$

where ϕ is the internal friction angle.

On the other hand, as is well known, Roscoe (1970) reported in his Rankine lecture that the rupture (failure) surface does not coincide with the stress characteristics, but with the zero-extension lines. This means that the angle of a shear band to the major principle stress plane θ (see Figure 8.1) is given by

$$\theta_R = \frac{\pi}{4} + \frac{\psi}{2} \tag{8.2}$$

where ψ is the dilatancy angle.

In addition, Roscoe (1970) pointed out that the thickness of a shear band is approximately ten times the average grain diameter.

For the angle of a shear band, Arthur et al. (1977) proposed a shear band angle based on experiments as

$$\theta_A = \frac{\pi}{4} + \frac{1}{4}(\phi + \psi) \tag{8.3}$$

since $\psi < \phi$ and $\theta_R < \theta_A < \theta_M$. The angles experimentally obtained by Arthur et al. (1977) were supported by theoretical works (Vardoulakis 1980; Vermeer 1982). Vardoulakis (1980) reported that the experimental results coincide with Equation (8.3). However, the experimental results for sand, obtained by Desrues (1990) and Lade (2003), are close to θ_M. Vardoulakis (1977) illustrated that the shear band angle depends on the particle diameter; in other words, the larger the soil particle size, the smaller the shear band angle. Tatsuoka et al. (1990) showed that the shear band angle depends on the b-value (see Equation 4.35) and the anisotropy. These indicate the other effects on the angle of the shear band. For clay, Oka et al. (2002) and Higo et al. (2006) numerically indicated that the angle of a shear band depends on the strain rates and the permeability, as shown in Sections 8.8 and 8.9.

8.2 LOCALIZATION ANALYSIS

It has been theoretically found that the onset conditions for strain localization can be captured by a bifurcation analysis (Hill 1962; Rice 1976). Rice (1976) showed that shear band formation is a problem because a shear band is the result of bifurcation from a homogeneous deformation. The onset conditions for localization are given by linear and nonlinear incremental constitutive models. For static problems, the conditions involve a loss of the elliptic onset condition, the thickness, the angle, and the distribution. Postlocalization behavior and the loss of ellipticity correspond to the loss of uniqueness, namely, the instability discussed within the framework of bifurcation and material instability.

Within the framework of a linear incremental model, the onset conditions of bifurcation are derived as the condition of equilibrium, which requires the time rate of change in stress traction $\dot{\sigma}_{ij}n_j$ at the shear band to be zero, namely,

$$[\dot{\sigma}_{ij}]n_j = 0 \tag{8.4}$$

where n_j is the unit normal to the shear band. The brackets [ ] of a quantity denote the difference in values of that quantity across the shear band.

The velocity is continuous, but the gradient of velocity is discontinuous.

$$[L_{ij}] = \lambda_i n_j \tag{8.5}$$

where $L_{ij} \equiv v_{i,j}$ is the velocity gradient.

The incremental constitutive model is given by

$$\dot{\sigma}_{ij} = D_{ijkl} L_{kl} \tag{8.6}$$

From Equations (8.4), (8.5), and (8.6),

$$H_{ik}\lambda_k = 0, \quad H_{ik} = D_{ijkl} n_j n_l \tag{8.7}$$

where D_{ijkl} is the stiffness tensor.

The onset conditions for nonzero λ_k are

$$\det[H_{ik}] = 0 \tag{8.8}$$

From these conditions, we can obtain shear band angle θ in a two-dimensional case, in which $n_1 = \cos\theta$ and $n_2 = \sin\theta$ in Equation (8.7) and θ is the angle between the unit normal to the shear band and the major principle stress plane.

Equation (8.8) can be written as

$$\det[H_{ik}] = a_0 n_1^4 + a_1 n_1^3 n_2 + a_2 n_1^2 n_2^2 + a_3 n_1 n_2^3 + a_4 n_2^4 = 0 \tag{8.9}$$

where

$$a_0 = L_{1111}L_{1212} - L_{1112}L_{1211},\ a_1 = L_{1111}L_{1222} + L_{1111}L_{2212} - L_{1112}L_{2211} - L_{1122}L_{1211},$$
$$a_2 = L_{1111}L_{2222} + L_{1112}L_{1222} + L_{1211}L_{2212} - L_{1122}L_{1212} - L_{1122}L_{2211} - L_{1212}L_{2211}$$
$$a_3 = L_{1112}L_{2222} + L_{1211}L_{2222} - L_{1122}L_{2212} - L_{1222}L_{2211},$$
$$a_4 = L_{1212}L_{2222} - L_{2212}L_{1222}$$

If we set $n_1 = \cos\theta$, $n_2 = \sin\theta$, we have to consider the symmetry with respect to θ.

$$g(\theta) = a_4 \tan^4\theta + a_2 \tan^2\theta + a_0 = 0 \tag{8.10}$$

For a two-dimensional infinitesimal strain model,

$$\begin{aligned} &\dot{\sigma}_{11} = D_{11}\dot{\varepsilon}_{11} + D_{12}\dot{\varepsilon}_{22},\ \dot{\sigma}_{22} = D_{21}\dot{\varepsilon}_{11} + D_{22}\dot{\varepsilon}_{22},\ \dot{\sigma}_{12} = 2G\dot{\varepsilon}_{12}, \\ &a_4 = GD_{11},\ a_2 = D_{11}D_{22} - D_{12}D_{21} - GD_{12} - GD_{21},\ a_0 = GD_{22} \end{aligned} \tag{8.11}$$

The character of Equation (8.9) is classified as elliptic, parabolic, and hyperbolic regions depending on the roots of Equation (8.10); the elliptic region for the four imaginary or four complex roots, the parabolic region for the two real and two imaginary roots, and the hyperbolic region for four real roots (Hill and Hutchinson 1975). It is worth noting that bifurcation never occurs in the elliptic region. On the contrary, bifurcation, such as shear bands, could develop in the parabolic and hyperbolic regions.

It is seen that Rice's bifurcation condition is the same as the singularity of the acoustic tensor (Mandel 1964). This means that the occurrence of a shear band coincides with the accelerated wave trapped in the narrow zone. For the propagation of an acceleration wave, that is, the propagation of the discontinuity of the acceleration, it is necessary for the eigenvalues to be positive and real. The acceleration is discontinuous across the wave front for the acceleration wave, although the displacement and the velocity are continuous. When the eigenvalues are negative and zero, discontinuity cannot propagate and it is referred to as stationary discontinuity, namely, shear bands or complex (flutter) instability.

The zero determinant of the acoustic tensor by Equation (8.7) corresponds to a change in the type of partial differential governing equations from elliptic to hyperbolic (Hill and Hutchinson 1975). Bigoni and Hueckel (1991) found that the loss of strong ellipticity coincides with the nullity of the determinant of the symmetric part of the acoustic tensor.

Strain localization has been well recognized as the development of shear bands. On the other hand, another type of strain localization has also been observed, namely, compaction bands (Mollema and Antonellini 1996; Olsson 1999). Compaction bands were experimentally found for sandstone as a tabular zone that exhibits closure, but no shear offset (Mollema and Antonellini 1996). The compaction bands comprise a kind of volumetric strain localization due to pore collapse or particle crushing. The compaction bands of porous sandstone exhibit a sharp reduction in permeability that causes difficulties in oil production. Olsson (1999) has shown that compaction bands can be described by the strain localization theory of Rudnicki and Rice (1975) as well as by shear localization problems. The re-examination of Rudnicki and Rice's theory comes from the correction of the Rudnick and Rice theory by Perrin and Leblond (1993). They showed that the possible range for the sum of the dilatancy factor and the coefficient of internal friction is wider than that shown in Rudnicki and Rice's paper; in other words, $\beta + \mu \leq \sqrt{3}/2$ should be $-\sqrt{3} \leq \beta + \mu \leq \sqrt{3}$.

Rudnicki and Olsson (1998) re-examined the fault angles predicted by the shear localization theory and obtained the following convenient form:

$$\theta = \frac{\pi}{4} + \frac{1}{2}\arcsin\alpha,$$

$$\alpha = \frac{(2/3)(1+\nu)(\beta+\mu) - N(1-2\nu)}{\sqrt{4-3N^2}} \tag{8.12}$$

$$N = \sigma'_{II} / \bar{\tau} \tag{8.13}$$

where $\bar{\tau} = \sqrt{s_{ij}s_{ij}/2}$, s_{ij} is the deviatoric stress tensor, ν is Poisson's ratio, β is the gradient of the plastic potential surface, σ'_{II} is the intermediate principal deviatoric stress, and μ is the gradient of the yield surface.

Using Equation (8.12), it has been realized that a smaller angle for the compaction bands can be described. Issen and Rudnicki (2000) obtained the conditions for compaction bands in porous rock.

In order to observe compaction bands for soil, Castellanza and Nova (2003) reported a compaction band forming in the cemented granular soil during odometer tests. They observed a stress–strain relation with a plateau that occurred after the peak stress had been reached, which is a typical stress–strain relationship for compaction bands. Oka and Kimoto (Kimoto and Oka 2004, 2005) pointed out that the large compression of structured clay is due to the development of compaction bands, and they numerically analyzed the unstable consolidation of clay with degradation. This compaction band phenomenon for a soft rock has been successfully simulated by an elastoviscoplastic model with volumetric strain softening by Oka et al. (2011).

8.3 INSTABILITY OF GEOMATERIALS

As mentioned earlier, strain localization can be described as bifurcation conditions by Equation (8.8), namely, a zero determinant of the acoustic tensor (Rice 1976; the loss of the positive definiteness of constitutive matrix D). The loci on the π plane for which the determinant of the acoustic tensor is zero, namely, the loci of the localization conditions, never intersect the stress axes. This means that shear banding cannot occur in conventional triaxial compression tests before the failure state. On the other hand, shear bands may easily develop in plane strain tests. The stress state reaches the surface given by the zero determinant of the acoustic tensor before it reaches the limit state. The limit state is characterized by the conditions under which the determinant of the constitutive matrix equals zero

(Tokuoka 1971) and it is called the failure condition. The failure condition is the limit of allowable stress space and is given by

$$det[D] = 0 \tag{8.14}$$

where $[D]$ is the constitutive stiffness matrix and $\{\dot{\sigma}\} = [D]\{\dot{\varepsilon}\}$. Since $\{\dot{\sigma}\} = \{0\}$ at the failure state, we have $[D]\{\dot{\varepsilon}\} = \{0\}$.

The basic definition of the stability is the stability in the sense of Lyapunov, as shown in Chapter 3, where the positive second-order work is a sufficient condition for the stability. The local form of the second-order work can be written in matrix form as

$$\Delta^2 W = \dot{\varepsilon}_{ij}\dot{\sigma}_{ij} = \{\dot{\varepsilon}\}^T[D]\{\dot{\varepsilon}\} > 0; \quad \forall\dot{\varepsilon} \neq 0 \tag{8.15}$$

From the fundamental theorem in linear algebra, the aforementioned condition is called positive definiteness. If the positive definiteness holds, $\det[D] > 0$ if and only if for the real symmetric $[D]$ is positive definite. Then, all minor determinants are positive.

On the contrary, if the $[D]$ is positive semidefinite, that is, $det[D] = 0$, many nonzero solutions for strain rate $\{\dot{\varepsilon}\}$ can exist.

$[D]$ is positive definite if and only if its symmetric part is positive definite, since for the skew-symmetric part of $[D]$, Equation (8.15) becomes zero. Nova (1994) pointed out that based on the theorem by Ostrowski and Taussky (1951):

$$\det[D] \geq \det[D_s] > 0 \tag{8.16}$$

Nova (1994) studied the other bifurcation criteria and found that the conditions for the loss of controllability are obtained as a zero determinant of the symmetric part of stiffness tensor $[D_s]$.

In general, for the nonassociated plasticity model, $\det[D_s] = 0$ even when $det[D] \neq 0$.

The constitutive matrix $[D]$ is decomposed into the symmetric part and the skew-symmetric part, namely,

$$[D] = [D_s] + [D_a] \tag{8.17}$$

And if $\det[D_s] = 0$ and $[D_s]\{\dot{\varepsilon}\} = \{0\}$ for a particular strain rate $\{\dot{\varepsilon}\}$, which is proportional to the eigenvector of $[D_s]$, then

$$\begin{aligned}\{\dot{\sigma}'\} &= [D]\{\dot{\varepsilon}\} = [D_s]\{\dot{\varepsilon}\} + [D_a]\{\dot{\varepsilon}\} \\ &= [D_a]\{\dot{\varepsilon}\}\end{aligned} \tag{8.18}$$

In this case, the second-order work becomes zero as

$$\begin{aligned}\Delta^2 W &= \{\dot{\varepsilon}\}^T\{\dot{\sigma}\} = \{\dot{\varepsilon}\}^T[D]\{\dot{\varepsilon}\}\\ &= \{\dot{\varepsilon}\}^T[D_a]\{\dot{\varepsilon}\} = 0\end{aligned} \tag{8.19}$$

This indicates the loss of Hill's stability condition (Equation 3.57). Nova (1994) indicated and discussed that the loss of controllability may occur at the same time as the loss of Hill's stability. The loss of controllability is explained in the following. The same discussion holds for the expression by compliance matrix $[C]$; $\{\dot{\varepsilon}\} = [C]\{\dot{\sigma}'\}$.

In the experiments on soil, we can choose the variables for controlling the system; for example, we control the axial strain rates and maintain the lateral stress constant in the strain control triaxial tests. When the conditions are satisfied, we cannot continue the experiments. For instance, we cannot execute triaxial tests for strain hardening–softening material after the peak stress by the axial stress control. Nova found that the loss of controllability is equal to the violation of the positive second-order work. This indicates that if the stress states satisfy the zero determinant of the symmetric part of the material stiffness matrix, $\det[D_s] = 0$, instability (bifurcation) may occur. Lade's instability condition (1992) is equivalent to the conditions under which the volumetric compliance is zero. This state comes before the failure state in the triaxial tests.

The loss of controllability is explained using compliance matrix $[C]$ based on an example following Imposimato and Nova (1998).

Under axisymmetric triaxial conditions, we can write the constitutive model as

$$\begin{Bmatrix} \dot{\varepsilon}_v \\ \dot{\varepsilon}_d \end{Bmatrix} = \begin{Bmatrix} C_{pp} & C_{pq} \\ C_{qp} & C_{qq} \end{Bmatrix} \begin{Bmatrix} \dot{p}' \\ \dot{q} \end{Bmatrix} \tag{8.20}$$

where $\dot{\varepsilon}_v$ is the volumetric strain rate, and $\dot{\varepsilon}_d$ is the shear strain rate, namely, $\dot{\varepsilon}_d = 2(\dot{\varepsilon}_a - \dot{\varepsilon}_r)/3$; $\dot{\varepsilon}_a$ is the axial strain rate and $\dot{\varepsilon}_r$ is the radial strain rate. $\dot{p}'$ is the mean effective stress rate, and $\dot{q}$ is the deviator stress rate; $\dot{q} = \dot{\sigma}_{11} - \dot{\sigma}_{33}$.

This expression indicates that the control parameters are the shear strain and the volumetric strain. For $\dot{\varepsilon}_v$ and $\dot{\varepsilon}_d$, see Chapter 4. In addition, we assume $\det[C] = C_{pp}C_{qq} - C_{pq}C_{qp} > 0$.

Consider the mixed control case in which we change the control parameters from two components of strain rates to the other set of control parameters, that is, shear strain rate $\dot{\varepsilon}_d$ and deviator stress rate $\dot{q}$.

In this case, we have

$$\begin{Bmatrix} \dot{\varepsilon}_v \\ \dot{q} \end{Bmatrix} = \frac{1}{C_{qq}} \begin{Bmatrix} C_{pp}C_{pq} - C_{pq}C_{qp} & C_{pq} \\ -C_{qp} & 1 \end{Bmatrix} \begin{Bmatrix} \dot{p}' \\ \dot{\varepsilon}_d \end{Bmatrix} \tag{8.21}$$

When the determent of the matrix is zero in the preceding equation,

$$C_{pp}C_{pq} = 0 \tag{8.22}$$

This condition holds when

$$C_{pp} = 0 \tag{8.23}$$

since $C_{pq} \neq 0$ for the existence of dilatancy. It is worth noting that the determinant of Equation (8.21) is zero, but that of Equation (8.20) is not zero.

Since $\dot{\varepsilon}_v = 0$ under undrained conditions, the constitutive model can be written as

$$\frac{1}{C_{qq}} \begin{Bmatrix} C_{pp}C_{pq} - C_{pq}C_{qp} & C_{pq} \\ -C_{qp} & 1 \end{Bmatrix} \begin{Bmatrix} \dot{p}' \\ \dot{\varepsilon}_d \end{Bmatrix} = \begin{Bmatrix} 0 \\ \dot{q} \end{Bmatrix} \tag{8.24}$$

When the stress rate is at the maximum q, $\dot{q} = 0$ under load-control conditions,

$$\det \begin{bmatrix} C_{pp}C_{pq} - C_{pq}C_{qp} & C_{pq} \\ -C_{qp} & 1 \end{bmatrix} = 0 \tag{8.25}$$

From Equation (8.25), we have $C_{pp} = 0$ (Equation 8.23), and in this case many solutions may exist for Equation (8.24). However, the solution indicates the ratio of the pore water pressure rate and the deviatoric strain rate such that

$$\frac{\dot{p}}{\dot{\varepsilon}_d} = \frac{1}{C_{qp}} \tag{8.26}$$

for $C_{pp} = 0$.

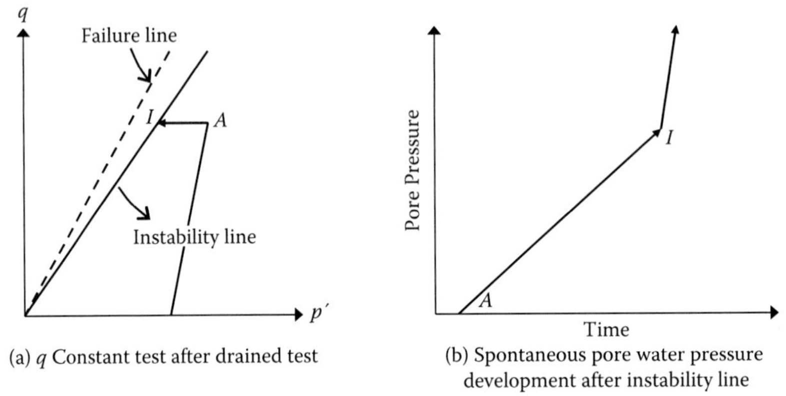

Figure 8.2 Schematic figure for the deviator stress constant undrained test. (a) q constant test after drained test. (b) Spontaneous pore water pressure development after instability line.

It is worth noting that only the ratio can be determined for the zero eigenvalue problem, but not for each value. Hence, at the peak value for the deviator stress, the pore water pressure and the deviatoric strain rate can be arbitrary for the constant total isotropic pressure, $\dot{p} = \dot{p}' + u_w = 0$, that is, $\dot{u} = -\dot{p}' = \text{arbitrary}$ and $\dot{\varepsilon}_d = -C_{qp}\dot{u}_w = \text{arbitrary}$. These results indicate the spontaneous pore water pressure development during the undrained test with stress control test (Imposimato and Nova 1998), schematically shown in Figure 8.2. Since the compliance matrix in Equation (8.20) is positive definiteness, Lade's instability condition (1992) corresponds, $C_{pp} = 0$, under undrained conditions, which manifests at the peak of the deviator stress shown in Figure 8.3.

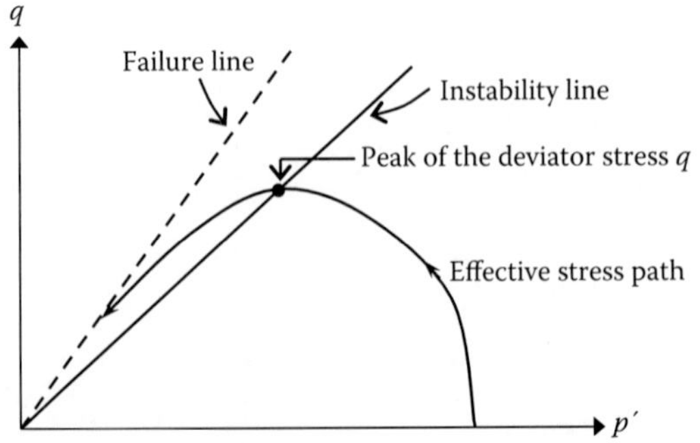

Figure 8.3 Effective stress path during the undrained triaxial test for loose sand.

Since the nonassociative flow rule of plasticity is not symmetric, a variety of unstable conditions may arise. They are the strain localization condition (the loss of the determinant of the acoustic tensor; Rice 1976), the zero second-order work condition (Hill 1957, 1958), and the loss of controllability condition (Nova 1994, 2009). The loss of positiveness of the secondary work condition and the loss of the controllability condition may lead to diffuse (homogeneous) bifurcation or instability that manifests as a barreling or a bulging type of phenomenon (Nova 1989; Darve 2001; Daouadji et al. 2010).

The surfaces that define these types of instability are inside the failure surface (Imposimato and Nova 2001). Benallal and Comi (2003) conducted a study using a perturbation approach and found that the failure mode of nonassociated material exhibits either a diffuse mode (long wavelength) or a localization mode (short wavelength).

Hill (1958) derived a sufficient condition for the uniqueness of the solution to boundary value problems using the virtual work theorem as

$$\int_V \Delta\dot{\sigma}_{ij}\Delta\dot{\varepsilon}_{ij}dv > 0 \tag{8.27}$$

where $\Delta\dot{\sigma}_{ij}, \Delta\dot{\varepsilon}_{ij}$ are the differences between the stress rates and the strain rates for two different solutions and zero at the appropriate boundary of the body in motion.

Equation (8.24) indicates that if the material is stable, that is, the second-order work is positive, and the uniqueness of the solution holds. This means that no localization solution exists. However, it is worth noting that if

$$\int_V \Delta\dot{\sigma}_{ij}\Delta\dot{\varepsilon}_{ij}dv = 0 \tag{8.28}$$

bifurcation, multiple solutions may occur, but bifurcation solutions include unstable and stable solutions.

For the incrementally strong nonlinear law, the onset conditions for a shear band have been obtained by Chambon et al. (2000) and Kolymbas and Rombach (1989). The incrementally nonlinear model has been proposed as a hypoplastic model by Kolymbas and Rombach (1989) and Chambon et al. (2000).

For the hypoplastic models,

$$\dot{\sigma}_{ij} = M_{ijkl}\dot{F}_{kl} + b_{ij}\left\|D_{kl}\right\| \tag{8.29}$$

where M_{ijkl} is the constitutive fourth order tensor, b_{ij} is the constitutive second-order tensor, $\dot{F}_{kl}$ is the rate of the deformation gradient, D_{ij} is the stretching tensor, and $\|\cdot\|$ denotes the Euclidian norm.

For the hypoplastic models, several criteria for bifurcation have been obtained based on the invertibility of $M_{ijkl}n_k n_l$. Namely, three types of bifurcation criteria were given as (1) the norm criteria, (2) the determinant criteria, and (3) the tangential criteria. The norm criteria are first met, while the determinant criteria are never met (Chambon et al. 2000; Tamaganini et al. 2001).

8.4 NONCOAXIALITY

It is known that the direction of the principle strain rates does not need to coincide with the direction of the stress rates for anisotropic materials. And even for isotropic materials, the direction of the principle strain rates does not need to coincide with the direction of the stress rates; this is called noncoaxiality. The yield vertex that comes from the micromechanical consideration leads to noncoaxiality. Rudnicki and Rice (1975) introduced this vertex effect into the model by a noncoaxial term called the Mandel–Spencer modulus term. The noncoaxial term was naturally derived into the planer model from the double slip theory by de Josselin de Jong (1971). Spencer (1964) proposed a double shearing model by assuming a micromechanical slip along the stress characteristic lines obtained from the equilibrium equations and Mohr–Coulomb failure criteria (Sokolovsky 1942). The double shearing model was generalized to be a model with dilatancy by Mehrabadi and Cowin (1978). The model has resulted in the noncoaxial term that is is workless, for example, purely deviatoric. Anand (1983) generalized it into the isotropic hardening model. This noncoaxial model has accounted for the vertex of the plastic potential surfaces (Rudnicki and Rice 1975). Yatomi et al. (1989) used the noncoaxial term in a Cam-clay model and showed that the noncoaxial term easily led to instability as shear banding or as accessibility to bifurcation from an elliptic to a hyperbolic type of governing equation. Papamichos and Vardoulakis (1995) developed a different type of noncoaxial theory by introducing a kinematical concept. Khojastehpour and Hashiguchi (2004) developed a tangential plasticity theory accounting for noncoaxiality. All of the models are rate-independent plasticity theories. Oka (1993b) developed a rate-dependent noncoaxial viscoplastic model using a transformed stress tensor with the current stress-induced quasi-anisotropy shown in the next paragraph. The other source of noncoaxiality is the anisotropy of the materials (Nemat-Nasser 1983).

8.5 CURRENT STRESS-DEPENDENT CHARACTERISTICS AND ANISOTROPY

As mentioned, the noncoaxial term, called the Mandel–Spencer modulus term, is derived from the characteristic plane. There are several constitutive models based on the characteristic plane. Matsuoka and Nakai (1974) proposed a model in which the failure criteria depend on the interim principal stress, although Mohr-Coulomb's criteria depend on the maximum and the minimum principal stresses. Oka (1993a, 1993b) proposed a transformed stress tensor to describe the current stress-dependent behavior of soil, by which Matsuoka and Nakai's failure criteria (1985), shown in Chapter 4, can be derived. In addition, the deviatoric flow rule can be reproduced in such a way that the direction of the strain rate is different from the direction of the stress path. The transformed stress tensor

$$\sigma_{ij}^{A} = F(\sigma_{ij}, A_{ij}) \tag{8.30}$$

is derived from Wang's representation theorem as a function of the stress tensor and the structural tensor (Spencer 1987). The structural tensor was first adopted by Boehler and Sawczuk (1977) as

$$A_{ij} = m_i m_j \tag{8.31}$$

where m_i is a component of unit normal vector to the characteristic plane.

Considering Matsuoka and Nakai's well-known failure criteria (1974), m_i can be taken as the unit normal to the spatial mobilized plane. If we take the sedimentation plane as the characteristic plane, we can construct a transversely isotropic theory (Oka 1993a, 1993b; Oka et al. 2002b). Nakai's t_{ij} theory can be classified as being similar to the transformed stress tensor (Nakai and Mihara 1984; Chowdhury and Nakai 2001), although they originally proposed the theory based on physical considerations.

8.6 REGULARIZATION OF ILL-POSEDNESS

The onset conditions for a shear band can be obtained analytically or numerically, but it is also necessary to predict the postlocalization behavior. After a shear band has occurred, the boundary value problem becomes ill-posed. Hence, ellipticity (well-posedness) must be retrieved. Well-posedness is the uniqueness of a solution and the continuous dependence on the boundary conditions. In the finite element analysis, this instability leads to a strong mesh-size dependency and, in turn, the thickness of the shear band becomes zero, although the thickness of shear bands is finite in the experiments. There are several methods for the regularization of the ill-posedness of governing equations.

8.6.1 Nonlocal formulation of constitutive models

Nonlocal models, such as the higher-order strain gradient-dependent model (Aifantis 1984; Mühlhaus and Aifantis 1991; Vardoulakis and Aifantis 1991), and micropolar models, such as the Cosserat model (Mühlhaus 1986; Teichman and Wu 1993; Swada et al. 1999, 2001; Bauer and Huang 2001), have been used for the postlocalization analysis. An integral type of nonlocal model has been developed to account for the nonlocal effect. The feature of nonlocal models is that they contain the material length scale. Mühlhaus and Oka (1996) and Oka et al. (1998) clarified that the higher strain gradient term naturally comes from the fact that the material is discrete and has an inherent length scale.

We will discuss the performance of the second-order strain gradient dependent model using the one-dimensional viscoelastic model. Let us consider a nonlocal model, for example, a viscoelastic model with a second order strain gradient as

$$\sigma = a\varepsilon + b\varepsilon_t - c\varepsilon_{xx} \tag{8.32}$$

where σ is the stress, ε is the strain, ε_t is the strain rate, ε_{xx} is the second-order gradient with respect to the space coordinate, and a,b,c are material parameters.

Incorporating Equation (8.32) into the equilibrium equation, $\sigma_x = 0$, we have

$$a\varepsilon_x + b\varepsilon_{tx} - c\varepsilon_{xxx} = 0 \tag{8.33}$$

Then, we consider the fluctuation in strain as

$$\varepsilon = A exp(ikx + \omega t) \tag{8.34}$$

in which A is a constant, k is the wave number, and ω is the growth rate of fluctuation. Introducing Equation (8.34) into Equation (8.33), we have

$$\omega = (-a - ck^2)/b \tag{8.35}$$

When a is positive, that is, the strain-hardening case, $\omega < 0$ for the usual viscous effect of $b > 0$ even if $c = 0$. On the other hand, when a is negative, that is, the strain-softening case, ω can be negative if and only if $c>0$. This means that the material is stable and the boundary value problem becomes well-posed even if the material exhibits strain softening. Otherwise, instability occurs in a finite time.

8.6.2 Fluid–solid two-phase formulation

For water-saturated soil, the Biot two-phase mixture theory can be applied to an elastoplastic model for the soil skeleton. In this type of formulation, Zhang and Schrefler (2001) found that the critical hardening moduli become smaller than those for a single material; the permeability is large. This indicates that the localization is delayed for the case of water-saturated soil. They also pointed out that there exists a domain of permeability where stability is lost, but the hyperbolicity remains in the dynamic analysis of water-saturated soil.

Vardoulakis (1996a, 1996b) reported that ill-posedness was encountered with the boundary value problem even when the two-phase formulation was adopted. However, from the numerical studies, it was found that the strain localization is depressed in the case of low permeability (Loret and Prévost 1991; Oka et al. 1995; Asaoka et al. 1997).

8.6.3 Viscoplastic regularization

An elastoviscoplastic formulation can retrieve the ill-posedness of the governing equations through the instantaneous elastic response. For example, Loret and Prévost (1991) adopted a viscoplastic regularization technique using the Duvaut–Lions theory (1976), presented in Chapter 3, by which the inviscid model can be transformed into a viscoplastic model with one viscous parameter.

The elastoviscoplastic model, which is transformed from an elastoplastic model, can maintain the well-posedness; namely, it can avoid the strong mesh-size dependency in the finite element analysis if the growth of the viscoplastic strain is bounded. The numerical example for the mesh-size dependency of the numerical analysis by an elastoviscoplastic model will be shown later in this chapter. It is worth noting that the elastoviscoplastic models can delay the instability, for example, shear bands, but cannot fully stabilize the catastrophic instability.

8.6.4 Dynamic formulation

The dynamic formulation of boundary value problems has been successfully used to solve static boundary value problems. This may be due to the fact that the type of governing equations remains hyperbolic. Vardoulakis (1996a, 1996b) proposed a regularization method that introduced microinertial and strain gradients into the nonassociated plasticity model called the second grade elastoplasticity model.

8.6.5 Discrete model and finite element analysis with strong discontinuity

Since discrete analyses such as the discrete element method (DEM; Cundall and Strack 1979; Kishino 1988; Thornton 1998) include an internal length scale, the instability could be avoided, although it needs many degrees of freedom. Oda and Kazama (1998) clearly showed that the shear band was able to be reproduced by DEM through the weak restriction of the moment between the particles. The finite element method with strong discontinuity has been developed to capture the slip surface without mesh-size dependency (Oliver et al. 1999; Regueiro and Borja 2001).

As mentioned earlier, many studies have been done on the strain localization problems of geomaterials, both experimentally and theoretically (Darve 1994; Vardoulakis and Sulem 1995). However, many of them are for cohesionless soils. On the other hand, comparatively fewer studies have been performed for cohesive soil, such as clay, although Palmer and Rice (1973) pointed out the strain localization of overconsolidated clay as a slip surface.

It is well known that cohesive soil, such as clay, exhibits strain–rate sensitivity. Hence, in the modeling of clayey soil, it is more natural to adopt elastoviscoplastic models instead of elastoplastic ones, although the elastoplastic models are extreme models for such materials. As was mentioned earlier, elastoviscoplastic models have been developed for clay to take into account the rate sensitivity.

Strain localization has been found to be the change in the type of partial differential equations of the governing equations of boundary value problems for the rate-independent modeling of materials such as sand. In other words, the type changes from elliptic to hyperbolic for static problems. On the other hand, for elastoviscoplastic modeling, such instability as strain localization can be treated as the exponential growth of fluctuation, that is, Lyapunov instability, and as the growth of kinetic energy and momentum for dynamic cases (see Table 8.1).

8.7 INSTABILITY AND EFFECTS OF THE TRANSPORT OF PORE WATER

In this section, we will discuss the instability of the viscoplastic water-saturated soil, and clarify the effects of permeability and initial heterogeneity on the strain localization of fluid-saturated cohesive soil modeled by a strain gradient-dependent poroviscoplastic constitutive model.

Table 8.1 Classification of plastic and viscoplastic instability

	Rate-Independent Model	*Rate-Dependent Model*
Body	Elastoplastic body	Elastoviscoplastic body
Static conditions	Loss of ellipticity Loss of controllability Zero secondary order work Loss of Lyapunov stability	Exponential growth of fluctuation Loss of Lyapunov stability
Dynamic conditions	Loss of hyperbolicity Loss of Lyapunov stability	Growth of kinetic energy and momentum Loss of Lyapunov stability

The effects of permeability and gradient parameters on the growth rate of the fluctuation were obtained by a linear instability analysis. The deformation behavior of the clay specimens modeled with a viscoplastic model with a second-order strain gradient during shear was numerically analyzed by a soil–water coupled FEM under both globally undrained and partially drained conditions. We found that the deformation pattern and the stress–strain curve greatly depend on the permeability, the drainage conditions, and the initial nonhomogeneous properties.

Rice (1976) and Rudnicki and Rice (1975) pointed out that the nature of this problem can be solved within the general framework of bifurcation problems and that localization problems should be studied within the wider framework of mechanics, including the rapid degradation of the material strength. In addition, Rice (1975) indicated the importance of local inhomogeneity and the behavior of pore fluid. The effect of pore fluid on localization problems has been analyzed by several researchers within the context of a two-phase mixture theory such as Biot's theory (1956) and de Boer (1996). Loret and Prévost (1991), Schrefler et al. (1995, 1996), and Ehlers and Volk (1998) numerically studied the localization problems of water-saturated geomaterials with the rate-independent constitutive model. Vardoulakis (1996a, 1996b) found that boundary value problems with a nonassociated rate-independent plastic model become mathematically ill-posed even if the pore water flow is included.

Oka, Adachi, and Yashima (1994) have been dealing with the localization problem of water-saturated clay through the use of viscoplastic constitutive equations due to the rate-dependent nature of cohesive soil. Zhang et al. (1999) and Zhang and Schrefler (2000) investigated the interaction between permeability and a gradient-dependent parameter with a one-dimensional instability analysis and a numerical simulation in the context of the dynamic strain localization of saturated and partially saturated porous media. As for the experimental study, Finno et al. (1998) discussed the effects of drainage conditions on strain localization in sand specimens. In these studies, many points have been clarified, such as the effect of dilatancy, permeability, and strain rates, for particular constitutive models. Loret and Prévost (1991) and Schrefler et al. (1995) showed that strain localizes in a narrow zone in the case of higher permeability levels. On the other hand, Oka, Adachi, and Yashima (1995) reported different results in which deformation was more localized in the case of low permeability levels compared with a material with absolutely very high permeability. Several problems remain to be studied. One of them is to clarify the roles of permeability and drainage conditions in the instability of the governing equations and the deformation patterning of nonlocal viscoplastic materials, such as a higher-order strain gradient-dependent model. The other problem is to clarify the effect of the initial heterogeneity.

8.7.1 Extended viscoplastic models for clay

Herein, we adopt the elastoviscoplastic model by Oka (1981), Adachi and Oka (1982), Adachi et al. (1987a), and Higo (2003), discussed in Chapter 5, and its generalized viscoplastic model with a higher-order strain gradient. The following formulation is based on the model presented in Chapter 5, which is an extended model of the original model (Adachi and Oka 1982) and takes shear softening into account.

The viscoplastic flow rule is given by

$$D_{ij}^{vp} = \gamma \langle \Phi_1(F) \rangle \Phi_2(\xi) \frac{\partial f}{\partial \sigma'_{ij}} \quad , \quad F = \frac{f - \kappa_s}{\kappa_s} \tag{8.36}$$

where D_{ij}^{vp} is the viscoplastic stretching, γ is the viscosity parameter, σ_{ij} is the total stress tensor, and σ'_{ij} ($\sigma'_{ij} = \sigma_{ij} - u_w \delta_{ij}$) is Terzaghi's effective stress tensor.

$$\Phi_2 = 1 + \xi \tag{8.37}$$

The internal variable ξ expresses the deterioration of the materials and obeys the following evolutional equation:

$$\dot{\xi} = \frac{{M_f^*}^2}{G_2^* (M_f^* - \eta^*)^2} \dot{\eta}^* \tag{8.38}$$

where M_f^* is the value of stress ratio η^* at the failure state, G_2^* is a material parameter, and η^* is the stress invariant ratio defined by $\eta^* = \sqrt{2J_2}/\sigma'_m$, where J_2 is the second invariant of deviatoric stress tensor s_{ij} and σ'_m is the mean effective stress.

It is worth noting that a more rigorous viscoplastic model with strain softening has been developed by Oka and Kimoto (2005) and was presented in Chapter 5. It has been experimentally found that the shear strength and the deformation characteristics of clay depend on the volumetric strain. The volumetric plastic strain is used as a hardening parameter in the well-known Cam-clay model (Roscoe et al. 1963). The volumetric inelastic strain associated with both consolidation and dilatancy is a measure of the deterioration of the granular materials. On the other hand, Mühlhaus and Oka (1995, 1996) demonstrated that the higher-order gradients may be attributed to the fact that the soil is discrete. Frantziskonis (1993) also showed that the material inhomogeneity can be described by the constitutive model with higher-order strain gradients. Thus, in the present paper, we have introduced the second-order gradient of the viscoplastic volumetric strain into the constitutive model to more accurately and more sufficiently describe the deformation of clay by considering the nonlocal and the viscoplastic effects of the material. In practice, the yield function includes the Laplacian of the viscoplastic volumetric strain and it is proposed as follows:

$$f - \kappa_s = \frac{\sqrt{2J_2}}{M^* \sigma_m} + \ln \frac{\sigma'_m}{\sigma'_{my}} - a_3 \nabla^2 v^p = 0 \tag{8.39}$$

in which $f - \kappa_s = 0$ indicates the static state $(f = f_s)$, κ_s is the hardening parameter, v^p is the viscoplastic volumetric strain $(= \int D_{kk}^{vp} dt)$, $a_3 \nabla^2 v^p$ is the gradient term with a_3, defined as a material constant, J_2 is the second invariant of deviatoric stress tensor S_{ij}, and σ'_m is the mean effective stress. σ'_{my} is a hardening parameter. The evolutional equation for σ'_{my} is

$$\frac{d\sigma'_{my}}{\sigma'_{my}} = \frac{1+e}{\lambda - \kappa} dv^p \tag{8.40}$$

where dv^p is an increment in v^p.

We assume that the dynamic yield function is the same as the static yield function. Following the experimental results, $\gamma\Phi_1(F)$ in Equation (8.36) is given by

$$\gamma\Phi_1(F) = C \exp\left\{ m' \left(\frac{\sqrt{2J_2}}{M_f^* \sigma'_m} + \ln \frac{\sigma'_m}{\sigma'_{me}} - \frac{1+e}{\lambda - \kappa} v^p - \alpha_3 \nabla^2 v^p \right) \right\} \tag{8.41}$$

$$C = C_0 \exp\left(m' \ln \frac{\sigma'_{me}}{\sigma'_{my0}} \right) \tag{8.42}$$

where m' and C are viscoplastic parameters, and the gradient coefficient a_3 is assumed to be constant. σ'_{me} is the initial value of σ'_m, σ'_{my0} is the initial value of the hardening parameter, λ is the consolidation index, κ is the swelling index, and e is the void ratio.

Elastic stretching D^e_{ij} (or strain rate tensor $\dot{\varepsilon}^e_{ij}$) is given by an isotropic Hooke's law, that is,

$$D^e_{ij} = \frac{1}{2G}\hat{s}_{ij} + \frac{\kappa}{3(1+e)\sigma'_m}\delta_{ij}\hat{\sigma}_m \tag{8.43}$$

where G is the elastic shear modulus, $\hat{s}_{ij}$ is the deviatoric part of the Jaumann rate of effective Cauchy stress rate tensor, $\hat{\sigma}_m$ is the Jaumann rate of mean effective stress, and δ_{ij} is Kronecker's delta.

The Jaumann rate of Cauchy's effective stress tensor is given by

$$\hat{\sigma}_{ij} = \dot{\sigma}'_{ij} - W_{ik}\sigma'_{kj} + \sigma'_{ik}W_{kj} \tag{8.44}$$

where W_{ij} is the spin tensor.

Total stretching D_{ij} (or the strain rate tensor for the small strain case of $\dot{\varepsilon}_{ij}$) can be broken down into elastic and viscoplastic parts as

$$D_{ij} = D^e_{ij} + D^{vp}_{ij} \quad \text{or} \quad \dot{\varepsilon}_{ij} = \dot{\varepsilon}^e_{ij} + \dot{\varepsilon}^{vp}_{ij} \tag{8.45}$$

8.7.2 Instability analysis of fluid-saturated viscoplastic models

8.7.2.1 Instability under locally undrained conditions

We can easily discuss the instability of the model given by Equation (8.36) to study the conventional undrained creep behavior under locally undrained conditions, that is, the permeability coefficient is zero. Oka, Adachi, and Yashima (1995) obtained the time rate of the second invariant of viscoplastic strain rate, I^{vp}_2, under the undrained creep conditions where the deviatoric stress levels are constant as

$$\dot{I}^{vp}_2 = a(\eta^*)[I^{vp}_2]^2 \tag{8.46}$$

where $I^{vp}_2 = \sqrt{\dot{\varepsilon}^{vp}_{ij}\dot{\varepsilon}^{vp}_{ij}}$, $\eta^* = \sqrt{2J_2}/\sigma'_m$.

When $a(\eta^*) \leq 0$, $\dot{I}^{vp}_2$ is negative or zero, the system is stable. On the other hand, when $a(\eta^*) > 0$, the system becomes unstable because the small fluctuation in the viscoplastic strain rates will grow. It is worth noting that

$\eta_c^* < M^*$ (M^* is the value of η^* at the critical state and η_c^* is the value of η^* when $a(\eta^*) = 0$). This indicates that the clay will be unstable before the critical state under the locally undrained conditions in which the permeability coefficient is zero in the normally consolidated region.

In addition, Oka, Adachi, and Yashima (1995) obtained the growth rate of small fluctuation, ω, under plane strain locally undrained conditions using the linear stability analysis as

$$\omega = \frac{-Z_2}{\mu Z_1 \cot^2 2\theta + H_1} \tag{8.47}$$

where θ is the angle of the shear band and $Z_1 > 0$, $\mu > 0$, and $H_1 > 0$.

Hence if $Z_2 < 0$, the material becomes unstable and the fluctuation will grow in the orientation of $\theta = 45°$. This means that the angle of the shear band is 45°. These results coincide with the prediction by the bifurcation analysis for an elastoplastic model (Vardoulakis 1996). The results of the aforementioned instability are limited in the normally consolidated region (Figure 8.4). However, the instability analysis for the viscoplastic model with degradation proposed by Kimoto (2002) and Kimoto and Oka (2005) (see Equation 5.87) showed that the material is more unstable in the overconsolidated region (Figure 8.5).

For the purpose of evaluating the instability when q equals the constant undrained creep condition, the numerical simulations of undrained creep tests under a triaxial compression stress state are conducted. The second

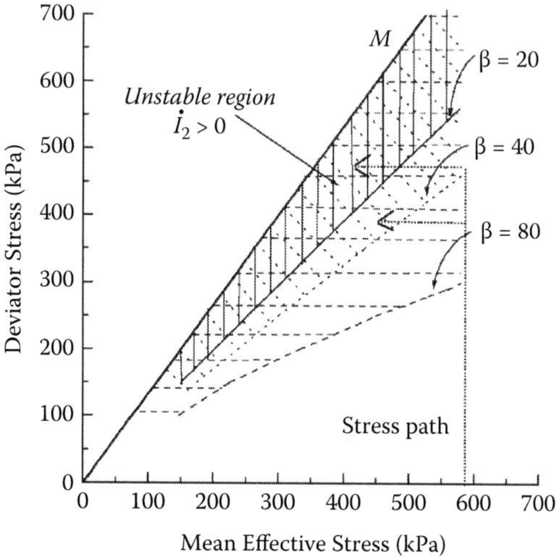

Figure 8.4 Unstable regions ($\eta^* < M^*$).

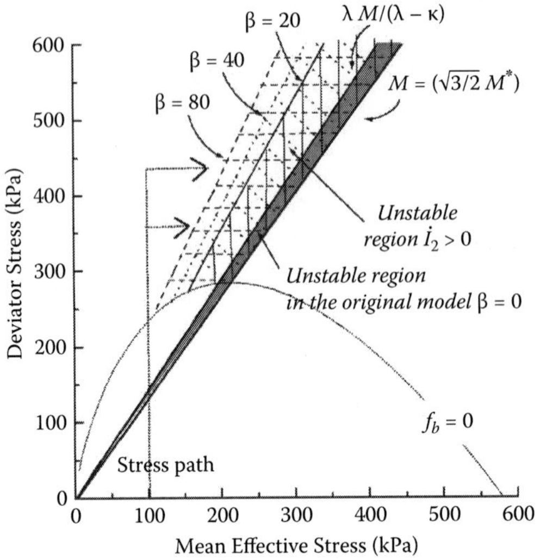

Figure 8.5 Unstable regions ($\eta^* > M^*$).

invariant of the deviatoric viscoplastic strain, I_2^{vp}, for the viscoplastic model with degradation, described in Chapter 5, is given as

$$I_2^{vp} = \sqrt{\dot{e}_{ij}^{vp}\dot{e}_{ij}^{vp}} = C\exp\left[m'\left\{\eta^* + M^*\left(\ln\frac{\sigma'_m}{\sigma'_{ma}} - y_m^*\right)\right\}\right] \tag{8.48}$$

$$dy_m^* = \frac{1+e_0}{\lambda-\kappa}d\varepsilon_v^{vp} \tag{8.49}$$

The differentiation of I_2^{vp}, with respect to time under the q equals constant undrained creep condition, is obtained as

$$\begin{aligned}\dot{I}_2^{vp} &= m'\left\{\dot{\eta}^* + M^*\left(\frac{\dot{\sigma}'_m}{\sigma'_m} - \frac{\dot{\sigma}'_{ma}}{\sigma'_{ma}} - \dot{y}_m^*\right)\right\}I_2^{vp} \\ &= m'\left\{-\frac{1+e_0}{\kappa}\left(M^* - \eta^*\right)\left(\frac{\lambda}{\lambda-\kappa}M^* - \eta^*\right)\right. \\ &\quad \left. + \frac{\beta\left(\sigma'_{mai} - \sigma'_{maf}\right)\exp(-\beta z)}{\sigma'_{ma}}M^*\sqrt{1+(M^*-\eta^*)^2}\right\}\left[I_2^{vp}\right]^2 \\ &= m'a_1(\eta^*, z)\left[I_2^{vp}\right]^2\end{aligned} \tag{8.50}$$

in which $\dot{z} = \sqrt{\dot{\varepsilon}_{ij}^{vp}\dot{\varepsilon}_{ij}^{vp}}$ and $z = \int_0^t \dot{z}dt$.

In the numerical simulations of undrained creep tests, the deviator stress levels are applied in a second for each case and kept constant until the sign for index $\dot{I}_2^{vp}$ becomes positive. Figures 8.4 and 8.5 show the unstable regions in which the sign for $\dot{I}_2^{vp}$ becomes positive during undrained creep for normally consolidated and overconsolidated clay, respectively. In the case of the original model with the material parameter of $\beta = 0$, $\dot{I}_2^{vp}$ becomes positive only in the region of $M^* < \eta^* < M^*\lambda/(\lambda - \kappa)$, whereas in the case of the model with degradation, the unstable region changes with the value of the material parameter β, which controls the degradation rate and expands around the failure line. This point will be numerically discussed in Sections 8.8 and 8.9.

8.7.2.2 Instability analysis considering the pore water flow

Let us consider the instability of the pore water flow in soil. Loret and Prévost (1991) and Schrefler et al. (1995) studied the effects of permeability on the dynamic strain localization analysis using a Mohr–Coulomb law with the associated flow rule. Loret and Prévost (1991) stated that in the case of low permeability levels, the instability may develop more slowly than in the case of high permeability levels. On the other hand, Oka et al. (1995) conducted a numerical analysis of strain localization under quasi-static conditions using an elastoviscoplastic model. From the numerical results, they pointed out that the distribution of pore water pressure is moderate with higher permeability levels and the strain localization is weaker. In their paper, the two results were compared and they suggested that the difference between the two results might be due to the different dilatancy characteristics. They pointed out that they took different loading conditions, such as quasi-static and dynamic conditions, and indicated the difference between rate-dependent and rate-independent models. However, the effects of permeability on the strain localization for a viscoplastic material under quasi-static conditions have not yet been fully studied.

Oka et al. (1995) stated only the effect of permeability on the strain localization obtained through the distribution of pore water pressure in their conclusions. Hence, it should be pointed out that the effect of permeability on the strain localization using an elastoviscoplastic model under quasi-static conditions has to be studied by both a numerical simulation and an instability analysis. To more clearly discuss the effects of permeability on strain localization, an instability analysis was carried out under two-dimensional conditions within the context of a small strain theory for simplicity (Oka et al. 1999; Higo et al. 2005). An instability analysis has been conducted by Oka et al. (1999) in which a simplified linear rigid-viscoplastic model was used. It was confirmed that the simplified rigid viscoplastic model neglecting the dilatancy model leads to the more instability with larger permeability in the strain softening. From the instability analysis, the effect of the permeability depends on the strain softening characteristics.

It is worth noting that the stability related to the permeability also depends on the dilatancy characteristics of the materials.

8.8 TWO-DIMENSIONAL FINITE ELEMENT ANALYSIS USING ELASTOVISCOPLASTIC MODEL

Using the elastoviscoplastic constitutive model by Oka et al. (1995) and the finite deformation numerical analysis method described in Chapters 5 and 6, a two-dimensional finite element analysis has been performed under plane strain conditions (Higo et al. 2005). From the numerical results, strain localization characters are discussed.

8.8.1 Effects of permeability

In Figure 8.6, the boundary conditions are shown for the plane strain problem used in the numerical analysis, while the parameters used in the computation are shown in Table 8.2. The gradient parameter, in principle, can be determined by the width of the shear band, namely, the wavelength of the localized pattern. The strain rate of the compression is 1.0%/min. The horizontal displacement at the top and the bottom of the specimen was fixed as a trigger of the localization. All of the boundaries were assumed to be impermeable, while the pore fluid was allowed to flow in the specimen.

The average vertical stress versus strain relations with different coefficients of permeability are shown in Figure 8.7. In the early stage of loading, that is, in the strain-hardening range, little difference can be seen among the three

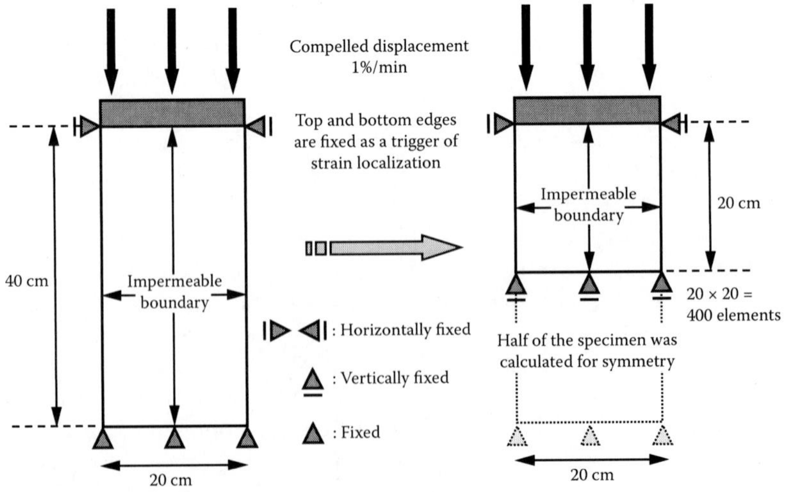

Figure 8.6 Boundary conditions and the size of the specimen.

Table 8.2 Material parameters for the strain localization analysis with different permeability coefficients

Compression index, λ	0.172
Swelling index, κ	0.054
Initial void ratio, e_0	1.28
Initial mean effective stress, σ'_{me}	200 (kPa)
Coefficient of earth pressure at rest, K_0	1.0
Viscoplastic parameter, m'	21.5
Viscoplastic parameter, C	4.5×10^{-8} (1/s)
Stress ratio at failure, M_f^*	1.05
Elastic shear modulus, G	5500 (kPa)
Softening parameter, G_2^*	100
Gradient parameter, a_3	0.0, 5.0, 30.0 ($\times 10^{-4}$m^2)
Coefficient of permeability, $k_x = k_y$(m/s)	$1.54 \times 10^{-6}, \times 10^{-8}, \times 10^{-12}$

cases. On the other hand, in the strain-softening range, the material with a low permeability level of 1.54×10^{-12} (m/s) is less unstable, that is, it is relatively stable because the average stress is larger than materials with higher permeability levels of $\times 10^{-6}$ (m/s) and $\times 10^{-8}$ (m/s) before an average axial strain of 5%. This behavior is consistent with the theoretical consideration mentioned in Section 8.7. After an axial strain of 5%, however, the average stress in the case of $\times 10^{-8}$ (m/s) is smaller than that in the case of $\times 10^{-6}$ (m/s). These results indicate that a material with a small permeability coefficient is not necessarily more stable than one with a larger permeability coefficient.

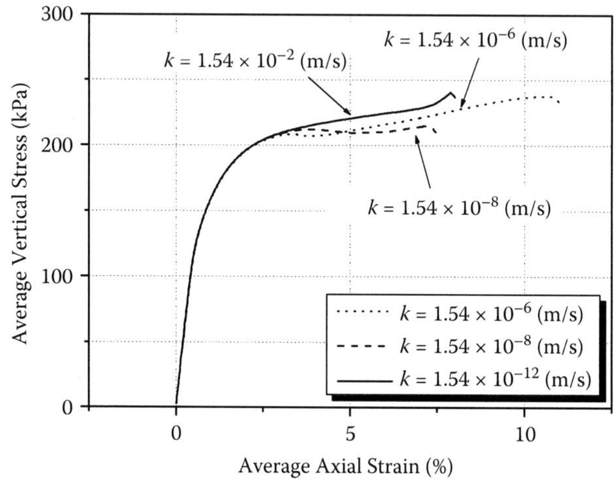

Figure 8.7 Average vertical stress–strain relations with different coefficients of permeability.

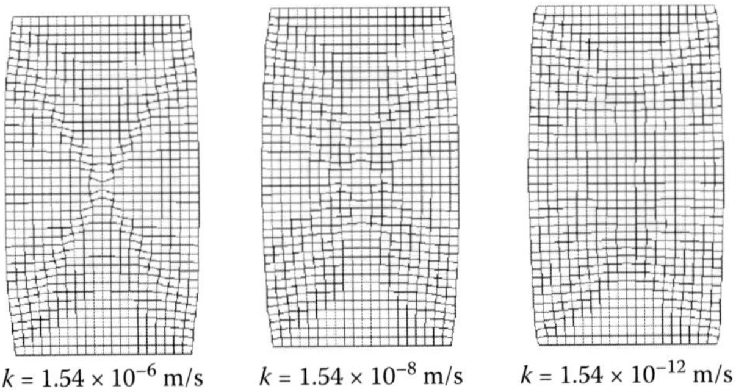

Figure 8.8 Deformed mesh at an average axial strain of 7%.

The calculations with coefficients of permeability of $k = 1.54 \times 10^{-8}$ (m/s) and 10^{-12} (m/s) diverged around 8% of the axial strain. The calculations with $k = 1.54 \times 10^{-6}$ (m/s) also diverged around 11% of the axial strain. This may be because the constraint conditions, that is, no lateral displacements at either the top or the bottom plates, induce numerical instability near the top and the bottom of the specimen. Figure 8.8 shows the deformed mesh at an average axial strain of 7% with different coefficients of permeability k. It is shown that the pore fluid has an apparent influence on the formation of shear bands. It is found that a symmetrical deformation can be seen in all cases, in particular, a clear shear band formed in the case of $k = 1.54 \times 10^{-6}$ (m/s). In Figure 8.9, velocity vectors are presented for half of the specimens at an average axial strain of 7% with different coefficients of permeability k. Discontinuous distributions of velocity fields are found in all cases due to the formation of shear bands. The patterns of distributions of the velocity vectors are consistent with the deformed mesh.

Figure 8.10 shows the distributions of accumulated viscoplastic shear strain, γ^p, at average axial strain levels of 3% and 7% with different coefficients of permeability k. γ^p is defined as follows:

$$\gamma^p = \int d\gamma^p, \quad d\gamma^p = (de_{ij}^p de_{ij}^p)^{1/2} \tag{8.51}$$

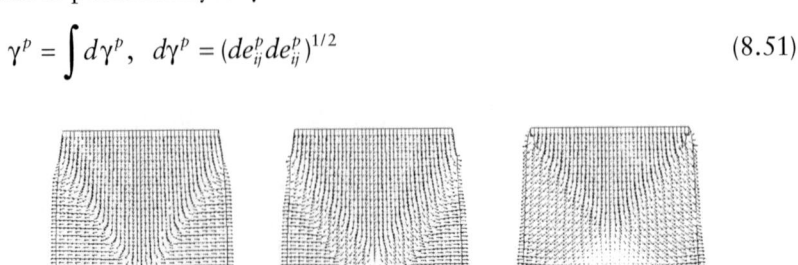

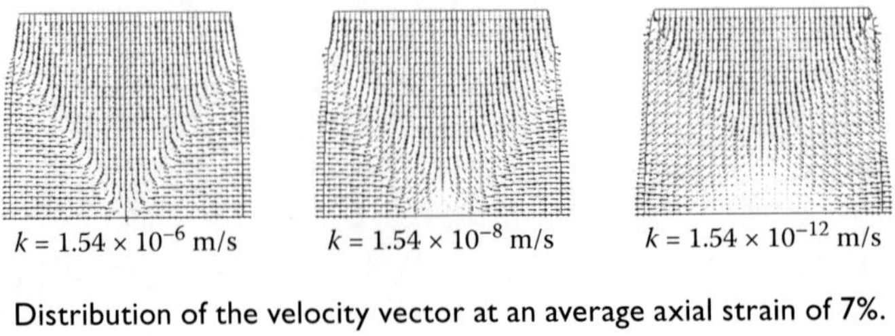

Figure 8.9 Distribution of the velocity vector at an average axial strain of 7%.

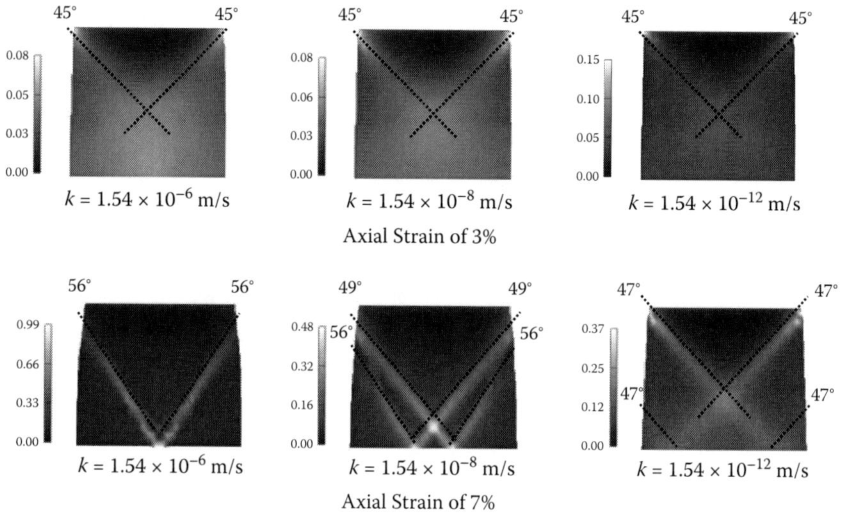

Figure 8.10 Distribution of γ^p at an average axial strain of 3% and 7%.

where de_{ij}^p is the viscoplastic deviatoric strain-increment tensor. The localized patterns of the figures for 3% are very similar to each other, but the maximum value for γ^p becomes larger as coefficient of permeability k decreases. On the other hand, at an average axial strain of 7%, the maximum value for γ^p is larger in the case of higher coefficients of permeability than that of lower coefficients of permeability. It can be said that materials with larger accumulated shear strain are more unstable than those with smaller shear strain. Following this point of view, when the average axial strain is small in the viscoplastic-hardening area, materials with lower permeability levels are rather unstable. On the other hand, when the average axial strain becomes large in the viscoplastic-softening area, materials with higher permeability levels are relatively unstable. This tendency is also consistent with the results obtained in Section 8.7. In addition, a larger difference between the maximum and the minimum values for γ^p is also seen in the case of higher permeability levels. This suggests that the strain localizes prominently when materials have high permeability levels. It is interesting, however, that two shear bands appear in the case of $k = 1.54 \times 10^{-6}$ (m/s), while the other cases have four shear bands and the distance between two shear bands is larger in the case of small permeability compared with the case of large permeability.

The inclination angles of the shear bands for all cases are 45° at the small axial strain of 3%. When the axial strain becomes 7%, the angles of the shear bands with higher permeability become larger than those with lower permeability. Oka et al. (1995) demonstrated that the preferred orientation

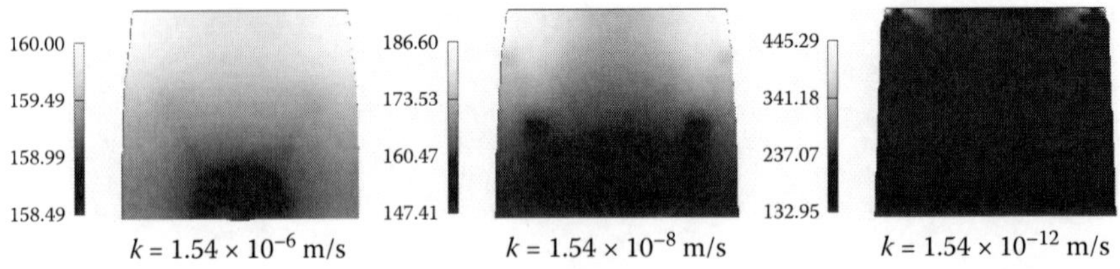

Figure 8.11 Distribution of pore water pressure at an average axial strain of 7% (unit: kPa).

for the shear bands is 45° under plane strain locally undrained conditions, that is, k=0, for Adachi and Oka's viscoplastic model introduced in Section 5.5. The reasons why the angles of the shear bands with lower strain are more proximate to 45° than materials with lower permeability levels are similar to the reasons for those under locally undrained conditions. In Figure 8.11, the distributions of pore water pressure are shown with different coefficients of permeability k. When permeability k is smaller, the pore water pressure is more localized. This trend is similar to the results obtained by Oka et al. (1995).

The distributions of mean effective stress, the second invariant of deviatoric stress, and the volumetric viscoplastic strain are all affected by the formation of shear bands and are inhomogeneous. The mean effective stress inside the shear band becomes smaller than that outside the shear band. The maximum values for the deviatoric stress and the volumetric viscoplastic strain become larger with higher permeability levels.

8.8.2 Strain localization analysis by the gradient-dependent elastoviscoplastic model

8.8.2.1 *Finite element formulation of the gradient-dependent elastoviscoplastic model*

For higher-order gradient-dependent constitutive equations, a second-order gradient of viscoplastic volumetric strain, v^p, is used in the constitutive equation. In order to obtain the second-order gradient by the finite element method, the discretization of v^p as an independent variable by the eight-node quadrilateral element is needed. Hence, we assume the weak form of the yield function and define viscoplastic volumetric strain, v^p, at each node in the same manner as that by Aifantis et al. (1999).

To obtain the weak form of the yield function, we adopt a Taylor series expansion around the current state and consider the first term. Introducing the second gradient of the viscoplastic volumetric strain, we have

$$\dot{v}^p = G(\sigma'_{ij}, v^p, \nabla^2 v^p) \tag{8.52}$$

Expanding the viscoplastic volumetric strain rate in a Taylor series and disregarding the second- and higher-order terms, we obtain a linearized evolutional equation in the form

$$\dot{v}^p = \dot{v}_0^p + \left([G_\sigma][\dot{\sigma}'] + G_{v^p}\dot{v}^p + G_\beta(\nabla^2\dot{v}^p)\right) \tag{8.53}$$

where

$$G_\sigma = \frac{\partial G}{\partial \sigma'_{ij}}, \quad G_{v^p} = \frac{\partial G}{\partial v^p}, \quad G_\beta = \frac{\partial G}{\partial(\nabla^2 v^p)} \tag{8.54}$$

$\dot{v}_0^p$ denotes the value of the volumetric strain rate at the current state.

Using a Taylor series expansion and truncating the first-order term, we obtain the following expression for the total strain rate tensor:

$$[\dot{\varepsilon}] = [L^*]^{-1}[\dot{\sigma}] + [\dot{\varepsilon}_0^p] + [A]\dot{v}^p + [A^*][\dot{\sigma}] + [A^{**}]\nabla^2\dot{v}^p \tag{8.55}$$

where $[L^*]^{-1}$ denotes the inversion of the elastic tensor and

$$A = \frac{\partial\dot{\varepsilon}_{ij}^{vp}}{\partial v^p}, \quad A^* = \frac{\partial\dot{\varepsilon}_{ij}^{vp}}{\partial\sigma_{kl}}, \quad A^{**} = \frac{\partial\dot{\varepsilon}_{ij}^{vp}}{\partial(\nabla^2 v^p)}$$

From Equation (8.55), the stress rate tensor, $[\dot{\sigma}]$, is obtained as

$$[\dot{\sigma}] = [L][\dot{\varepsilon}] - [L][\dot{\varepsilon}_0^p] - [L][A]\dot{v}^p - [L][A^{**}]\nabla^2\dot{v}^p, [L]^{-1} = [L^*]^{-1} + [A^*] \tag{8.56}$$

where $[\dot{\varepsilon}]$ is the total strain rate tensor and $[\dot{\varepsilon}_0^p]$ is the viscoplastic strain rate tensor at the current state. In addition, for the two-phase material we use the effective stress tensor for the stress.

8.8.2.2 *Effect of the strain gradient parameter*

Strain gradients, in principle, can describe the thickness of shear bands. In addition, it is found in the instability analysis of Section 8.7 that strain gradients act as stabilizers. In this section, the effects of the strain gradient parameter on the strain localization analysis are investigated. The boundary conditions and the material parameters are the same as those

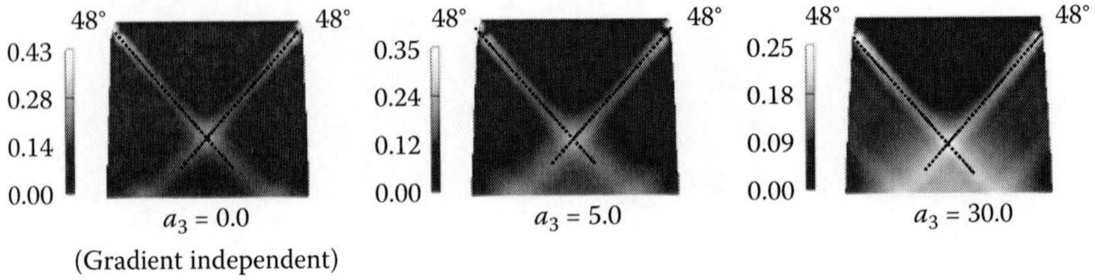

Figure 8.12 Distribution of γ^p at an average axial strain of 7% (k = 1.54 × 10^{-8} cm/s).

mentioned in the last section. Figure 8.12 depicts the distributions of accumulated viscoplastic shear strain, γ^p, at an average axial strain of 7% with different gradient parameters. In this case, $k = 1.54 \times 10^{-10}$ (m/s). It can be found from the figure that the thickness, the spacing of the shear bands, and the extent of the strain localization also depend on the gradient parameter, a_3. The accumulated strain is more localized when the gradient parameter, a_3, is rather small. This suggests that the gradient term makes the system more stable. The distance between shear bands will decrease with the gradient parameter, a_3, while the angles of the shear bands are consistently 48°.

8.8.2.3 Effect of the heterogeneity of the soil properties

The boundary conditions are shown in Figure 8.13, and Table 8.3 shows the material parameters used in the analysis in this section. We assumed three cases of distribution for the stress ratio at critical state, M^*_m, as shown in Figure 8.14. The perturbation of M^*_m was obtained using a pseudo-random number by the linear congruential method.

In Figure 8.15, the stress–strain relations for cases 1 to 3 are obtained by the different ranges in perturbation of M^*_m, R; R = 0.5%, 1.0%, and 3.0%. R = 0% means a homogeneous clay sample. The applied strain rate is 1.0/min. The effects of heterogeneity on the stress–strain relations are dependent on the initial distribution of M^*_m. It can be seen in case 1 that the average vertical stress of the nonhomogeneous clay is a little larger than that of the homogeneous clay, but that it becomes smaller in the failure state. On the other hand, the heterogeneous clay in case 3 shows softening behavior and the average stress is smaller than that of the homogeneous clay. The axial strain at the failure state is consistently smaller as the range of perturbation, R, is larger. Figure 8.16 shows the deformed mesh and the distributions of γ^p for a homogeneous one and for case 1 with different R at an average axial strain of 10%. Regarding the homogeneous case, the deformation and the distribution of γ^p are uniform in the specimen. On the

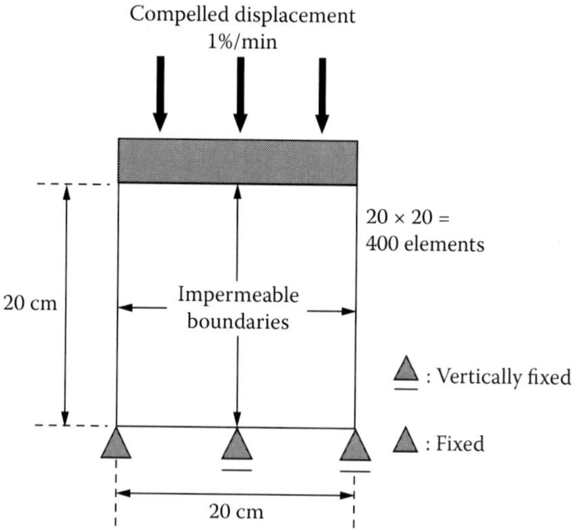

Figure 8.13 Boundary conditions and the size of the specimen (undrained plane strain condition).

contrary, localized deformations are seen in the nonhomogeneous clay and the shear band of $R = 3\%$ is clearer than the others. Similar results have been obtained by Ehlers and Volk (1998). They showed that a random distribution of the Lamé constants within local deviations of ±0.5% provided clear shear bands under plane strain conditions, although the homogeneous specimens do not.

Table 8.3 Material parameters for the strain localization analysis of homogeneous and heterogeneous types of clay

Compression index, λ	0.372
Swelling index, κ	0.054
Initial void ratio, e_0	1.28
Initial mean effective stress, σ'_{me}	600 (kPa)
Coefficient of earth pressure at rest, K_0	1.0
Viscoplastic parameter, m'	21.5
Viscoplastic parameter, C	4.5×10^{-8} (1/s)
Stress ratio at critical state, M^*_m	1.05
Elastic shear modulus, G	13210 (kPa)
Softening parameter, G^*_2	100
Gradient parameter, a_3	0.0 (m^2)
Coefficient of permeability, $k_x = k_y$	1.16×10^{-14} (m/s)

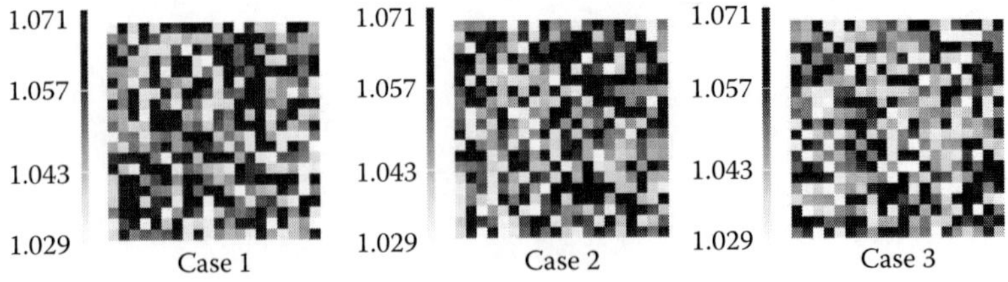

Figure 8.14 Initial distribution of M_f^* (R = 2.0%).

8.8.2.4 Mesh-size dependency

We should discuss the mesh-size dependency of the present analysis. It is well known that the finite element analysis has an inherent mesh-size dependency and that many researchers have studied regularization methods. The first method is to introduce the rate dependency of the material through

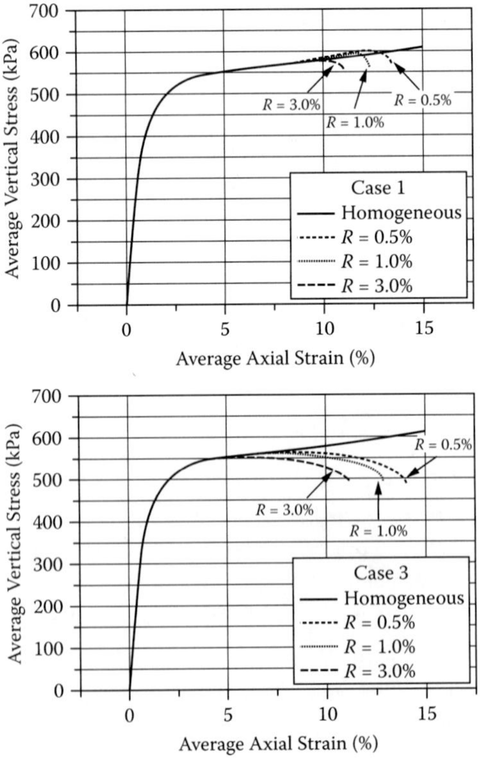

Figure 8.15 Stress–strain curves for Case 1 and Case 3 obtained by different *R*.

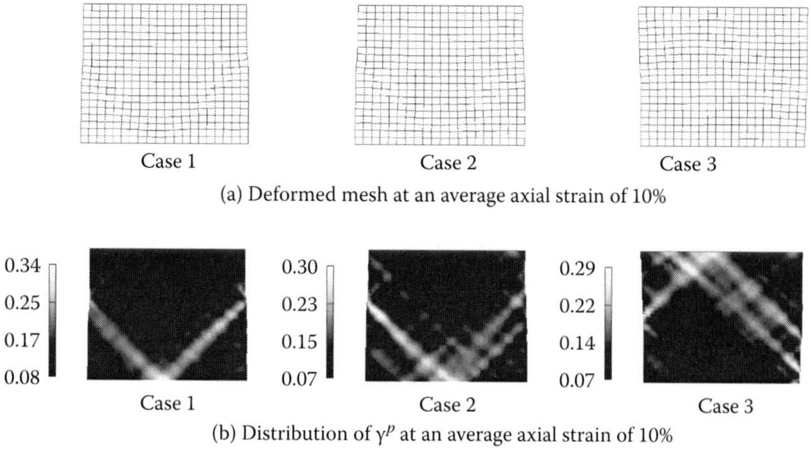

Figure 8.16 Deformed mesh and the distribution of γ^p for cases 1, 2, and 3 (R = 2.0%). (a) Deformed mesh at an average axial strain of 10%. (b) Distribution of γ^p at an average axial strain of 10%.

the use of an elastoviscoplastic model or regularization in the numerical analysis (Hughes and Taylor 1978). The second method is to introduce higher-order strain gradients into the constitutive model (Aifantis 1984). The third approach is to incorporate Darcy's law of soil–fluid interaction, which can alleviate the problem of instability by delaying the onset of material instability (Rice 1975). Herein, the mesh-size dependency is small since an elastoviscoplastic model with the second-order strain gradient and a solid–fluid mixture theory was applied. Oka et al. (2002b) found that the analysis method has no significant mesh-size dependency. It is worth noting that Zhang and Schrefler (2000) have shown that gradient dependence and permeability can regularize the finite element solution and that the second sometimes prevails over the first.

8.9 THREE-DIMENSIONAL STRAIN LOCALIZATION ANALYSIS OF WATER-SATURATED CLAY

Strain localization has been numerically analyzed for geomaterials by many researchers. Many of them, however, were treated as two-dimensional problems although the phenomena are generally three dimensional. To investigate the strain localization behavior of geomaterials under three-dimensional conditions, undrained triaxial compression tests using rectangular specimens and their simulation by a finite element analysis using an elastoviscoplastic model by Kimoto and Oka (2005) have been conducted

(Higo et al. 2006). In the experiments, both normally consolidated and overconsolidated clay samples were tested with different strain rates (Kodaka et al. 2001; Oka et al. 2005; Higo et al. 2006). Using the distribution of shear strain obtained by the image analysis of digital photographs taken during deformation, the effects of the strain rates, dilatancy, and overconsolidation on strain localization have been studied in detail. The method of numerical simulation is a soil–water coupled finite element method that is based on the finite deformation theory described in Chapter 6, using an elastoviscoplastic model for water-saturated clay considering structural changes, as shown in Chapter 5.

8.9.1 Undrained triaxial compression tests for clay using rectangular specimens

8.9.1.1 Clay samples and the testing program

The clay used in the experiment is Fukakusa clay, which is a Pleistocene marine clay produced in the southeastern part of the Kyoto Basin. The liquid limit was $w_L = 62\%$, plasticity index $I_p = 33$, and the density of soil solid $\rho_s = 2.69$ g/cm. Reconstituted clay samples were prepared by remolding them in slurry and then preconsolidating them. The specimens were consolidated one dimensionally at a preconsolidation pressure of 98 kPa. The preconsolidated specimens were covered with paraffin and not to be disturbed. The scale of the transverse section is 4 cm × 4 cm and the height is 8 cm (see Figure 8.17). The test cases are listed in Table 8.4. All the

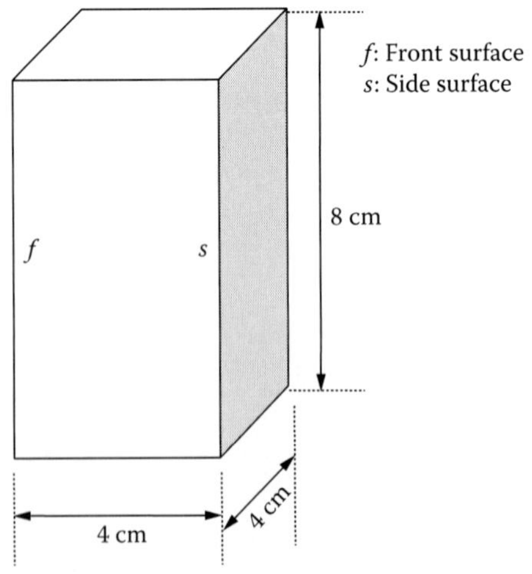

Figure 8.17 Sizes of the specimens.

Table 8.4 Test cases

No.	Sizes (c_m)	Strain rate (%/min)
Normally Consolidated Clay		
σ'_{m0} = **200 *kPa (Consolidation pressure)***		
NC-1	4 × 4 × 8	1
NC-2	4 × 4 × 8	0.1
NC-3	4 × 4 × 8	0.01
Overconsolidated clay		
σ'_{mc} = **300 *kPa (Pre-consolidation pressure)***		
σ'_{m0} = **50 *kPa (Swelling pressure), OCR = 6***		
OC-1	4 × 4 × 8	1
OC-2	4 × 4 × 8	0.1
OC-3	4× 4 × 8	0.01

specimens used were saturated by the double vacuum method and were acted upon by 200 kPa of back pressure. The normally consolidated clay specimens were isotropically consolidated to 200 kPa. The overconsolidated clay specimens were isotropically consolidated to 300 kPa, and then isotropically swelled to 50 kPa. Therefore, the overconsolidation ratio (OCR) is 6.

After the consolidation or the swelling procedure, axial pressure was applied under undrained conditions by an axial loading device with an axial strain or displacement control system. The three axial strain rates monotonically applied in the tests were 1%/min, 0.1%/min, and 0.01%/min. The tests were stopped at an axial strain of 20%.

8.9.1.2 Image analysis

We drew 2 mm square meshes on the rubber membranes covering the specimen. A digital camera was used to take photographs of two surfaces of the specimens during the tests. The following photograph and Figure 8.18 show a sample of the digital photographs taken through the triaxial cell and a schematic figure of the photography, respectively. After correcting the effects of the refraction, we digitized the nodal coordinates of the meshes. Using the coordinates at the initial state, that is, before the undrained loading, and those of each axial strain level, the nodal displacements were calculated. Adopting the B matrix for the four-node isoparametric

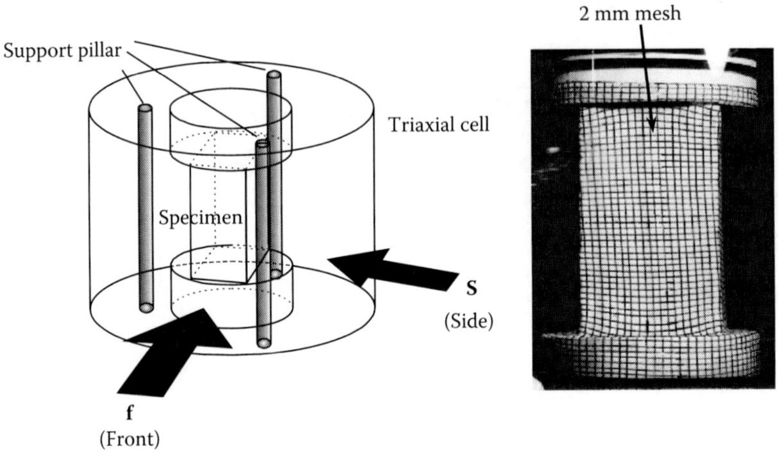

(a) (b)

Figure 8.18 (a) Schematic figure of the triaxial cell. (b) An Example of the photographs taken through the acrylic cell.

finite elements provided the strain for each element (Kodaka et al. 2001; Oka et al. 2005).

8.9.2 Three-dimensional soil–water coupled finite element analysis method

We used the three-dimensional finite element method presented in Chapter 6, based on Biot's two-phase mixture theory and the finite deformation theory, to simulate the three-dimensional strain localization tests for rectangular-shaped clay specimens. The strain localization phenomenon is a geometrically nonlinear problem since the deformation in a shear band is large. In addition, the constitutive equation for clay used in this study is nonlinear and is defined in an incremental form. In order to deal with such a nonlinear large deformation problem using an incremental constitutive model, an updated Lagrangian method with the objective Jaumann rate of Cauchy stress is used for the weak form of the rate type of equilibrium equations. As for the element type, a 20-node quadrilateral isoparametric element with a reduced Gaussian four-point integration is used for the displacement to eliminate shear locking as well as to reduce the appearance of a spurious hourglass mode. The pore water pressure is defined by an eight-node quadrilateral isoparametric element. The formulation of the three-dimensional soil–water coupled finite element analysis method is the same as those presented in the last section. Detailed formulations are in the references (Oka et al. 2002b; Higo 2003; Higo et al. 2005, 2006).

Table 8.5 Material parameters used in the numerical simulation

Parameter		*NC Clay*	*OC Clay*
Compression index, λ		0.191	0.191
Swelling index, κ		0.043	0.043
Initial void ratio, e_0		1.10	1.11
Initial elastic shear modulus, G_0	0.01%/min	16300 (kPa)	9190 (kPa)
	0.1%/min	17700 (kPa)	9920 (kPa)
	1%/min	23400 (kPa)	13080 (kPa)
Initial mean effective stress, σ'_{m0}		200 (kPa)	50 (kPa)
Compression yield stress, σ'_{mbi}		200 (kPa)	300 (kPa)
Coefficient of earth pressure at rest, K_0		1.0	1.0
Stress ratio at maximum compression, M^*_m		1.14	1.14
Viscoplastic parameter, m'		24.3	20.5
Viscoplastic parameter, C		5.8×10^{-10} (1/s)	2.7×10^{-9} (1/s)
Structural parameter, σ'_{maf}		170 (kPa)	270 (kPa)
Structural parameter, β		10	5
Coefficient of permeability, k		1.63×10^{-9} (m/s)	2.86×10^{-9} (m/s)

8.9.3 Numerical simulation of triaxial tests for rectangular specimens

8.9.3.1 Determination of the material parameters

The material parameters required by the constitutive model introduced in the last section are listed in Table 8.5. We determined λ to be 0.191 and κ to be 0.043 using the isotropic consolidation and the swelling test results for Fukakusa clay. For initial void ratio, e_0, we used the average obtained in each test, that is, 1.10 for normally consolidated clay and 1.11 for overconsolidated clay, since calculating with different void ratios is not appropriate for a comparison of the simulation results.

The initial elastic shear modulus, G_0, is determined by the initial slope of the undrained triaxial compression tests, namely, $G_0 = \Delta q/(3\Delta\varepsilon_{11})$, in which Δq is the increment in deviator stress and $\Delta\varepsilon_{11}$ is the increment in axial strain. In this study, $\Delta\varepsilon_{11}$ was determined to be 0.1%. G_0 is dependent on the strain rate because of the viscoelastic properties. The compression yield stress, σ'_{mbi}, is assumed to be the preconsolidation stress. Therefore, that of normally consolidated clay is 200 kPa and that of overconsolidated clay is 300 kPa. The stress ratio M^*_m, is defined as the stress ratio whereby maximum compression occurs in the drained compression tests. Herein, M^*_m is assumed to be determined from the stress ratio at the critical state in the undrained triaxial compression tests.

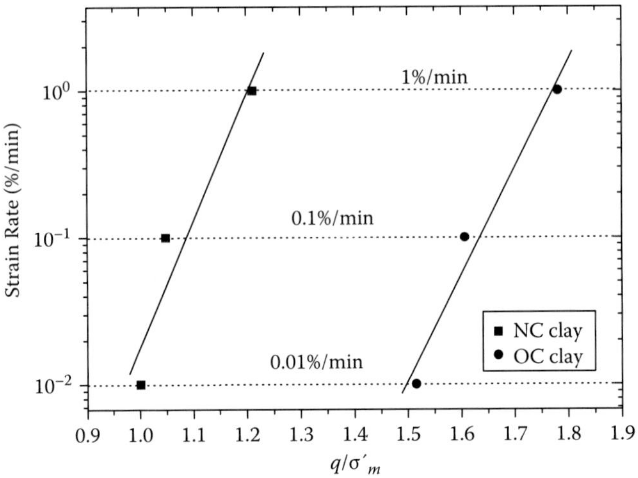

Figure 8.19 Relations between the logarithm of the strain rate and the stress ratio (NC clay: $\sigma'_m = 0.7\ \sigma'_{m0}$, OC clay: $\sigma'_m = 2.0\ \sigma'_{m0}$).

The viscoplastic parameters m' and C can be determined from undrained triaxial tests with different strain rates. Adachi and Oka (1982) noted that viscoplastic parameter m' is estimated from the slope of the relation between the stress ratio and the logarithm of the strain rate. In principle, material parameters are determined from elemental tests. In this study, however, undrained triaxial compression tests using cylindrical specimens were conducted only for a strain rate of 1.0%/min, since we could not prepare a sufficient number of specimens for the tests using cylindrical specimens that had been consolidated under the completely same conditions. Hence, we applied the test results using rectangular specimens (4 cm × 4 cm × 8 cm) to estimate the viscoplastic parameter, m' (Higo 2003).

Figure 8.19 shows the relations between the applied strain rates and stress ratio q/σ'_m in which q is the deviator stress and σ'_m is the mean effective stress. For normally consolidated clay, we plotted the stress ratios at $\sigma'_m = 0.7\sigma'_{m0}$, and estimated m' to be 24.3. As for the overconsolidated clay, m' is estimated to be 20.5 from the stress ratios at $\sigma'_m = 2.0\sigma'_{m0}$. After m' is fixed, the viscoplastic parameter C is determined by the peak stress. The structural parameter, σ'_{maf}, can be obtained by the deviator stress at the residual stress state, while β dominates the decreasing rate of deviator stress.

8.9.3.2 Boundary conditions

Figure 8.20 shows the boundary conditions, which are set up according to the same boundary conditions as those of the undrained triaxial

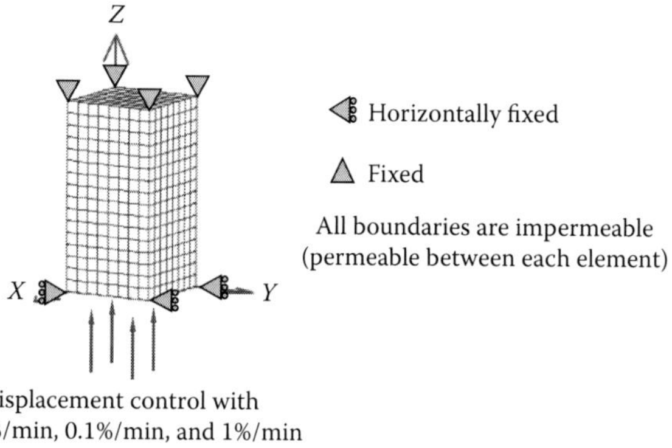

Figure 8.20 Boundary conditions for three-dimensional finite element analysis.

compression tests with displacement control. All the boundaries are assumed to be impermeable, however, the transport of pore water between each element is allowed. Constant displacement rates (z-direction) of 0.01%/min, 0.1%/min, and 1%/min are applied to the nodes on the bottom surface. The time increment is determined by the increment of average strain $\Delta\varepsilon = 0.05\%$. As for the top and the bottom surfaces, frictional force occurs between the top and the bottom surfaces and the top cap and the pedestal. Hence, the top and the bottom surfaces deform. However, it is difficult to accurately estimate the friction force. In addition, the horizontal displacements of the top and the bottom surfaces, which were measured after the tests (see Higo 2003), are rather small. Consequently, we assumed that the horizontal (x-direction and y-direction) displacement of the nodes on both the top and the bottom surfaces is constrained.

8.9.3.3 Comparison between experimental and simulation results

Undrained triaxial compression tests for normally consolidated clay and overconsolidated clay with different axial strain rates have been simulated. Figures 8.21 and 8.22 show the experimental results and the simulation results for normally consolidated clay and overconsolidated clay, respectively.

8.9.3.3.1 Stress–strain relations and effective stress paths

Figures 8.21a and 8.22a illustrate the stress–strain relations for both the simulation and the experiment. The deviator stress, the mean effective stress, and the pore water pressure used in the stress–strain curves and the effective

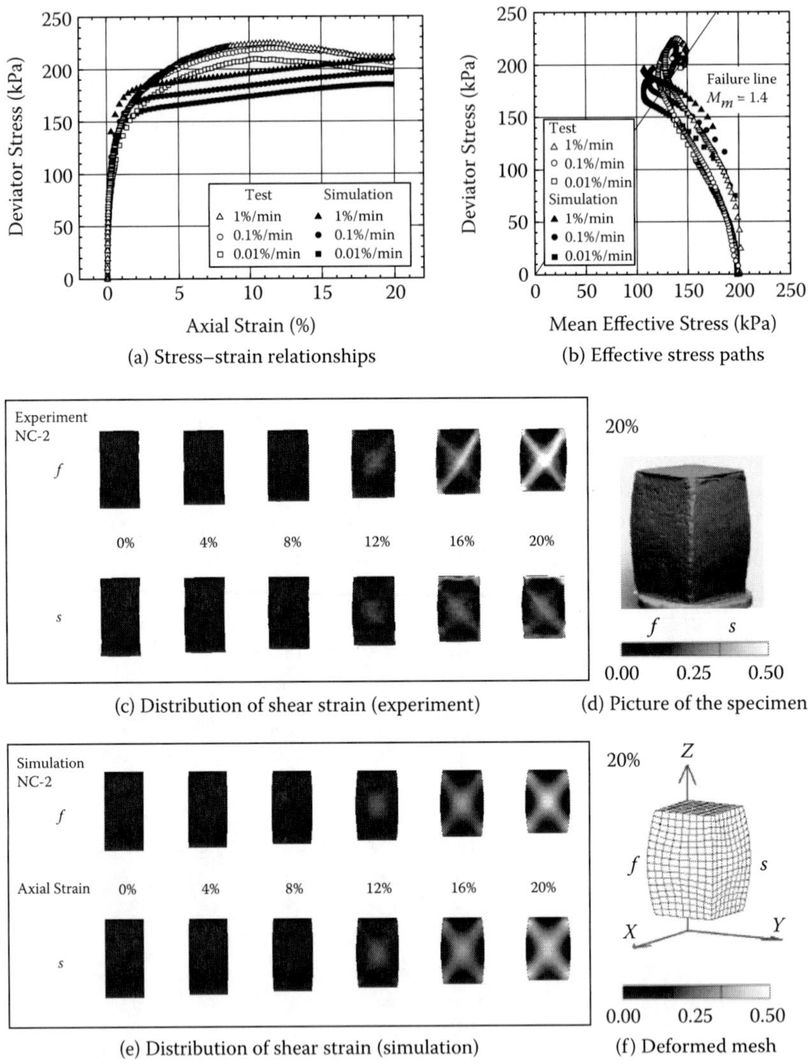

Figure 8.21 Comparison between the experimental results and simulation results (NC clay, 0.1%/min).

stress paths for simulation results are calculated in the same manner as the experimental data. The deviator stress and the pore water pressure are obtained using the average of those nodal values of the top surface.

We can see that the stress–strain relations for both the experiment and the simulation are greatly dependent on the strain rate and the dilatancy characteristics. In the experimental results, we can also observe

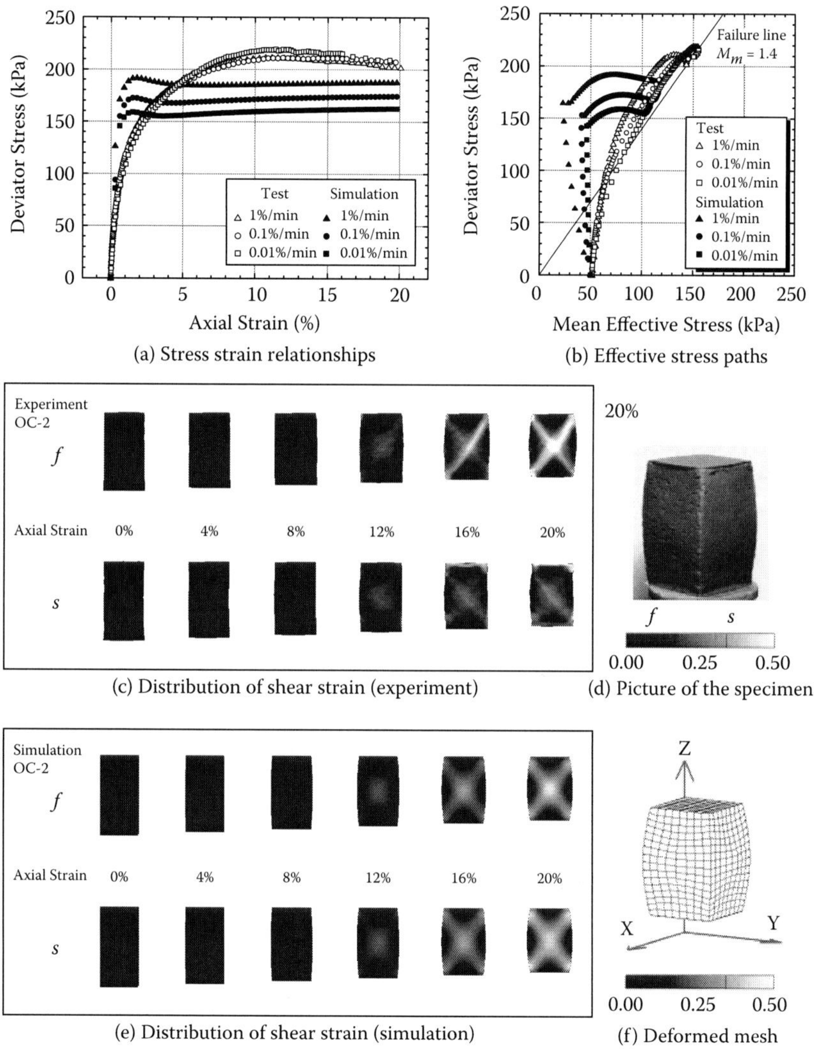

Figure 8.22 Comparison between the experimental results and simulation results (OC clay, 0.1%/min).

gradual strain-softening behavior for both normally consolidated clay and overconsolidated clay. On the other hand, in the case of the simulation, it is seen that the stress–strain relations for normally consolidated clay consistently show strain-hardening behavior, while those for overconsolidated clay show strain-softening behavior just after the peak stress around an axial strain of 2%, and then they show a gradual hardening.

It is seen in Figures 8.21b and 8.22b that the effective stress paths for normally consolidated clay exhibit negative dilatancy, that is, a decrease in mean effective stress, whereas those for overconsolidated clay exhibit positive dilatancy, that is, an increase in mean effective stress.

We can say that the stress–strain curves and the effective stress paths for normally consolidated clay are well reproduced by the presented analysis method. However, there are some differences between the experiment and the simulation for both cases. We would say that the differences can be improved by the inverse analysis technique to accurately determine the input parameters.

8.9.3.3.2 Distribution of shear strain

Figures 8.21c and 8.22c show the distributions of shear strain, γ, for the experiment, whereas Figures 8.21e and 8.22e are those for the simulation. Figures 8.21d and 8.22d are the pictures taken after the tests, and Figures 8.21f and 8.22f indicate the deformed meshes at an axial strain of 20%. In these figures, s and f indicate the side surface and the front surface of the specimens, respectively (see Figure 8.17). To obtain the distributions of shear strain, γ, on the surfaces of the specimens for the simulation results, we used a special method, which is the same method as that for the experiment.

We can see in the experimental results (Figures 8.21c and 8.22c) that strain localization starts at an axial strain of 8% and that shear bands are clearly seen at an axial strain of 12%. In addition, shear bands develop from the edges of the top and the bottom of the specimens since the friction force generated between the specimen and the top cap or the pedestal acts as a trigger of strain localization. As the axial strain becomes large, clear shear bands appear on the side surface and the front surface, and develop with increases in the thickness of the shear bands.

As shown in Figures 8.21e and 8.22e, the simulation results can well reproduce the strain localization behavior observed in the experiment. Although homogeneous deformations can be seen until an axial strain of 4% is reached, the strain starts to localize at an axial strain of 8%, and then four or two shear bands appear at an axial strain of 12% and develop with an increased thickness on both surfaces. The generating process of the shear bands is simulated well in both cases for normally consolidated clay and overconsolidated clay.

8.9.3.3.3 Strain localization pattern

In the experiment and the simulation, we can see a deformation pattern in which two or four shear bands develop from the edges of the top and the bottom of the specimens. This mode is due to the material instability induced by the frictional boundary conditions between the clay specimens

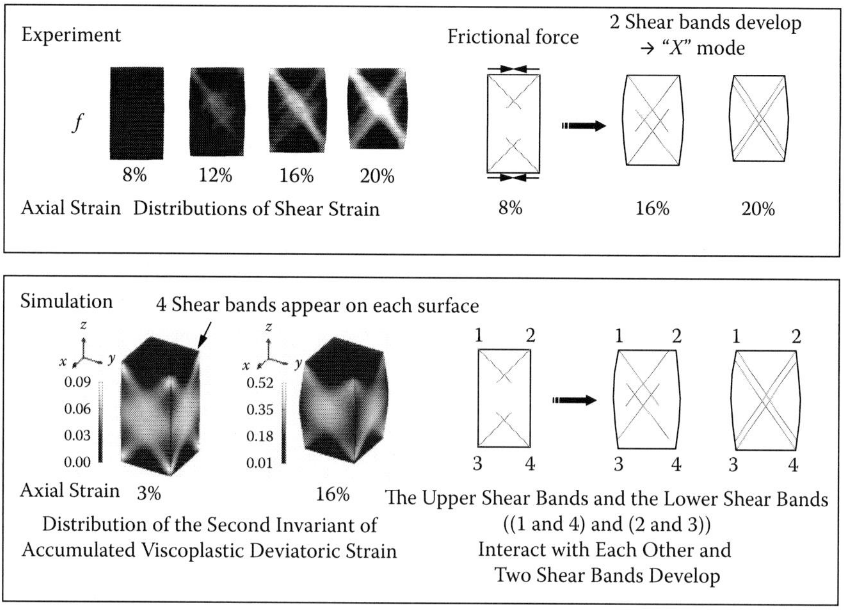

Figure 8.23 Schematics of the estimated process of the X mode for the experiment and simulation (OC-1, 1%/min).

and the top caps and the pedestals. The two shear bands intercrossing each other are just like an X, thus, we call it the X mode. Figure 8.23 shows the schematics of the estimated process generating the X mode. The four generated shear bands finally develop two clear and thick shear bands. In the case of the finite element analysis, we depict the distribution of the second invariant of accumulated viscoplastic deviatoric strain, γ^p, for all cases. In the distributions, by disregarding smaller values of γ^p, we can see the localized strain, that is, three-dimensional shear bands. Note that γ^p is obtained at the Gaussian integration points of the finite element method and is different from the shear strain γ used in the previous figures.

8.9.3.4 Three-dimensional shear bands

Estimated shear bands for case NC-2 (normally consolidated clay, 0.1%/min), tested and simulated, are depicted in Figure 8.24. In the contour for the simulation results, the accumulation of the second invariant of viscoplastic deviatoric strain, γ^p, is illustrated if γ^p is more than 0.32. Since the shear bands observed on the front surface are clearer than those on the side surface, two shear planes are estimated. It is found, however, that the X mode appears just on the surface and that higher levels of shear strain are distributed in the center of the specimen.

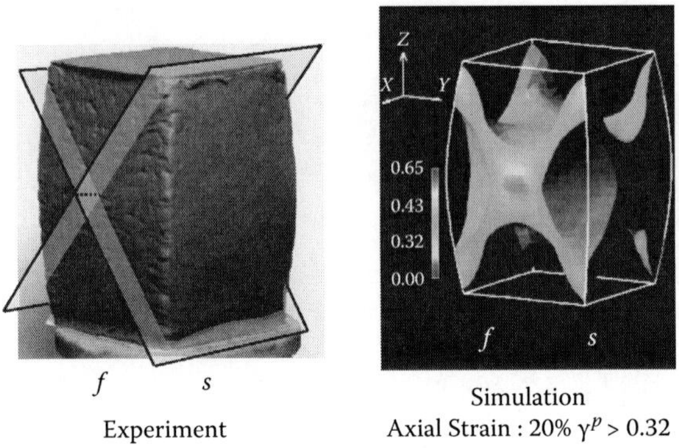

Figure 8.24 Estimated three-dimensional shear bands for experiment and simulation (NC-2, 0.1%/min).

8.9.3.5 *Effects of the strain rates*

8.9.3.5.1 Strain rate sensitivity

It is well known that clay exhibits strain rate sensitivity. Oka et al. (2003) reported the strain rate sensitivity of Fukakusa clay under undrained triaxial compression conditions. Figure 8.21a,b and Figure 8.22a,b show the rate sensitivity of the stress–strain relations and the stress paths in both cases of normally consolidated clay and overconsolidated clay can be observed. It can be seen that the strain rate sensitivity observed in the experimental results is smaller than that in the simulation results.

8.9.3.5.2 Strain localization pattern

The shear-strain distributions and the inclination angles of the shear bands for both the simulation and the experiment with different strain rates at an axial strain of 20% are shown in Figure 8.25. The numerical simulation very well reproduces the experimental results with respect not only to the X mode but also to the effects of the strain rates on the strain localization pattern. Shear bands develop from the top and the bottom edges in the case of higher strain rates, while those with lower strain rates develop beneath the top and the bottom edges. Due to this tendency, the angles of the shear bands become smaller as the strain rate decreases. If we could assume that the effect of the strain rate is equal to the effect of permeability, that is, the higher strain rates correspond to the lower permeability levels, and vice versa, we would then obtain a different trend from the one in the present study, that is, angles with higher strain rates would be smaller than those with lower strain rates.

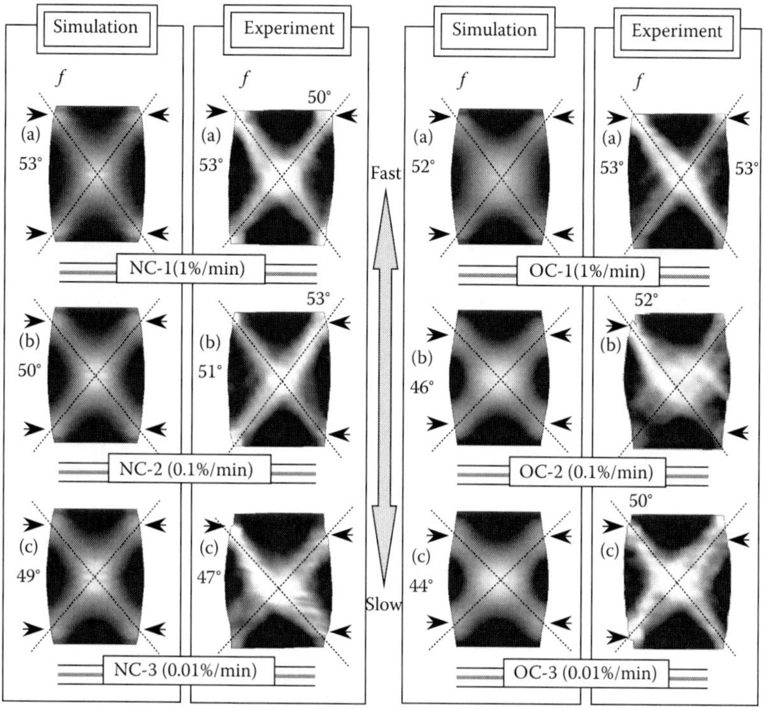

Figure 8.25 Comparison of distributions of shear strain and inclination angles of shear bands for specimen B with different strain rates between simulation and experimental results (axial strain: 20%, front surface). (a) Shear bands develop from the top edge. (b–c) Shear bands develop beneath the top edge.

Note that the constitutive equation used in Oka, Kimoto, et al. (2002) is slightly different than the one used in the present study. In addition, the maximum thickness of the shear bands with lower strain rates is larger than that of the shear bands with higher strain rates. It should be noted that these types of behavior are more clearly seen in the case of overconsolidated clay.

8.9.3.5.3 Effect of the sample shape

The effect of the sample shape has been studied by Kodaka et al. (2001) for normally consolidated clay. For a tall specimen with a high height over width ratio of H/B = 3.0, the bucking type of deformation mode is predominant for normally consolidated clay. On the contrary, for overconsolidated clay with H/B = 3.0, the X type of shear bands was experimentally observed and well simulated by the model (Kodaka et al. 2001; Oka, Kodaka, et al. 2005).

8.9.3.5.4 Mesh-size dependency

Figure 8.26 indicates the numerical results for the deformation of a quarter of the specimen using different mesh sizes under the globally undrained conditions using the same technique for the three-dimensional analysis mentioned earlier. As is shown in Figure 8.26, the effect of the mesh-size dependency is very weak in these numerical examples.

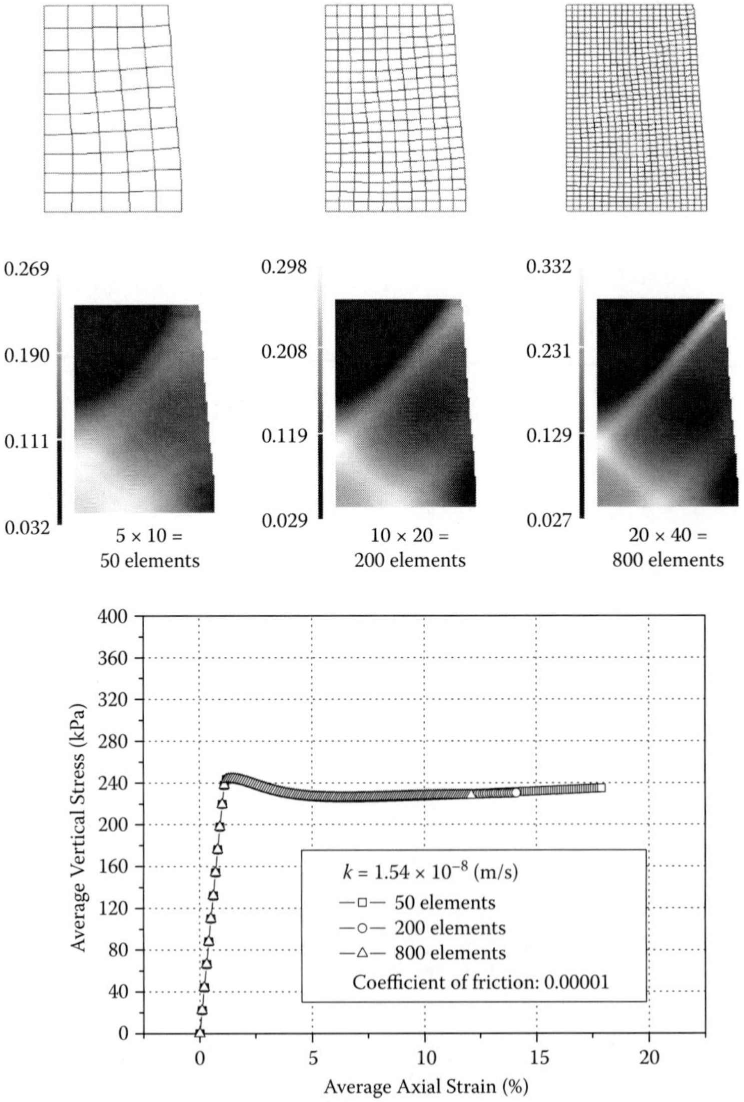

Figure 8.26 Average stress–strain relations for different numbers of finite element mesh.

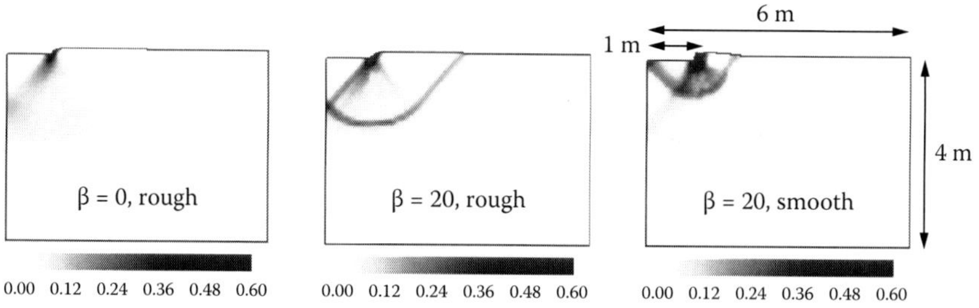

Figure 8.27 Distribution of viscoplastic deviatoric strain at a footing displacement of 10 cm for homogeneous soil cases.

8.10 APPLICATION TO BEARING CAPACITY AND EARTH PRESSURE PROBLEMS

A series of analyses for the footing of a clay deposit with different microstructure parameters was carried out (Siribumrungwong et al. 2004). The results of the analyses show that strain localization can be predicted during the loading of a footing on highly structured soil. This strain localization acts as a "slip line" and it affects the bearing behavior of the strip footing. The effects of footing roughness on the failure mechanism were also discussed.

For a smooth strip footing on clay soil with a constant level of undrained shear strength, both the Hill and the Prandtl failure mechanisms are theoretically possible. For a rough footing, however, the Hill mechanism is not appropriate as it implies horizontal soil movement at the soil-footing interface. It can be seen in Figures 8.27 and 8.28 that the Hill mechanism (Hill 1950) is predicted for a smooth footing and the Prandtl mechanism (1920) is predicted for a rough footing. Through a plasticity solution, both mechanisms yield the same value of NC = 5.14. However, a comparison shows that with a large footing displacement, the reaction

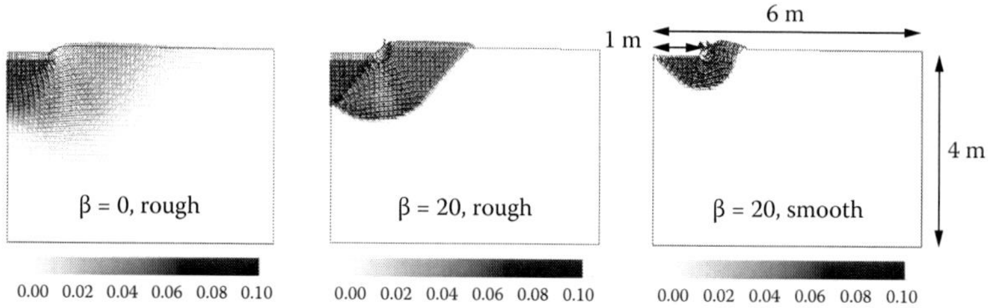

Figure 8.28 Vectors of incremental nodal displacements at a footing displacement of 10 cm for homogeneous soil cases.

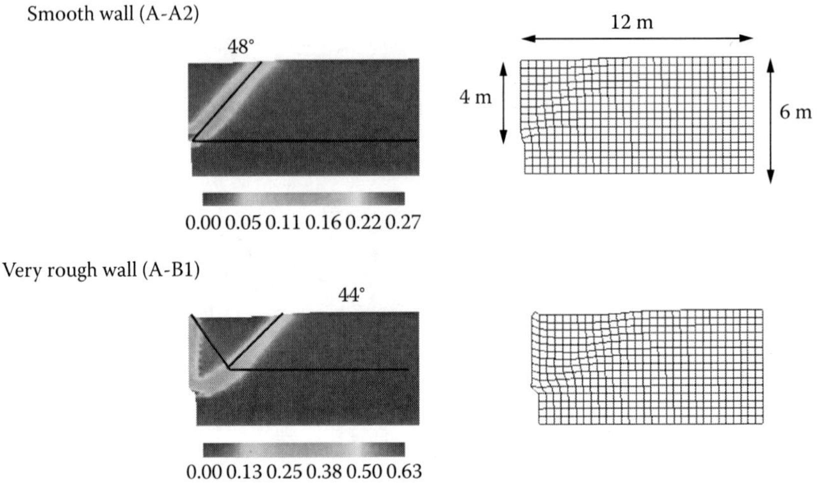

Figure 8.29 Comparison of deformed mesh and the accumulated viscoplastic shear strain from soil with different friction conditions between the soil and the wall in the case of active earth pressure.

force predicted for the smooth footing is lower than that for the rough footing.Figure 8.29 illustrates the behavior of backfill due to the movement of the wall analyzed by the same elastoviscoplastic model for clay as that mentioned earlier. It is well simulated that the strain localization depends on the friction of the wall.

8.11 SUMMARY

In the present chapter, the strain localization problem in geomaterials was described. First, strain localization phenomenon and related theoretical problems were reviewed and discussed. Then, the instability of viscoplastic geomaterials was discussed. The instability of viscoplastic geomaterials is rather complex compared with the rate-independent case. Namely, the effect of strain rate, permeability, rate sensitivity and strain softening interrelates each other depending on the boundary conditions. Finally, numerical simulations of viscoplastic geomaterials were performed and it was shown that strain localization features, such as the formation of shear bands in the material, can be reproduced well by the viscoplastic softening model. In addition, it has been proven that the method is applicable to practical problems such as bearing capacity and earth pressure.

REFERENCES

Adachi, T., and Oka, F. 1982. Constitutive equations for normally consolidated clay based on elasto-viscoplasticity, *Soils and Foundations*, 22(4):55–70.

Adachi, T., Oka, F., and Mimura, M. 1987a. An elasto-viscoplastic theory for clay failure, Proc. 8th Asian Reg. Conf. Soil Mech. Found. Eng., Kyoto, 1, JSSMFE:5–8.

Adachi, T., Oka, F., and Mimura, M. 1987b. Mathematical structure of an overstress elasto-viscoplastic model for clay, *Soils and Foundations*, 27(3):31–42.

Aifantis, E.C. 1984. On the microstructural origin of certain inelastic models, *J. Eng. Mater. Tech.*, ASME, 106:326–330.

Aifantis, E.C., Oka, F., Yashima, A. and Adachi, T. 1999. Instability of gradient dependent elasto-viscoplasticity for clay. *Int. J. Numer. Anal. Meth. Geomech.*, 23:973–994.

Anand, L. 1983. Plane deformations of ideal granular materials, *J. Mech. Phys. Solids*, 31(2):105–122.

Anandarajah, A., Rashidi, H., and Arulanandan, K. 1995. Elasto-plastic finite element analysis of a soil-structure system under earthquake excitation, *Comput. Geotech.*, 17:301–325.

Arthur, J.R.F., Dustan, T., Al-Ani, Q.A.J. and Assadi, A. 1977. Plastic deformation and failure of granular media, *Géotechnique*, 27(1):53–74.

Asaoka, A., Nakano, M., and Noda, T. 1997. Soil-water coupled behavior of heavily overconsolidated clay near/at critical state, *Soils and Foundations*, 37(1):13–28.

Bauer, E., and Huang, W. 2001. Evolution of polar quantities in a granular Cosserat material under shearing, Proceedings of 5th International Workshop on Bifurcation, Localization Theory in Geomechanics, H.B. Mühlhaus, A.V. Dyskin, and E. Pasternak, eds., Balkema, Rotterdam, 227–238.

Benallal, A., and Comi, C. 2002. Material instabilities in inelastic saturated porous media under dynamic loadings, *Int. J. Solids Struct.*, 39:3693–3716.

Benallal, A., and Comi, C. 2003. Perturbation growth and localization in fluid-saturated inelastic porous media under quai-static loadings, *J. Mech. Phys. Solids*, 51:851–899.

Bigoni, D., and Hueckel, T. 1991. Uniqueness and localization–I, Associative and non-associative plasticity, *Int. J. Solids Struct.*, 28:197–213.

Biot, M.A. 1956. Theory of propagation of elastic waves in a fluid-saturated porous solid, *J. Acoust. Soc. Am.*, 28(2):168–178.

Boehler, J.P., and Sawczuk, A. 1977. On yielding of oriented solids, *Acta Mechanica*, 27:185–206.

Bolzon, G., Schrefler, B.A., and Zienkiewicz, O.C. 1996. Elasto-plastic soil constitutive laws generalized to partially saturated states, *Géotechnique*, 46(2):279–289.

Borja, R.I. 2004. Computational modeling of deformation bands in granular media, II Numerical simulations, *Meth. Appl. Mech. Eng.*, 193:2699–2718.

Borja, R.I., and Aydin, A. 2004. Computational modeling of deformation bands in granular media, I Geological and mathematical framework, *Comput. Meth. Appl. Mech. Eng.*, 193:2667–2698.

Castellanza, R., and Nova, R. 2003. Compaction bands in oedometeric tests on cemented granular soils, Procedings of International Workshop on Prediction and Simulation Methods in Geomechanics, Athens, F. Oka, I. Vardoulakis, A. Murakami, and T. Kodaka, eds., Japanese Geotechnical Society, 37–40.

Chambon, R., Crochepeyre, S., and Desrues, J. 2000. Localization criteria for non-linear constitutive equations of geomaterials, *Mech. Cohesive-Frictional Mater.*, 5:61–82.

Chowdhury, E.Q., and Nakai, T. 2001. An unconventional approach of elastoplastic modeling of soils, In Bifurcation and Localization Theory in *Geomechanics*, H.-B. Mühlhaus, A.V. Dyskin, and E. Pasternak, eds., Balkema, 193–200.

Cleall, P.J., Thomas, H.R., Seetharam, S.C., and Melhuish, T.A. 2004. Proceedings of 9th Numerical Models in Geomechanics, G.N. Pande and S. Pietruszczak, eds., Balkema, 311–317.

Coulomb, C.A. 1773. Essai sur application des regles des maiximas et minimis a quelques problems de statique relatifs a l'architecture, Memoires de l'Academie Royale pres Divers Savants, 7:342–382.

Cundall, P.A., and Strack, O.D.L. 1979. A discrete numerical model for granular assemblies, *Géotechnique*, 29(1):47–65.

Dafalias, Y. 1982. Bounding surface elasto-viscoplasticity for particulate cohesive media, Proceedings of IUTAM Conference on Deformation and Failure of Granular Materials, P.A. Vermeer and H.J. Luger, eds., Balkema, 97–107.

Daouadji, A., Darve, F., Al Gali, H., Hicher, P.Y., Laouafa, F., Lignon, S., Nico, F., et al. 2010. Diffuse failure in geomaterials: Experiments, theory and modeling, *Int. J. Numer. Anal. Meth. Geomech.* DOI:10.1002/nag.975.

Darve, F. 1994. Stability and uniqueness in geomaterials constitutive modeling, *Localization and bifurcation theory for soils and rocks*, R. Chambon, J. Desrues and I. Vardoulakis eds., Balkema, Rotterdam: 73–88.

Darve, F. 2000. Instabilities in granular materials and application to landslides, *Mech. Cohesive-Frictional Mater.*, 5(8):627–652.

Darve, F. 2001. Modes of rupture in geomaterials, Proceedings of International Workshop on Deformation of Earth Materials, TC34, ISSMGE, Sendai, F. Oka, ed., 39–50.

de Boer, R. 1996. *Theory of Porous Media, Highlight in the Historical Development and Current State*, Springer, Berlin.

De Josselin de Jong, G. 1971. The double sliding, free rotating model for granular assemblies, *Géotechnique*, 21(3):155–162.

De Josselin de Jong, G. 1977. Mathematical elaboration of the double-sliding, free-rotating model, *Archs Mech.*, 29:561–591.

Desrues, J. 1990. Shear band initiation in granular materials: experiments and theory, *Geomaterials Constitutive Equations and Modeling*, F. Darve, ed., Elsevier Applied Science, Oxford, 283–310.

Desrues, J., Chambon, R., Mokni, M., and Mazerolle, F. 1996. Void ratio evolution inside the shear band in triaxial sand specimen studied by computer tomography, *Géotechnique*, 46(3):529–546.

Duvaut, G., and Lions, J.L. 1976. *Inequalities in Mechanics and Physics*, Springer-Verlag, Berlin.

Earthquake Engineering Committee, JSCE, 2003. Soil liquefaction under level II earthquake motion subcommittee, Report on the Soil Liquefaction under Level II Earthquake Motion (by F. Oka, K. Furuya, and R. Uzuoka), 157–201 (in Japanese).

Ehlers, W. and Volk, W. 1998. On the theoretical and numerical methods in the theory of porous media based on polar and non-polar elasto-plastic solid materials, *Int. J. Solids and Structures*, 35:4597–4617.

Finno, R.J., Viggiani, G., Harris, W.W., and Mooney, M.A. 1998. Localization of strains in plane strain compression of sand, Proceedings of 4th International Workshop on Localization and Bifurcation Theory for Soils and Rocks, Gifu, T. Adachi, F. Oka and A. Yashima, eds., Balkema, Rotterdam, 249–258.

Frantziskonis, G. 1993. Crack pattern related universal constants, *Probabilities and Materials, Tests, Models and Applications*, D. Breysse, ed., Kluwer Academic, New York, 361–376.

Gudehus, G., and Nübel, K. 2004. Evolution of shear bands in sand, *Géotechnique*, 54(3):187–201.

Hashiguchi, K. 1980. Constitutive equations of elastoplastic materials with elastic-plastic transition, *J. Appl. Mech.*, ASME, 47:266–272.

Hashiguchi, K. 1989. Subloading surface model in unconventional plasticity, *Int. J. Solids Struct.*, 25:917–945.

Hashiguchi, K., and Tsutsumi, S. 2003. Shear band formation analysis in soils by the subloading surface model with tangential stress rate effect, *Int. J. Plasticity*, 19:1651–1677.

Higo, Y. 2003. Instability and strain localization analysis of water-saturated clay by elasto-viscoplastic constitutive models, PhD thesis, Kyoto University.

Higo, Y., Oka, F., Jiang, M., and Fujita, Y. 2005. Effects of transport of pore water and material heterogeneity on strain localization of fluid-saturated gradient-dependent viscoplastic geomaterial, *Int. J. Numer. Anal. Meth. Geomech.*, 29:495–523.

Higo, Y., Oka, F., Kodaka, T., and Kimoto, S. 2006. Three-dimensional strain localization of water-saturated clay and numerical simulation using an elasto-viscoplastic model, *Philosoph. Mag.*, 86(21)–(22):3205–3240.

Hill, R. 1950. *The Mathematical Theory of Plasticity*, Oxford Clarendon Press.

Hill, R. 1957. On uniqueness and stability in the theory of finite elastic strain, *J. Mech. Phys. Solids*, 5:229–241.

Hill, R. 1958. A general theory of uniqueness and stability in elastic-plastic solids, *J. Mech. Phys. Solids*, 6:239–249.

Hill, R. 1962. Acceleration waves in solids. *J. Mech. Phys. Solids*, 10:1–16.

Hill, R., and Hutchinson, J.W. 1975. Bifurcation phenomena in the plane strain test, *J. Mech. Phys. Solids*, 23:239–264.

Hughes, T.J.R., and Taylor, R.J. 1978. Unconditionally stable algorithms for quasi-static elasto-viscoplastic finite element analysis. *Comput. Struct.*, 8:169–173.

Imposimato, S., and Nova, R. 1998. Instability of loose sand specimens in undrained tests, Proceedings of International Workshop on Localization and Bifurcation Theory for Soils and Rocks, T. Adachi, F. Oka, and A. Yashima, eds., Balkema, 313–322.

Issen, K.A., and Rudnicki, J.W. 2000. Conditions for compaction bands in porous rock, *J. Geophys. Res.*, B9, 105(21):529–521, 536.

Khojastehpour, M., and Hashiguci, K. 2004. Plane strain bifurcation analysis of soils by the tangential-subloading surface model, *Int. J. Solids Struct.*, 41:5541–5563.

Khojastehpour, M. and Hashiguchi, K. 2004. Axisymmetric bifurcation analysis in soils by the tangential-subloading surface model, *J. Mech. Phys. Solids*, 52:2235–2261.

Kimoto, S. 2002. Constitutive models for geomaterials considering structural changes and anisotropy, PhD thesis, Kyoto University.

Kimoto, S., and Oka, F. 2004. Numerical study on the effects of strain rate and structure on the consolidation process of soft clay deposits, Proceedings of International Symposium on Engineering Practice and Performance of Soft Deposits, T. Matsui, Y. Tanaka, and M. Mimura, eds., IS-Osaka, Japanese Geotechnical Society, 105–108.

Kimoto, S., and Oka, F. 2005. An elasto-viscoplastic model for clay considering destructuration and consolidation analysis of unstable behavior, *Soils and Foundations*, 45(2):29–42.

Kimoto, S., Oka, F., and Higo, Y. 2004. Strain localization analysis of elasto-viscoplastic soil considering structural degradation, *Int. J. Comput. Meth. Appl. Mech. Eng.*, 193(27–29):2845–2866.

Kishino, Y. 1988. Discrete model analysis of granular media, in *Poromechanics and Granular Media*, M. Satake and J.T. Jenkins, eds., Elsevier, 285–292.

Kodaka, T., Higo, Y., and Takyu, T. 2001. Deformation and failure characteristics of rectangular clay specimens under three-dimensional condition, Proc. 15th ICSMGE, Istanbul, 1:167–170.

Kolymbas, D., and Rombach, G. 1989. Shear band formation in generalized hypoplasticity, *Ingenieur-Archive*, 59:177–186.

Lade, P.V. 1992. Static instability and liquefaction on loose fine sandy slopes, *J. Geotech. Eng.*, ASCE, 118(1):51–71.

Lade, P.V. 2003. Analysis and prediction of shear banding under 3D conditions in granular materials, *Soils and Foundations*, 43(4):161–172.

Leroy, Y. 1991. Linear stability analysis of rate dependent discrete systems, Int. *J. Solids Struct.*, 27(6):783–808.

Loret, B., and Prévost, J.H. 1991. Dynamic strain localization in fluid-saturated porous media. *J. Eng. Mech.*, ASCE, 117(4):907–922.

Mandel, J. 1964. Conditions de stabilite et postulat de Drucker, Proceedings of IUTAM Symposium on Rheology and Soil Mechanics, J. Kravtchenko and P.M. Sirieys, eds., Springer, Berlin: 58–67.

Matsuoka, H., and Nakai, T. 1974. Stress-deformation and strength characteristics of soil under three different principal stresses, *Proc. JSCE*, 232:59–70.

Matsuoka, H., and Nakai, T. 1985. Relationship among Tresca, Mieses, Mohr-Coulomb and Matsuoka-Nakai failure criterion, *Soils and Foundations*, 25(4):123–128.

Mehrabadi, M.M., and Cowin, S.C. 1978. Initial planar deformation of dilatant granular materials, *J. Mech. Phys.* Solids, 26:269–284.

Mollema, P.M., and Antonellini, M.A. 1996. Compaction bands: A structural analog for anti-mode I cracks in aeolian sandstone, *Techtonophysics*, 267:209–228.

Mühlhaus H.-B. 1986. Shear band analysis in granular materials by Cosserat theory, *Ing. Archiv.*, 56:389–399.

Mühlhaus, H.-B., and Aifantis, E.C. 1991. A variational principle for gradient plasticity, *Int. J. Solids Struct.*, 28(7):845–857.

Mühlhaus, H.-B., and Oka, F. 1995. A continuum theory for random packings of elastic spheres, Fracture of Brittle and Disordered Materials, Proceedings of IUTAM Symposium on Fracture of Brittle and Disordered Materials, Concrete, Rocks and Ceramics, Brisbane, Australia, G. Baker and B.L. Karihaloo eds., E & FN Spon, 285–298.

Mühlhaus, H.-B., and Oka, F. 1996. Dispersion and wave propagation in discrete and continuous models for granular materials. *Int. J. Solids Struct.*, 33(19):2841–2858.

Mühlhaus, H.-B., and Vardoulakis, I. 1987. The thickness of shear bands in granular materials, *Géotechnique*, 37:271–283.

Muir Wood, D. 2002. Some observations of volumetric instabilities in soils, *Int. J. Solids Struct.*, 39:3429–3449.

Nakai, T., and Mihara, Y. 1984. A new mechanical quantity for soils and its application to elasto-plastic constitutive models, *Soils and Foundations*, 24(2):82–94.

Nemat-Nasser, S. 1983. On finite plastic flow of crystalline solids and geomaterials, Trans. ASME, *J. Appl. Mech.*, 50:1114–1126.

Nova, R. 1989. Liquefaction, stability, bifurcation of soil via strain-hardening plasticity, Numerical Methods for the Liquefaction and Bifurcation of Granular bodies: Proceedings of the International Workshop, Gdansk, E. Dembicki, G. Gudehus, and Z. Sikora, eds., 117–132.

Nova, R. 1994. Controllability of the incremental response of soil specimens subjected to arbitrary loading programmes, *J. Mech. Behav. Mater.*, 5(2):193–201.

Nova, R. 2009. Controllability of geotechnical tests and their relationship to the instability of soils, in *Micromechanics of Failure in Granular Geomaterials*, F. Nicot and R. Wan, eds., ISTE Wiley, London, 1–34.

Oda, M. and Kazama, H. 1998. Micro-structure of shear band and its relation to the mechanism of dilatancy and failure of granular soils, *Géotechnique*, 48(1):1–17.

Oka, F. 1981. Prediction of time-dependent behaviour of clay, Proc. 10th Int. Conf. Soil Mech. Found. Eng., Stockholm, Balkema, Rotterdam, 1:215–218.

Oka, F. 1993a. An elasto-viscoplastic constitutive model for clay using a transformed stress tensor, *Mech. of Mater.*, 16:47–53.

Oka, F. 1993b. Anisotropic and pseudo-anisotropic elasto-viscoplastic constitutive models for clay, Modern Approaches to Plasticity, Proceedings of a Workshop held in Horton, D. Kolymbas, ed., Elsevier Science, 505–526.

Oka, F. 2000. *Elasto-viscoplastic constitutive models of geomaterials*, Morikita Shuppan, Japan 103 (in Japanese).

Oka, F. 2005. Computational modeling of large deformations and the failure of geomaterials, Proc. 16th ICSMGE, Millpress Science, Rotterdam, 1:95–122.

Oka, F., Adachi, T., and Yashima, A. 1994. Instability of an elasto-viscoplastic constitutive model for clay and strain localization, *Mech. Mater.*, 18:119–129.

Oka, F., Adachi, T., and Yashima, A. 1995. A strain localization analysis using a viscoplastic softening model for clay, *Int. J. Plasticity*, 11(5):523–545.

Oka, F., Adachi, T., Yashima, A., and Chu, L.L. 1994. Strain localization analysis by elasto-viscoplastic softening model for frozen sand, *Int. J. Numer. Anal. Meth. Geomech.*, 18:813–832.

Oka, F., Higo, Y., and Kimoto, S. 2002a. Effect of dilatancy on the strain localization of water-saturated elasto-viscoplastic soil, *Int. J. Solids Struct.*, 39(13–14):3625–3647.

Oka, F., Jiang, M., and Higo, Y. 1999. Effect of transport of water on strain localization analysis of fluid-saturated strain gradient dependent viscoplastic geomaterial, Proceedings of 5th International Workshop on Bifurcation and Localization, Perth, Australia, H.-B. Mühlhaus, A.V. Dyskin, and E. Pasternak, eds., Swetst Zeitlinger, B.V. Lisse, 77–83.

Oka, F., Kimoto, S., and Adachi, T. 2005. Calibration of elasto-viscoplastic models for cohesive soils, Proceedings of 11th IACMAG, G. Barla and M. Barla, eds., AGI, Patron Editore, Bologna, 1:449–456.

Oka, F., Kimoto, S., Higo, Y., Ohta, H. Sanagawa, T., and Kodaka, T. 2011. An elasto-viscoplastic model for diatomaceous mudstone and a numerical simulation of compaction bands, *Int. J. Numer. Anal. Meth. Geomech.*, 35(2):244–263.

Oka, F., Kimoto, S., Kobayashi, H., and Adachi, T. 2002b. Anisotropic behavior of soft sedimentary rock and a constitutive model, *Soils and Foundations*, 42(5):59–70.

Oka, F., Kodaka, T., Kimoto, S., Ichinose, T., and Higo, Y. 2005. Strain localization of rectangular clay specimen under undrained triaxial compression conditions, Proc. 16th ICSMGE, Osaka, Millpress Science, Rotterdam, 2:841–844.

Oka, F., Kodaka, T., Kimoto, S., Ishigaki, S., and Tsuji, C. 2003. Step-changed strain rate effect on the stress-strain relations of clay and a constitutive modeling, *Soils and Foundations*, 43(4):189–202.

Oka, F., Muhlhaus, H.B., Yashima, A., and Sawada, M. 1998. Quasi-static and dynamic characteristics of strain gradient dependent non-local constitutive models, in *Material Instabilities in Solids*, R. de Borst and E. Van der Giessen, eds., John Wiley & Sons, Chichester, 387–404.

Oka, F., Yashima, A., and Kohara, I. 1992. A finite element analysis of clay foundation based on finite elasto-viscoplasticity, Proceedings of 4th International Symposium on Numerical Models in Geomechanics, Swansea, UK, G.N. Pande, and S. Pietruszczak, eds., Balkema, Rotterdam, 915–922.

Oka, F., Yashima, A., Sawada, K., and Aifantis, E.C. 2000. Instability of gradient-dependent elastoviscoplastic model for clay and strain localization, *Comput. Meth. Appl. Mechan. Eng.*, 183:67–86.

Oliver, J., Cervera, M., and Manzoli, O. 1999. Strong discontinuities and continuum plasticity models: the strong discontinuity approach. *Int. J. Plasticity*, 15:319–351.

Olsson, W.A. 1999. Theoretical and experimental investigation of compaction bands in porous rock, *J. Geophys. Res.*, 104, B4:7219–7228.

Ostowski, A. and Taussky, O. 1951. On the variation of the determinant of a positive definite matrix, Neder. Akadem. *Wet. Proc.* A54:333–351.

Palmer, A.C., and Rice, J. 1973. The growth of slip surfaces in the progressive failure of overconsolidated clay, Proc. Roy. Soc. London, A269:500–527.

Papamichos, E., and Vardoulakis, I. 1995. Shear band formation in sand according to non-coaxial plasticity model, *Géotechnique*, 45(4):649–661.

Peirce, D., Shih, C., and Needleman, A. 1984. A tangent modulus method for rate dependent solids, *Comput. Struct.*, 18(5):845–887.

Perrin, G., and Leblond, J.B. 1993. Rudnicki and Rice's analysis of strain localization revisited, *J. Appl. Mech.*, 60:842–846.

Perzyna, P. 1963. The constitutive equations for work-hardening and rate sensitive plastic materials, *Proc. Vib. Probl.*, Warsaw, 4(3):281–290.

Prandtl, L. 1920. Über die Harte plashecher Körper, Nachr. Math-phys. KL. Gottingen, 74.

Regueiro, R., and Borja, R. 2001. Plane strain finite element analysis of pressure sensitive plasticity with strong discontinuity, *Int. J. Solids Struct.*, 38:3647–3672.

Rice, J.R. 1975. On the stability of dilatant hardening for saturated rock masses, *J. Geophys. Res.*, 80(11):1531–1536.

Rice, J.R. 1976. The localization of plastic deformation, Proceedings of 14th International Congress on Theoretical and Applied Mechanics, W.T. Koiter, ed., North-Holland, 207–220.

Roscoe, K.H. 1970. The influence of strains in soil mechanics, *Géotechnique*, 20(2):129–170.

Roscoe, K.H., and Burland, J.B. 1968. On the generalised stress-strain behaviour of "wet" clay, In Engineering Plasticity, J. Heyman and F.A. Leckie, eds., Cambridge, 535–609.

Roscoe, K.H., Schofield, A.N., and Thrairajah, A. 1963. Yielding of clays in states wetter than critical, *Géotechnique*, 13(3):211–240.

Rudnicki, J.W, and Rice, J.R. 1975. Condition for the localization of deformation in pressure-sensitive dilatant material, *J. Mech. Phys. Solids,* 23:371–394.

Rudnicki, J.W., and Olsson, W.A. 1998. Reexamination of fault angles predicted by shear localization theory, *Int. J. Rock Mech. Min. Sci.*, 35(4/5):512–513.

Sawada, K., Yashima, A. and Oka, F. 1999. Numerical analysis of deformation of saturated clay based on elasto-viscoplastic Cosserat type model, Proc. 5th Int. Workshop on Bifurcation and Localization, Perth, Australia, H.-B. Mühlhaus, A.V. Dyskin, and E. Pasternak eds., Swets & Zeitlinger B.V., Lisse: 155–160.

Sawada, K., Yashima, A., and Oka, F. 2001. Numerical analysis of deformation of saturated clay based on Cosserat type elasto-viscoplastic model, *J. Soc. Mater. Sci.* Japan, 50, 6:585–592 (in Japanese).

Schrefler, B.A., Majorana, C.E., and Sanavia, L. 1995. Shear band localization in saturated porous media. *Arch. Mech.*, 47(3):577–599.

Schrefler, B.A., Sanavia, L., and Majorana, C.E. 1996. A multiphase medium model for localization and postlocalization simulation in geomaterials. *Mech. Cohesive-Frictional Mater.*, 1:95–114.

Scott, R.F. 1987. Failure, *Géotechnique*, 37(4):423–466.

Siribumrungwong, B. 2005. Elasto-viscoplastic numerical analysis of soil-foundation interaction in soft clay ground, PhD thesis, Kyoto University.

Siribumrungwong, B., Oka, F., and Kimoto, S. 2004. Numerical study of the effect of degradation on bearing capacity of clay using an elasto-viscoplastic model, Proc. 17th KKCNN Symp. Civil Eng., Thailand, 589–594.

Siribumrungwong, B., Oka, F., Kimoto, S., Kodaka, T., and Higo, Y. 2007. Elasto-viscoplastic finite element study of the effect of degradation on bearing capacity of footing on clay ground, *Geomech. Geoeng.*, 2(4):235–251.

Sokolovsky, V.V. 1942. *Statics of earthy media* (Russian), Moscow, Izdatelstiv Akademii Nauk.

Spencer, A.J.M. 1964. A theory of the kinematics of ideal soils under plane strain conditions, *J. Mech. Phys. Solids*, 12:337–351.

Spencer, A.J.M. 1987. Isotropic polynominal invariants and tensor functions, *Applications of tensor functions in solid mechanics*, J.P. Boehler, ed., Springer, 141–164.

Tamaganini, R., Viggiani, G., and Chambon, R. 2001. Some remarks on shear band analysis in *Hypoplasticity, Bifurcation and localization theory in geomechanics*, H.B. Mühlhaus, A.V. Dyskin, and E. Pasternak eds., Balkema, Rotterdam, 85–93.

Tatsuoka, F., Nakamura, S., Huang, C.C., and Tani, K. 1990. Strength anisotropy and shear band direction in plane strain tests of sands, *Soils and Foundations*, 30(1):35–54.

Taylor, D.W. 1948. Fundamentals of Soil Mechanics, New York: John Wiley & Sons.

TC34 of ISSMGE. 2005. The state-of-the-art report, Prediction methods in large strain geomechanics, Proceedings of 16th ICSMGE, Osaka, JGS, Tokyo.

Teichman, J., and Wu, W. 1993. Numerical study of patterning of shear bands in a Cosserat continuum, *Acta Mechanica*, 69:61–74.

Thornton, C. 1998. Numerical simulations of deviatoric shear deformation of granular media, *Géotechnique*, 50(2):43–53.

Tokuoka, T. 1971. Yield conditions and flow rules derived from hypo-elasticity, *Arch. Rat. Mech. Anal.*, 42:239–252.

Valdoulakis, I. 1977. Scherfugenbildung in Sandkörpern als Verzweiguns problem, Dissertation, Universität Karlsruhe, Veröffentlichungen IBF, Heft Nr. 70.

Vardoulakis, I. 1980. Shear band inclination and shear modulus of sand in biaxial tests, *Int. J. Numer. Anal. Meth. Geomech.*, 4:103–119.

Vardoulakis, I. 1996a. Deformation of water-saturated sand: I uniform undrained deformation and shear banding, *Géotechnique*, 46(3):441–456.

Vardoulakis, I. 1996b. Deformation of water-saturated sand: II effect of pore water flow and shear banding, *Géotechnique*, 46(3):457–472.

Vardoulakis, I. 2002. Dynamic thermo-poro-mechanical analysis of catastrophic landslides, *Géotechnique*, 52(3):157–171.

Vardoulakis, I., and Aifantis, E.C. 1991. A gradient flow theory of plasticity for granular materials, *Acta. Mech.*, 87:197–217.

Vardoulakis, I., and Sulem, J. 1995. *Bifurcation Analysis in Geomechanics*, Glasgow, Blackie.

Vermeer, P. 1982. A simple shear band analysis using compliance, Proceedings of IUTAM Conference on Deformation and Failure of Granular Materials, P.A. Vermeer and H.J. Luger, eds., Balkema, Rotterdam, 493–499.

Wan, R.G., Chan, D.H., and Morgenstern, N.R. 1990. A finite element method for the analysis of shear bands in geomaterials, *Finite Elements Anal. Design*, 7(2):129–143.

Wang, R., and Guo, P.J. 1997. Strain localization of granular materials with special reference to Barotropy and Pyknotropy, Proceedings 4th Interanational Symposium on Deformation and Progressive Failure in Geomechanics, A. Asaoka, T. Adachi, and F. Oka, eds, Pergamon, 45–50.

Yashima, A., Leroueil, S., Oka, F., and Guntoro, I. 1998. Modelling temperature and strain rate dependent behavior of clays: One-dimensional consolidation, *Soils and Foundations*, 38(2):63–73.

Yatomi, C., Yashima, A., Iizuka, A., and Sano, I. 1989. General theory of shear bands formation by a non-coaxial Cam-clay model. *Soils and Foundations*, 29(3):41–53.

Zhang, H.W., Sanavia, L., and Schrefler, B.A. 1999. An internal length scale in dynamic strain localization of multiphase porous media. *Mech. Cohesive-Frictional Mater.*, 4:443–460.

Zhang, H.W., and Schrefler, B.A. 2000. Gradient-dependent plasticity model and dynamic strain localization analysis of saturated and partially saturated porous media: One dimensional model, *Eur. J. Mechs. A/Solids*, 19(3):503–524.

Zhang, H.W., and Schrefler, B.A. 2001. Uniqueness and localization analysis of elasto-plastic saturated porous media, *Int. J. Num. Anal. Meth. Geomech.*, 25:29–48.

Zienkiewicz, O.C., Humpheson, C., and Lewis, R.W. 1975. Associated and non-associated viscoplasticity and plasticity in soil mechanics, *Géotechnique*, 25(4):671–689.

Chapter 9

Liquefaction analysis of sandy ground

9.1 INTRODUCTION

The aim of this chapter is to present a cyclic elastoplastic constitutive model for sand and its application to an effective-stress-based fully coupled liquefaction analysis. The model is a cyclic elastoplastic model for sand based on the nonlinear kinematical hardening theory considering the strain dependency of the plastic modulus (Oka et al. 1999). The model has already been applied to liquefaction analyses, in particular, after the 1995 Hyogo-ken Nambu earthquake in Japan. First, an outline of the model for sand is described. Second, a possible modification for the strain dependency of the plastic modulus is presented. Finally, the numerical application of the model to a liquefaction analysis of a ground and a ground–structure system is shown.

9.2 CYCLIC CONSTITUTIVE MODELS

During the past three decades, many elastoplastic constitutive models have been proposed. The elastoplastic theory is known as a representative tool for modeling the nonlinear and hysteretic behavior of soil. In the realm of soil mechanics, an associated flow rule has been used to model the behavior of soil, in particular, soil with negative dilatancy such as normally consolidated clay. In contrast, the nonassociated flow rule is frequently adopted to model sand with positive–negative dilatancy. One of the well-known models for soil is the Drucker–Prager model (1952). The Drucker–Prager model includes a dependency of the mean effective stress. There are, however, several shortcomings, namely, an unreasonable amount of dilatancy. In the 1970s, DiMaggio and Sandler (1971) proposed an elastoplastic model for sand called the cap model to generalize the Drucker–Prager model. The model contains two yield functions, the first function being a modified Drucker–Prager model and the second function intersecting the mean effective stress axis in the space of a second invariant of the deviatoric stress tensor and the mean effective stress. This model improved the behavior of the soil in the process with an increase in the mean effective stress. Although this cap model is a simple model, it has been used for a variety of geological materials, including rocks. Nishi and Esashi (1978) and Vermeer (1978) proposed a double-hardening model based on the nonassociated flow rule. These models are consistent with the experimental results obtained by Poorooshasb et al. (1966).

A number of constitutive models for sand have been proposed and then applied to design practice with appropriate numerical methods, such as the finite element method. In order to analyze the dynamic behavior of a liquefiable ground, it is necessary to predict the cyclic behavior of saturated sand.

For the cyclic behavior of sand, many models have been proposed, including those from Ghaboussi and Momen (1979), Pastor and Zienkiewicz (1986), Nishi and Kanatani (1990), Ishihara and Kabilamany (1990), and Prévost and Keane (1990). Others have proposed cyclic elastoplastic models for sand. They have tried to incorporate their models into computer programs for two-dimensional liquefaction analyses. Hashiguchi and Ueno (1977) and Hashiguchi (1980) proposed a cyclic plasticity model called a subloading surface model, which considers the expansion and the contraction of the loading surface.

One of the authors proposed a constitutive model based on the rotational hardening rule (Adachi and Oka 1982; Oka 1982) that can incorporate the Masing rule (Masing 1926). This model was incorporated into a two-dimensional liquefaction analysis and some practical problems were solved by it (Oka, Yashima, Kato, et al. 1994). Furthermore, the nonlinear kinematic hardening rule originally proposed by Armstrong and Frederick (1966) for metal, and later modified by Chaboche and Rousselier (1983), was introduced (Oka et al. 1992) into the rotational hardening model (Oka 1982) for generalizing kinematic hardening. This nonlinear kinematic hardening rule is described by an evolutional differential equation that is a generalized Prager's kinematic hardening rule and can take into account initial anisotropy as well as the Bauschinger effect. The model by Oka et al. (1992) is based on the nonlinear kinematic hardening rule and has shown a better performance for describing the behavior of sand under cyclic loading than the model based on the simple rotational hardening model (Oka and Washizu 1981). It has already been applied to several types of liquefaction problems.

The effectiveness of the cement mixing column method and the gravel drain method as countermeasures against liquefaction has been clarified by numerical work and a two-dimensional liquefaction analysis (Shibata et al. 1992; Kato et al. 1994). The liquefaction of a seabed due to ocean waves (Oka et al. 1993) and the seepage failure of loose sand layers (Oka, Yashima, Shibata, et al. 1994) were also studied with the original kinematic hardening model. In the model proposed by Taguchi et al. (1995, 1997a,b) and Oka et al. (1999) a generalized flow rule was incorporated to accurately describe the cyclic behavior of sand and the liquefaction strength curve that relates the number of loading cycles to the shear stress amplitude.

9.3 CYCLIC ELASTOPLASTIC MODEL FOR SAND WITH A GENERALIZED FLOW RULE

Oka et al. (1999) developed an elastoplastic constitutive model based on the nonlinear kinematic hardening rule. Several modifications to the original kinematic hardening model (Oka et al. 1992) were proposed based on a comparison between the experimental results and the numerical predictions by the model. The first feature of the model is the flow rule and the second feature is the introduction of the cumulative plastic strain dependence of the plastic shear modulus. The third feature is the introduction of a fading memory of the initial anisotropy. The nonlinear kinematic hardening variables are used in both the yield function and the plastic potential function.

9.3.1 Basic assumptions

The basic assumptions are taken as follows:

- Infinitesimal strain theory
- Elastoplastic theory
- Nonassociated flow rule
- Overconsolidated boundary surface
- Nonlinear kinematic hardening rule

9.3.2 Overconsolidation boundary surface

An overconsolidation boundary surface is adopted that distinguishes the overconsolidated region from the normally consolidated region. The overconsolidation boundary surface, $f_b = 0$, is defined as

$$f_b = \overline{\eta}^*_{(0)} + M^*_m ln \frac{\sigma'_m}{\sigma'_{mb}} = 0 \tag{9.1}$$

$$\bar{\eta}^*_{(0)} = \left\{ \left(\eta^*_{ij} - \eta^*_{ij(0)} \right) \left(\eta^*_{ij} - \eta^*_{ij(0)} \right) \right\}^{1/2} \tag{9.2}$$

$$\eta^*_{ij} = s_{ij} / \sigma'_m \tag{9.3}$$

where σ'_m is the mean effective stress, s_{ij} is the deviatoric stress tensor, M^*_m is the value of the stress ratio expressed by $\sqrt{\eta^*_{ij}\eta^*_{ij}}$ when the maximum volumetric strain during shearing takes place and that can be called the phase transformation stress ratio, and $\eta^*_{ij(0)}$ denotes the value of η^*_{ij} at the end of consolidation.

The condition $f_b < 0$ means that the stress state stays in an overconsolidated (OC) region, whereas $f_b \geq 0$ means that the stress state stays in a normally consolidated (NC) region. Herein, σ'_{mb} in Equation (9.1) is given as follows:

$$\sigma'_{mb} = \sigma'_{mbi} \exp\left(\frac{1+e}{\lambda - \kappa} \upsilon^P \right) \tag{9.4}$$

where σ'_{mbi} is the initial value of σ'_{mb}, κ is the swelling index, λ is the compression index, e is the void ratio, and υ^P is the plastic volumetric strain.

σ'_{mbi} is determined by considering the volume change characteristics of sand, although it is usually thought that for isotropically consolidated soils, σ'_{mbi} is equal to the preconsolidation pressure in the context of the conventional concept of overconsolidation. σ'_{mbi} of the sand samples is not always equal to σ'_{m0} (mean effective stress at the end of consolidation) due to the material anisotropy, the method of sample preparation, the degree of compaction, aging, and so forth. $OCR^* = \sigma'_{mbi} / \sigma'_{m0}$ should be called the quasi-overconsolidation ratio (quasi-OCR). Furthermore, σ'_{mc}, which is the mean effective stress at the intersection of the overconsolidated boundary surface and the σ'_m axis, is defined as

$$\sigma'_{mc} = \sigma'_{mb} \exp\left(\frac{\eta^*_{(0)}}{M^*_m} \right). \tag{9.5}$$

The overconsolidated boundary surface, that is, the surface of isotropically consolidated sand in the $\sqrt{2J_2} - \sigma'_m$ plane, is shown in Figure 9.1. J_2 is the second invariant of deviatoric stress tensor, s_{ij}, that is, $J_2 = \frac{1}{2} s_{ij} s_{ij}$.

$$f_b = 0 \left(\sqrt{\frac{2}{3}} \left| \frac{q}{\sigma'_m} - x \frac{q_0}{\sigma'_{m0}} \right| + M^*_m \ln \left(\sigma'_m / \sigma'_{mb} \right) = 0 \right)$$

$$g = 0 \left(\sqrt{\frac{2}{3}} \left| \frac{q}{\sigma'_m} - x \right| + \tilde{M}^*_m \ln \left(\sigma'_m / \sigma'_{mb} \right) = 0 \right)$$

$q = \sigma_1 - \sigma_3$

x = kinematic hardening parameter under triaxial conditions

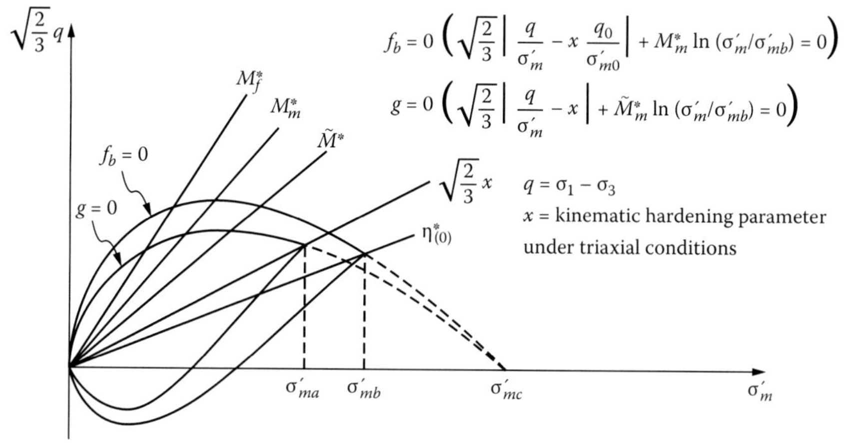

Figure 9.1 Plastic potential and overconsolidation boundary surface.

9.3.3 Fading memory of the initial anisotropy

During cyclic loading in soil, the effect of the initial anisotropy decreases. In the original model, the overconsolidation boundary surface depends on the initial anisotropy of the soil, as shown in Equation (9.5), so that the existence of the initial anisotropy influences the shape of the overconsolidation boundary surface. Herein, the initial anisotropy is assumed to fade during cyclic loading in soil. To take this into account, the following coefficient, $\zeta = \zeta(\gamma^{P*})$, is used in Equation (9.5), that is,

$$\sigma'_{mc} = \sigma'_{mb} \exp \left(\zeta \frac{\eta^*_{(0)}}{M^*_m} \right) \qquad (9.6)$$

$$\zeta = -C_d \gamma^{P*} = -C_d \int d\gamma^{P*} \qquad (9.7)$$

$$d\gamma^{P*} = \left(de^P_{ij} de^P_{ij} \right)^{1/2} \qquad (9.8)$$

in which, γ^{P*} is the cumulative or accumulated plastic shear strain from the initial condition, de^P_{ij} is the plastic deviatoric strain increment tensor, and C_d is a constant that controls the rate of disappearance of anisotropy. When sand is initially at the isotropic state, Equation (9.6) is equivalent to Equation (9.5) because the value of $\eta^*_{(0)}$ is zero.

9.3.4 Yield function

The yield function for changes in stress ratio f_{y1} is denoted as

$$f_{y1} = \left\{ \left(\eta^*_{ij} - \chi^*_{ij} \right) \left(\eta^*_{ij} - \chi^*_{ij} \right) \right\}^{1/2} - k = 0 \tag{9.9}$$

where k is a numerical parameter that controls the size of the elastic region and χ^*_{ij} is the nonlinear kinematic hardening parameter. χ^*_{ij} has the same dimensions as the stress ratio η^*_{ij} and is the so-called back stress parameter. The evolution equation for the hardening parameter is defined by Equation (9.10), namely,

$$d\chi^*_{ij} = B^* \left(A^* de^P_{ij} - \chi^*_{ij} d\gamma^{P*} \right) \tag{9.10}$$

in which A^* and B^* are material parameters. A^* and B^* are related to the stress ratio at failure, M^*_f, and the initial plastic shear modulus normalized with the mean effective stress, G^P, respectively, as follows:

$$A^* = M^*_f, \qquad B^* = \frac{G^P}{M^*_f}. \tag{9.11}$$

In general, parameter B^* can be considered to follow the evolution equation as

$$dB^* = C_f (B^*_1 - B^*) d\gamma^{P*} \tag{9.12}$$

in which C_f and B^*_1 are material parameters. $\chi^*_{ij} d\gamma^{P*}$ in Equation (9.10) is the nonlinear term that depends on the magnitude of the plastic shear strain increment. If $\chi^*_{ij} d\gamma^{P*}$ is negligible, Equation (9.10) will lead to the well-known Prager's linear kinematical hardening rule given by $d\chi^*_{ij} = B^* A^* de^P_{ij} = G^P de^P_{ij}$.

In the following Equation (9.13), the accumulated plastic shear strain between two sequential stress reversal points in the previous cycle is given by $\gamma^{P*}_{(n)} = \int_{(n)} d\gamma^{P*}$.

Cycle n can be easily indebted. In reality, a half cycle is defined as the process between two sequential stress reversal points. The stress reversal point is judged by the change in the sign of df_{y1}, increment of f_{y1} in Equation (9.9), in the general stress cognition. This automatic identification of a cycle is an advantage of using the nonlinear kinematic hardening rule. G^P decreases with an increase in $\gamma_{(n)}^{P*}$ (Oka et al. 1999). After the stress state earnest cyclic mobility, the rate of decrease in G^P, due to the accumulation of plastic strain, is accelerated. Therefore, a decrease in G^P with an increase in $\gamma_{(n)}^{P*}$ is introduced as

$$G^P = (G_{\max}^P - G_{\min}^P)\exp(-C_f \gamma_{(n)}^{P*}) + G_{\min}^P \tag{9.13}$$

$$\gamma_{(n)}^{P*} = \int_{(n)} d\gamma^{P*} \tag{9.14}$$

where $G_{\max}^P$ is the initial value for G^P, $G_{\min}^P$ is the lower limit of G^P, and C_f is a constant. The regression curves are found to be in good agreement with the experimental results shown by the symbols. Using Equation (9.13), three parameters, namely, $G_{\max}^P, G_{\min}^P$, and C_f, can be identified in each cycle. $\gamma_{(n)\max}^{P*}$ is the maximum accumulated plastic shear strain between sequential stress reversal points in the previous nth cycle, $\gamma_{(n)\max}^{P*}$, after entering into a condition of cyclic mobility. Herein, the onset of the cyclic mobility condition is judged when stress ratio η^* reaches M_m^*. It is found that during cyclic mobility, the value of $G_{\max}^P$ decreases with an increase in $\gamma_{(n)\max}^{P*}$. The remaining parameters, $G_{\min}^P$ and C_f, are also generally found to be dependent on $\gamma_{(n)\max}^{P*}$. In the modeling, however, only $G_{\max}^P$ is assumed to be dependent on $\gamma_{(n)\max}^{P*}$. The two remaining parameters are assumed not to be dependent on $\gamma_{(n)\max}^{P*}$, because $G_{\max}^P$ was found to have a stronger effect on the change in G^P than the other two parameters. The relationship between $G_{\max}^P$ and $\gamma_{(n)\max}^{P*}$ is assumed as follows:

$$G_{\max}^P = \frac{G_{\max 0}^P}{1 + \gamma_{(n)\max}^{P*} / \gamma_{(n)r}^{P*}} \tag{9.15}$$

where $G_{\max 0}^P$ is the initial value of $G_{\max}^P$ and $\gamma_{(n)r}^{P*}$ is the reference value of $\gamma_{(n)\max}^{P*}$ when the value of $G_{\max}^P$ is the half value of $G_{\max 0}^P$. The lower limit of $G_{\max}^P$ is $G_{\min}^P$. The same type of equation is also applied to model the reduction in the elastic shear modulus in which $\gamma_{(n)r}^{E*}$ is used as a reference value instead of $\gamma_{(n)r}^{P*}$.

$$G^E = \frac{G_0^E}{1 + \gamma_{(n)\max}^{E*} / \gamma_{(n)r}^{E*}} \tag{9.16}$$

where $\gamma^{E*}_{(n)\max}$ is the maximum accumulated elastic shear between sequential stress reversal points in past cycles. G_0^E is the initial value of G^E, the elastic shear modulus normalized with the mean effective stress.

Changes in yield function f_{y1} in Equation (9.9) are governed by changes in the stress ratio. As for the analysis under the general stress conditions, the second yield function, f_{y2}, which is described based on changes in the mean effective stress, should be taken into account (Oka 1992).

9.3.5 Plastic–strain dependence of the shear modulus

So far, three methods have been proposed by which the plastic–strain dependence of the plastic shear modulus is described as follows.

9.3.5.1 *Method 1*

Oka et al. (1993) proposed the plastic–strain dependency of B^* as

$$B^* = (B_0^* - B_1^*)exp(-C_f\gamma_0^{P*}) + B_1^* \tag{9.17}$$

in which B_1^* is the lower bound of B^* and γ_0^{P*} is the accumulated value of the second invariant of the deviatoric plastic strain from the initial state. In method 1, the elastic modulus is assumed to not be dependent on the plastic strain.

9.3.5.2 *Method 2*

Oka et al. (1999) proposed a method using Equations (9.13), (9.15), and (9.16). Method 2 is the most general and the most complicated method among the three methods. In method 2, $B^* = G^P/M_f^*$ is updated with the accumulated plastic shear strain between the two sequential stress reversal points in the previous nth cycle when the stress reverses. Equation (9.13) can be rewritten with B^* as

$$B^* = G^P / M_f^* = (B^*_{\max} - B_1^*)\exp(-C_f\gamma^{P*}_{(n)}) + B_1^* \tag{9.18}$$

Moreover, $B^*_{\max}$ is reduced with the maximum accumulated plastic shear strain between the sequential stress reversal points in the previous nth cycle after the stress ratio η^* has reached M_m^*. Equation (9.15) can be rewritten with B^* as

$$B^*_{\max} = G^P_{\max} / M_f^* = \frac{B_0^*}{1+\gamma^{P*}_{(n)\max}/\gamma^{P*}_{(n)r}} \tag{9.19}$$

The elastic modulus is also reduced by using Equation (9.16) in a similar manner as the plastic modulus.

9.3.5.3 Method 3

Uzuoka (2000) proposed a simplified method 3 in which B^* is determined by the following equation only when the stress ratio has reached the phase transformation line:

$$B^* = \frac{B_0^*}{1+\gamma_{ap}^{P*}/\gamma_r^{P*}} \tag{9.20}$$

in which γ_{ap}^{p*} is the accumulated value of the second invariant of the deviatoric plastic strain tensor after it has reached the phase transformation line and γ_r^{P*} is the reference strain. The lower limit of B^* is given by B_1^*. As for elastic shear modulus G^E, a similar relation is adopted for updating it in method 3.

9.3.5.4 Method 4

Oka et al. (2004) proposed a method 4 that is similar to method 2, except for using Equation (9.20) in place of Equation (9.19). B^* is updated with the accumulated plastic shear strain between the two sequential stress reversal points with Equation (9.18). After the stress path reaches the phase transformation line, B^* and the elastic shear modulus decrease with Equation (9.20).

In general, methods 2 and 4 have a high potential to reproduce the cyclic behavior of various soils. However, methods 2 and 4 require more parameters than the other methods. From a practical point of view, it is not always easy to determine many parameters. In contrast, when we adopt method 1 or method 3, the soil parameters can be determined more simply. With the knowledge gained from numerical studies (Oka et al. 2003), it has been found that method 3 is effective for modeling the plastic–strain dependence of the shear modulus even though the accuracy of the method is low; method 4 provides an effective method for determining both the plastic modulus and the elastic modulus.

Changes in yield function f_{y1} in Equation (9.9) are governed by changes in the stress ratio. As for the analysis under general stress conditions, the second yield function, f_{y2}, which is defined on the basis of changes in the mean effective stress, should be taken into account (Oka 1992), namely,

$$f_{y2} = M_m^* \left| \ln\left(\frac{\sigma'_m}{\sigma'_{m0}}\right) - y_m^* \right| - R_d = 0 \tag{9.21}$$

where y_m^* is the scalar kinematic hardening parameter, σ'_{m0} is the unit value of the mean effective stress, and R_d is the scalar variable. Since the strains brought about by changes in the mean effective stress are small in the overconsolidated region, it can be assumed that the second static yield function can be disregarded in the overconsolidated region.

The kinematic hardening parameter y_m is broken down into y_{m1}^* and y_{m2}^*. The evolutional equations for these kinematic hardening parameters are given by

$$dy_m^* = dy_{m1}^* + dy_{m2}^* \tag{9.22}$$

$$dy_{m1}^* = B_2^*(A_2^* dv^p - y_{m1}^* |dv^p|) \tag{9.23}$$

$$dy_{m2}^* = H_2^* dv^p \tag{9.24}$$

where $v^P = \varepsilon_{kk}^P$.

9.3.6 Plastic potential function

The nonlinear kinematic hardening variable is used in the plastic potential function as well as in the yield function. The plastic potential function, g, is denoted as follows:

$$g = \left\{ \left(\eta_{ij}^* - \chi_{ij}^* \right) \left(\eta_{ij}^* - \chi_{ij}^* \right) \right\}^{1/2} + \tilde{M}^* ln \left(\frac{\sigma'_m}{\sigma'_{mk}} \right) = 0 \tag{9.25}$$

where σ'_{mk} is a constant and $\tilde{M}^*$ is defined by

$$\tilde{M}^* = \begin{cases} -\dfrac{\eta^*}{ln(\sigma'_m/\sigma'_{mc})} & \text{(OC region in which } f_b < 0) \\ M_m^* & \text{(NC region in which } f_b \geq 0). \end{cases} \tag{9.26}$$

M_m^* is the stress ratio when a maximum contraction of the material takes place. $\tilde{M}^*$, which is a variable depending on the stress state, controls the direction of the plastic strain increment. When the stress state is inside the overconsolidated region, $\tilde{M}^*$ takes a value that is less than that of M_m^*. In addition, it is assumed that $\tilde{M}^*$ becomes equal to M_m^* once $\tilde{M}^*$ has reached M_m^*. The plastic potential function and the overconsolidated boundary surface are shown in Figure 9.1.

The stress–dilatancy characteristic, which is normally derived from Equation (9.25), sometimes gives a rather steeper slope to the liquefaction strength curve than that obtained from laboratory tests. To counteract this shortcoming in the original model, the flow rule is generalized using the fourth rank isotropic tensor, H_{ijkl} (Naghdi and Trapp 1975), as

$$d\varepsilon_{ij}^{P} = H_{ijkl}\frac{\partial g}{\partial \sigma'_{kl}} \tag{9.27}$$

$$H_{ijkl} = a\delta_{ij}\delta_{kl} + b(\delta_{ik}\delta_{jl} + \delta_{il}\delta_{jk}) \tag{9.28}$$

Namely,

$$de_{ij}^{P} = 2b\frac{\partial g}{\partial s_{ij}}, \qquad d\varepsilon_{kk}^{P} = (3a+2b)\frac{\partial g}{\partial \sigma'_{m}} \tag{9.29}$$

Desai and Siriwardane (1980) adopted a similar approach to the general description of $d\varepsilon_{ij}^{P}$ by introducing a correction factor into the flow rule. Coefficients a and b in Equation (9.28) generally depend on the state parameters, for example, stress and strain. From Equations (9.25), (9.27), and (9.28), a stress–dilatancy characteristic relation of the generalized model can be derived by the generalized flow rule, Equation (9.27), as follows:

$$\frac{dv^{P}}{d\gamma^{P*}} = D^{*}(\tilde{M}^{*} - \eta_{x}^{*}) \tag{9.30}$$

where $v^{P} = \varepsilon_{kk}^{P}$ and

$$D^{*} = \frac{3a}{2b} + 1 \tag{9.31}$$

D^{*} is a so-called coefficient of dilatancy, which controls the proportion of the plastic deviatoric strain increment to the plastic volumetric strain increment. In cases where coefficient a is zero, D^{*} is equal to one and the stress-dilatancy relation is equivalent to that of the original model. As for the functional form of D^{*}, two types of functions are used in this study. The first one is as follows:

$$D^{*} = D_{0}^{*} = const. \tag{9.32}$$

A constant coefficient of dilatancy has been often used, for example, by Pradhan and Tatsuoka (1989). In the model, D^* is the nonlinear function:

$$D^* = D_0^* \left(\frac{\eta^*}{\eta_r^*} \right)^n = D_0^* \left(\frac{\eta^*}{M_m^* ln(\sigma'_{mc}/\sigma'_m)} \right)^n = D_0^* \left(\frac{\tilde{M}^*}{M_m^*} \right)^n \tag{9.33}$$

where η_r^* is the value of the stress ratio on the overconsolidation boundary surface with the current mean effective stress. When η^* is larger than η_r^*, η^* is kept to the value of η_r^*. In cases where the stress state is inside the overconsolidated region, Equation (9.33) is the nth power of the stress ratio invariant. When the value of n is larger than one and the stress conditions are inside the overconsolidated region, the value of the coefficient of dilatancy drastically changes due to changes in the stress ratio.

9.3.7 Stress–strain relation

The total strain increment tensor, $d\varepsilon_{kl}$, is given by adding the elastic strain increment tensor, $d\varepsilon_{kl}^E$, and the plastic strain increment tensor, $d\varepsilon_{kl}^P$, as

$$d\varepsilon_{kl} = d\varepsilon_{kl}^E + d\varepsilon_{kl}^P \tag{9.34}$$

The plastic part of the deviatoric strain increment is derived from Equation (9.29) and the consistency condition $df = 0$. The elastic parts of the deviatoric strain increment are given as

$$de_{ij}^E = \frac{1}{2(G^E \sigma'_m)} ds_{ij} \tag{9.35}$$

where G^E is the elastic shear modulus normalized to the mean effective stress, σ'_m. The elastic volumetric strain increment is also obtained with the swelling index, κ, and void ratio, e, as

$$d\varepsilon_{kk}^E = \frac{\kappa}{1+e} \frac{d\sigma'_m}{\sigma'_m} \tag{9.36}$$

In the proposed constitutive model, the independent parameters are as follows: void ratio, e; compression index, λ; swelling index, κ; elastic shear modulus, G_E; the stress ratio at the phase transformation, M_m^*; the stress ratio at failure, M_f^*; anisotropic consolidation ratio, K_0; fading memory parameter, C_d; kinematic hardening parameters, B_0^*, B_1^*, and C_f; dilatancy parameters D_0^* and n; and strain parameters $\gamma_{(n)r}^{P*}$ and $\gamma_{(n)r}^{E*}$.

All parameters can be theoretically determined by conventional tests, physical properties tests, or triaxial compression tests. As for parameters D_0^*, n, B_0^*, B_1^*, C_f, $\gamma_{(n)r}^{P*}$, and $\gamma_{(n)r}^{E*}$, the curve-adjusting technique is available for easier identification. Through the internal friction angle, ϕ_f, and the phase transformation angle, ϕ_m, for drained triaxial tests, M_f^* and M_m^* are expressed as

$$M_f^* = \frac{2\sqrt{6}\sin\phi_f}{3-\sin\phi_f}, \quad M_m^* = \frac{2\sqrt{6}\sin\phi_m}{3-\sin\phi_f} \tag{9.37}$$

9.4 PERFORMANCE OF THE CYCLIC MODEL

A numerical simulation of the sand behavior in undrained cyclic shear tests has been conducted using a torsional hollow cylinder test apparatus (Oka et al. 2003). The specimen was consolidated under an isotropic pressure of 98 kPa. The minimum void ratio was 0.605 and the maximum void ratio was 0.977. The material parameters are listed on Table 9.1. Figures 9.2 through 9.9 show the numerical and the experimental results. From these figures, it is seen that the proposed model can reproduce the cyclic behavior of sandy soil.

9.4.1 Determination of material parameters

Parameters e_0, λ, κ, M_m^*, and M_f^* were determined directly from physical property tests and undrained monotonic shear tests. λ can be determined by the slope of the compression curve ($e - ln\sigma'_m$) and κ by the slope of the swelling curve ($e - ln\sigma'_m$). The quasi-overconsolidation ratio, OCR^*, was set to be 1.0 based on the experimental conditions. M_m^* and M_f^* were calculated by Equation (9.37) through monotonic shear tests. Laboratory tests were not performed to determine the initial shear modulus. The initial shear modulus of the specimen was estimated using the following equation (Iwasaki and Tatsuoka 1977):

$$G_0 = 14092\frac{(2.17-e)^2}{1+e}\sigma_m'^{\,0.4} \quad (kPa) \tag{9.38}$$

The normalized elastic shear modulus, G^E, was calculated by G_0/σ'_m.

Although, in principle, the remaining parameters could be determined by physical property tests and undrained monotonic and cyclic shear tests, the

Table 9.1 Material parameters

Parameters	*Dr = 60%*	*Dr = 70%*
Initial void ratio,e_0	0.754	0.716
Swelling index, κ	0.00052	0.00052
Compression index, λ	0.0091	0.0091
G_0/σ'_{m0}	2023.6	1980.0
OCR*	1.2	1.2
M_m^*	0.707	0.707
M_f^*	0.990	0.990
B_0^*	4089	4001
B_1^*	54.5	61.5
C_f	0	0
Referential strain, γ_r^{P*}	0.002	0.003
Referential strain, γ_r^{E*}	0.012	0.015
D_0^*	0.60	0.60
n	5.1	7.0
Parameters	*Dr = 80%*	*Dr = 90%*
Initial void ratio, e_0	0.683	0.672
Swelling index, κ	0.00052	0.00052
Compression index, λ	0.0091	0.0091
G_0/σ'_{m0}	1941.0	1877.3
OCR*	1.2	1.2
M_m^*	0.707	0.707
M_f^*	0.990	0.990
B_0^*	3924.0	3793.0
B_1^*	65.4	108.4
C_f	0	400
Referential strain, γ_r^{P*}	0.005	0.030
Referential strain, γ_r^{E*}	0.025	0.36
D_0^*	0.52	0.22
n	8.5	10.0

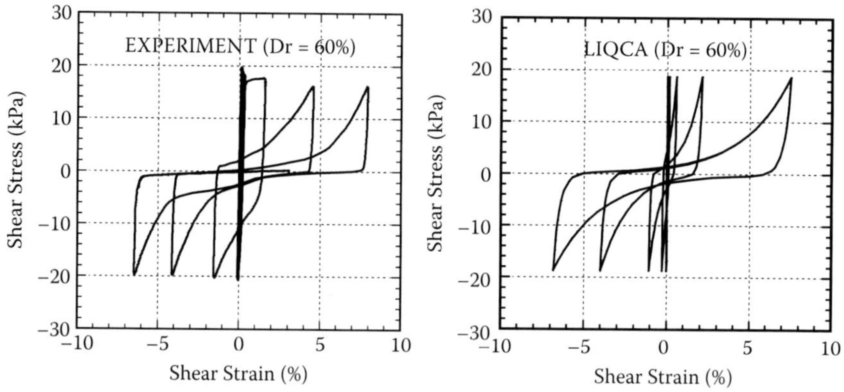

Figure 9.2 (a) Stress–strain relation (Dr = 60%). (b) Simulated stress–strain relation (Dr = 60%).

"data-adjusting method" is more practical for determining the soil parameters. In this method, the values of the material parameters are selected to provide a good description of the stress–strain relations under monotonic and cyclic loading conditions. The data-adjusting method was used to determine the dilatancy parameters, D_0^* and n, the hardening parameters, and the reference strain parameters. These parameters basically influence the effective stress path, the stress–strain curve, and the liquefaction strength curve in the undrained tests. The dilatancy parameters, D_0^* and n, control the decrease in the mean effective stress before the cyclic mobility condition and adjust the slope of the liquefaction strength curve. The hardening parameters and the reference strain parameters, on the other hand,

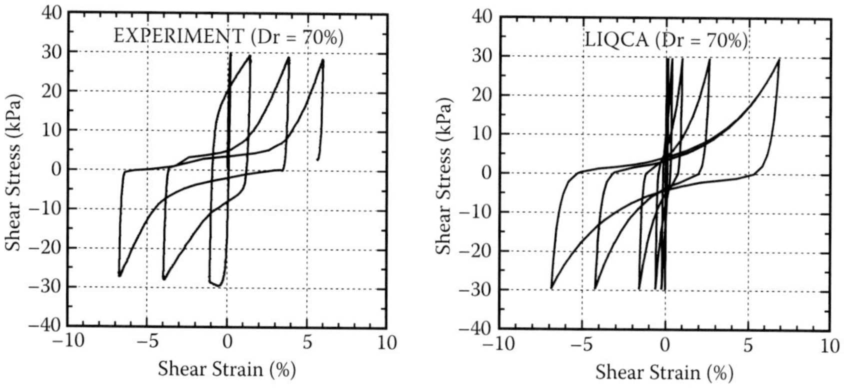

Figure 9.3 (a) Stress–strain relation (Dr = 70%). (b) Simulated stress–strain relation (Dr = 70%).

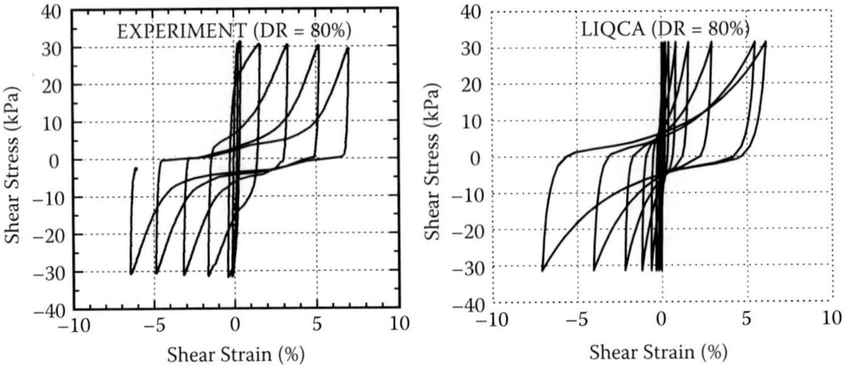

Figure 9.4 (a) Stress–strain relation (Dr = 80%). (b) Simulated stress–strain relation (Dr = 80%).

mainly control the behavior under cyclic mobility. They have an effect on the effective stress path and stress–strain relations in the range of lower mean effective stress.

9.5 LIQUEFACTION ANALYSIS OF A LIQUEFIABLE GROUND

In this section, an application of the proposed elastoplastic model to a liquefaction analysis (Oka et al. 2003) is presented. An effective-stress-based fully coupled method for the dynamic analysis of a ground considering soil liquefaction has been developed by many researchers. Herein, we adopt

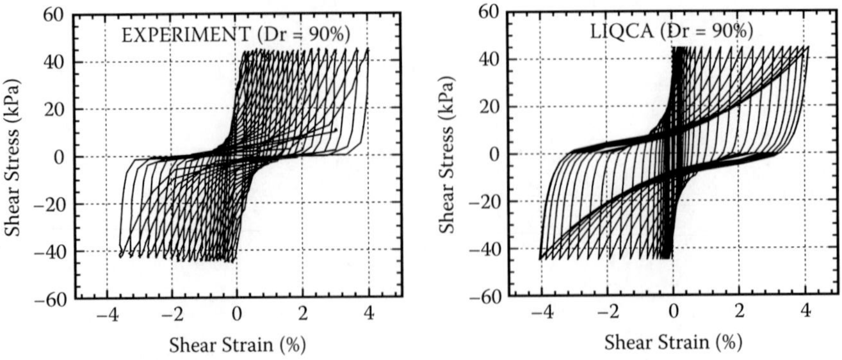

Figure 9.5 (a) Stress–strain relation (Dr = 90%). (b) Simulated stress–strain relation (Dr = 90%).

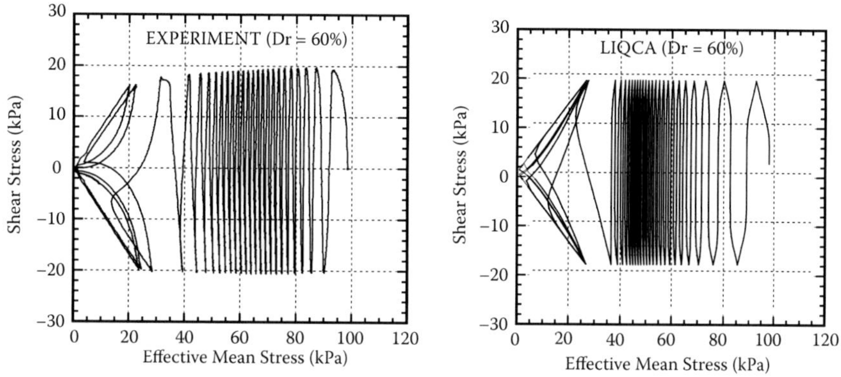

Figure 9.6 (a) Stress path (Dr = 60%). (b) Simulated stress path (Dr = 60%).

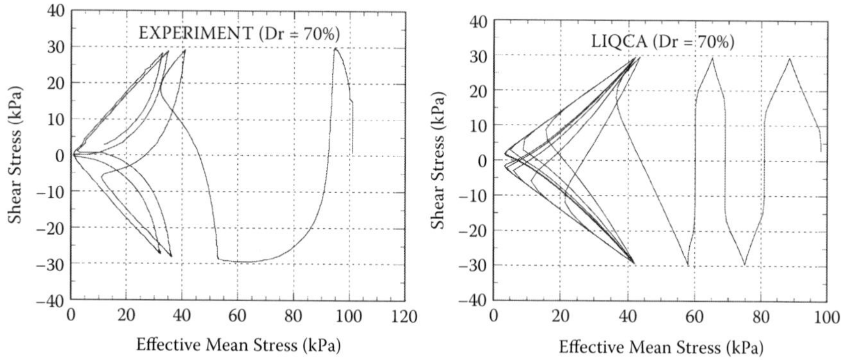

Figure 9.7 (a) Stress path (Dr = 70%). (b) Simulated stress path (Dr = 70%).

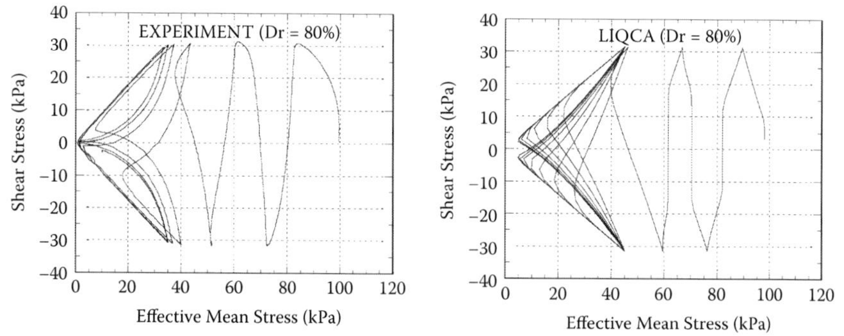

Figure 9.8 (a) Stress path (Dr = 80%). (b) Simulated stress path (Dr = 80%).

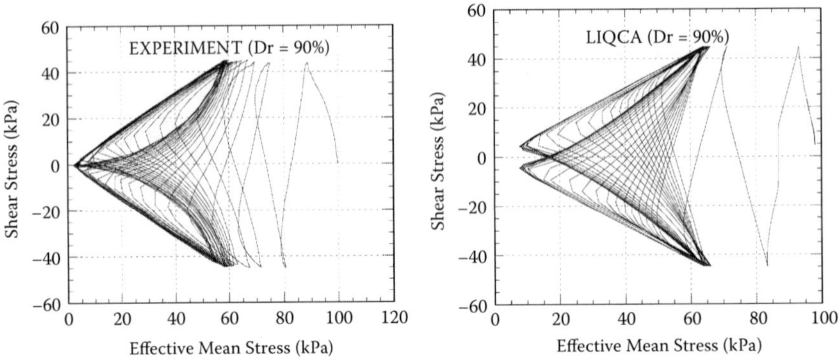

Figure 9.9 (a) Stress path (Dr =90%). (b) Simulated stress path (Dr = 90%).

the method presented in Chapter 6 based on Biot's mixture theory. The numerical analysis of a three-dimensional liquefaction analysis of Port Island in Kobe City, Japan, during the 1995 Hyogo-ken Nambu earthquake is presented, and we discuss the applicability of the proposed constitutive model through a comparison of the simulated results and the vertical array records.

9.5.1 Vertical array records on Port Island

Borehole strong motion vertical array observations were conducted on Port Island by the Development Bureau of Kobe City. Three components of accelerations, consisting of N-S, E-W, and U-D directions, were recorded at four levels and at depths of GL-83 m, GL-32 m, GL-16 m, and GL-0 m. The depth of GL-83 m corresponds to the base layer that is on the top of the diluvial dense sand: Ds. The depth of GL-32 m is under the alluvial medium-dense sand: As. The depth of GL-16 m is under the reclaimed gravelly sand: B3.

The acceleration histories of the main shock during the 1995 Hyogo-ken Nambu earthquake of January 17, 1995, were obtained. The epicenter was the northern edge of Awaji Island and the Japanese Meteorological Agency's (JMA) magnitude was 7.2 (later modified to 7.3 by JMA). The obtained acceleration histories will be shown later with the simulated results.

9.5.2 Numerical models

In the analysis, a computer code called LIQCA3D was used. This program was developed based on the cyclic elastoplasticity model for sand, mentioned earlier, and a cyclic elastoviscoplasticity model developed by Oka and Yashima (1995). The cyclic elastoplasticity model for sand was applied

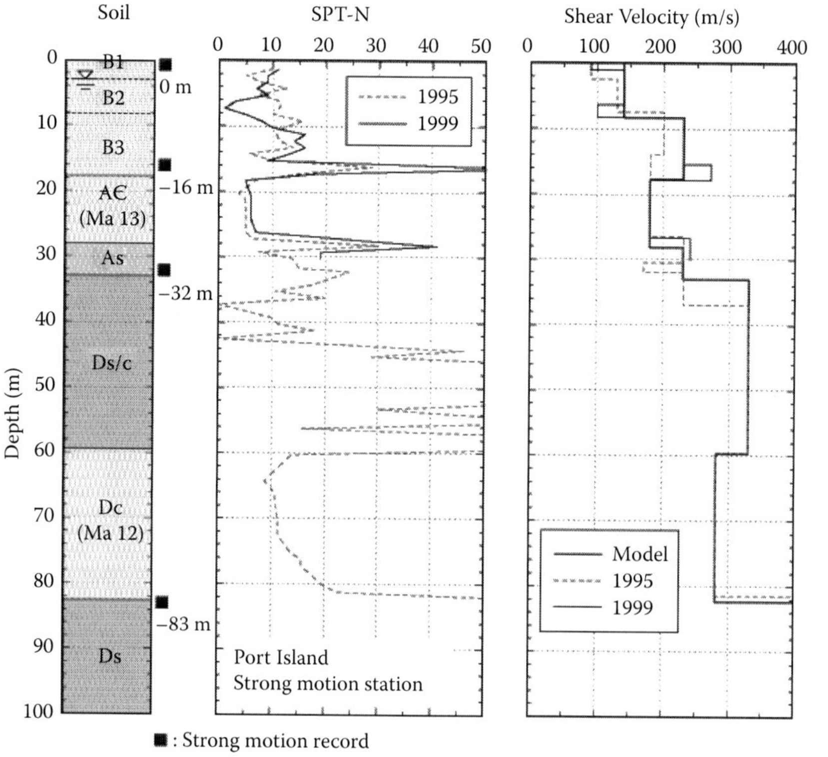

Figure 9.10 Soil profile at the observation site.

to the B1, B2, B3, and As layers shown in Figure 9.10. In this simulation, we used method 3 for the shear modulus reduction depending on the plastic shear strain history. The cyclic elastoviscoplasticity model for clay was applied to the Ac layer in Figure 9.10. To model the stiff sandy ground and sandy gravel, a generalized Ramberg–Osgood model was used in which the second invariant of the deviatoric stress tensor is incorporated. The Ramberg–Osgood model was applied to the Ds/c, Dc, and Ds layers in Figure 9.10

In Table 9.2, the material parameters used in the analysis are listed. These parameters are determined by the existing field and laboratory testing data.

9.5.3 Common parameters

The soil profile and the groundwater table were determined from Figure 9.10, which was obtained by the boring at the observation site. The density and the void ratio were determined from the density measurement of the undisturbed

Table 9.2 Material parameters

Name of Soil Profile		*B1*	*B2*	*B3*	*Ac*	*As*	*Ds/c*	*Dc*	*Ds*
Soil type		Gravelly sand	Gravelly sand	Gravelly sand	Marine clay	Sand	Sand with clay	Marine clay	Sand
Model type		E-P	E-P	E-P	E-VP	E-P	R-O	R-O	R-O
Density	ρ (t/m^3)	1.90	2.10	2.10	1.67	2.00	2.00	2.00	2.00
Initial void ratio	e_0	0.42	0.42	0.42	1.41	0.50	0.50	1.20	0.50
Coefficient of permeability	k (cm/s)	—	0.003	0.003	0.0002	0.002	0.002	0.0001	0.002
Compression index	λ	0.01	0.01	0.01	0.331	0.01	—	—	—
Swelling index	k	0.001	0.001	0.001	0.0425	0.001	—	—	—
Poisson's ratio	ν	—	—	—	—	—	0.35	0.35	0.35
Initial shear velocity	V_s (m/s)	140	140	230	180	230	330	280	450
Initial shear modulus ratio	G_o/σ'_m	2002	730	1019	328	516	—	—	—
Failure stress ratio	M^*_f	1.34	1.34	1.34	1.23	1.26	—	—	—
Phase transformation stress ratio	M^*_m	0.91	0.91	0.91	1.03	0.91	—	—	—
Internal friction angle	φ' (deg)	—	—	—	—	—	45	45	45
R-O parameter	α	—	—	—	—	—	1.00	1.17	0.87
	r	—	—	—	—	—	2.13	2.06	2.13
	a	—	—	—	—	—	12414	6466	17971
	b	—	—	—	—	—	0.5	0.5	0.5
Hardening parameter	B^*_0	6000	1500	2100	55	5000	—	—	—
	B^*_1	0	150	140	—	100			
Control parameter of anisotropy	C_d	2000	2000	2000	—	2000	—	—	—

Reference strain parameter	$\gamma_r^{P^*}$	1000	0.005	0.004	—	0.01	—	—	—
	$\gamma_r^{E^*}$	1000	0.005	0.004	—	0.1	—	—	—
Dilatancy parameter	D_0^*	0.0	1.0	1.0	—	0.0	—	—	—
	n	0.0	4.0	4.0	—	0.0	—	—	—
Viscoplastic parameter	m_0'	—	—	—	14.0	—	—	—	—
	C_1 (1/s)	—	—	—	5.54E–06	—	—	—	—
	C_2 (1/s)	—	—	—	7.76E–07	—	—	—	—

sample obtained after the earthquake. The coefficient of permeability was determined from laboratory permeability tests on the undisturbed sample obtained after the earthquake. The shear velocity was estimated from the PS logging tests at the observation site in 1995 and 1999.

9.5.4 Parameters for elastoplasticity model

It is important to estimate an appropriate shear modulus in an earthquake response analysis. The initial shear modulus of the elastoplastic model is given by the elastoplastic shear modulus, G^{EP}, in the analysis. When there exists an elastic region, the initial shear modulus $G_0^{EP} = G_0^E$ and can be obtained as

$$G_0^E = \rho V_s^2 \tag{9.39}$$

where G_0^E is the initial elastic shear modulus, ρ is the density, and V_s is the shear velocity. It is worth mentioning that the elastic modulus is normalized by the mean effective stress.

The failure stress ratio was determined by undrained triaxial tests using the undisturbed samples. The phase transformation stress ratio was determined from past laboratory tests by Pradhan et al. (1989).

The remaining parameters were determined by the data-adjusting method for the undrained cyclic shear tests. In this method, the values for the material parameters were selected in order to provide a good description of the stress–strain relations under cyclic loading conditions and liquefaction strength curves. The undrained cyclic shear tests for the reclaimed soil were conducted with the frozen in situ sample by Suzuki et al. (1997). The experimental and the simulated liquefaction strength, which is the cyclic shear stress that produces a double amplitude strain of 7.5% with a particular number of cycles, is shown in Figure 9.11. The simulated liquefaction strength for the reclaimed B2 layer agrees well with the experimental one.

9.5.5 Parameters for elastoviscoplasticity model

For clay layers we have used a cyclic elastoviscoplastic model presented in Section 5.9.1. All remaining parameters were determined by undrained triaxial tests using the clay samples obtained at the observation site. In particular, the viscoplastic parameters were determined from monotonic shear tests with two different loading rates. We confirmed the applicability of the determined parameters through a simulation of the

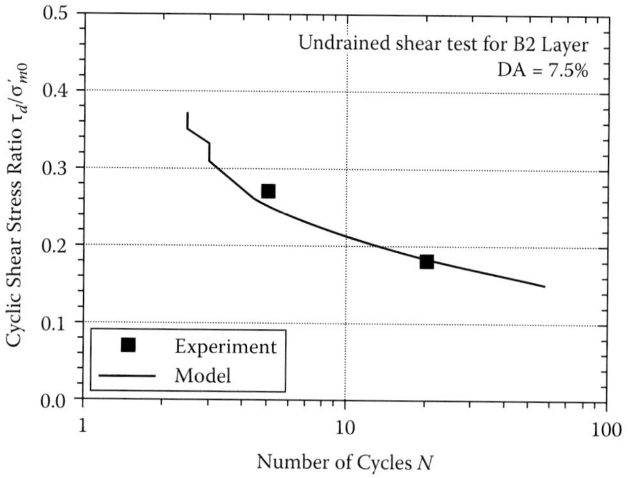

Figure 9.11 Liquefaction strength for reclaimed soil.

dynamic strength obtained by the cyclic shear tests (Oka 1992; Oka et al. 2004).

9.5.6 Parameters for Ramberg–Osgood model

The remaining parameters were determined by the strain-dependent shear modulus and the damping ratio, which were obtained by the undrained cyclic shear tests with multi-step loading. The Ds/c, Dc, and Ds layers modeled by the Ramberg–Osgood model have no dilatancy; hence, these layers generate no buildup of excess pore water pressure (LIQCA Res. Development Group 2005).

The elastic shear modulus is given by

$$G_{\max} = a\sigma'^{b}_{m0} \text{ (unit: kPa)} \tag{9.40}$$

The skeleton curve of the stress–strain relation is given by

$$\gamma = \frac{\tau}{G_{\max}}\left\{1+\alpha\left|\frac{\tau}{\tau_f}\right|^{r-1}\right\} \tag{9.41}$$

The shear strength follows Mohr–Coulomb failure criterion: $\tau_f = \sigma'_{m0}\sin\phi + c\cos\phi$.

9.5.7 Finite element model and numerical parameters

The single-column ground model, composed of three-dimensional solid finite elements, was used for the analysis. The number of finite elements is 31 and the number of nodes is 128. The model displacements are fixed at the base. Four nodes at the same depth were assumed to move coincidentally. The lateral and the bottom boundaries were assumed to be impermeable, while the ground water surface was assumed to be permeable. As an input earthquake motion at the base rock, the three components (NS, EW, and UD) of the acceleration records obtained at the depth of –83 m at the vertical array recording site on Port Island were used.

Rayleigh damping proportional to the initial stiffness, which was determined by assuming that the damping factor was 1%, was used as the convenient method in this study. A time integration step of 0.002 seconds was adopted to obtain sufficient accuracy. β and γ in the Newmark β method were set to be 0.3025 and 0.6, respectively, to ensure numerical stability.

9.5.8 Numerical results

A comparison of the simulated and the observed absolute acceleration histories is presented in Figure 9.12. The amplitudes as well as the phases were reproduced in the simulation results, although the peak accelerations at the depths of GL-16 m and GL-32 m were underestimated. The simulated peak acceleration was affected by the numerical parameters related to damping. In this simulation, we overestimated the damping of the lower layers more than the reclaimed layer. A comparison of the simulated and the observed relative velocity histories is presented in Figure 9.13. The relative velocity is given by subtracting the base velocity at the depth of GL-83 m. The observed velocity was obtained by integrating the acceleration histories in the time domain. The simulation reproduced the observed velocity very well for the amplitudes and the phases. The simulated time histories of the effective stress decreasing ratio (ESDR), which is given by $1.0 - \sigma'_m/\sigma'_{m0}$, in reclaimed layers B2 and B3 are shown in Figure 9.14. Complete liquefaction occurred after a few strong motion cycles of about 8 seconds in the reclaimed layer. The changes in the frequency property of the acceleration and the velocity at the ground surface are due to the liquefaction of the reclaimed layer. It is well known that extensive soil liquefaction was observed on Port Island, a man-made island in Kobe City (Shibata et al. 1996). It is seen that the proposed liquefaction method, based on the cyclic elastoplasticity model, can simulate the dynamic response of liquefiable grounds well.

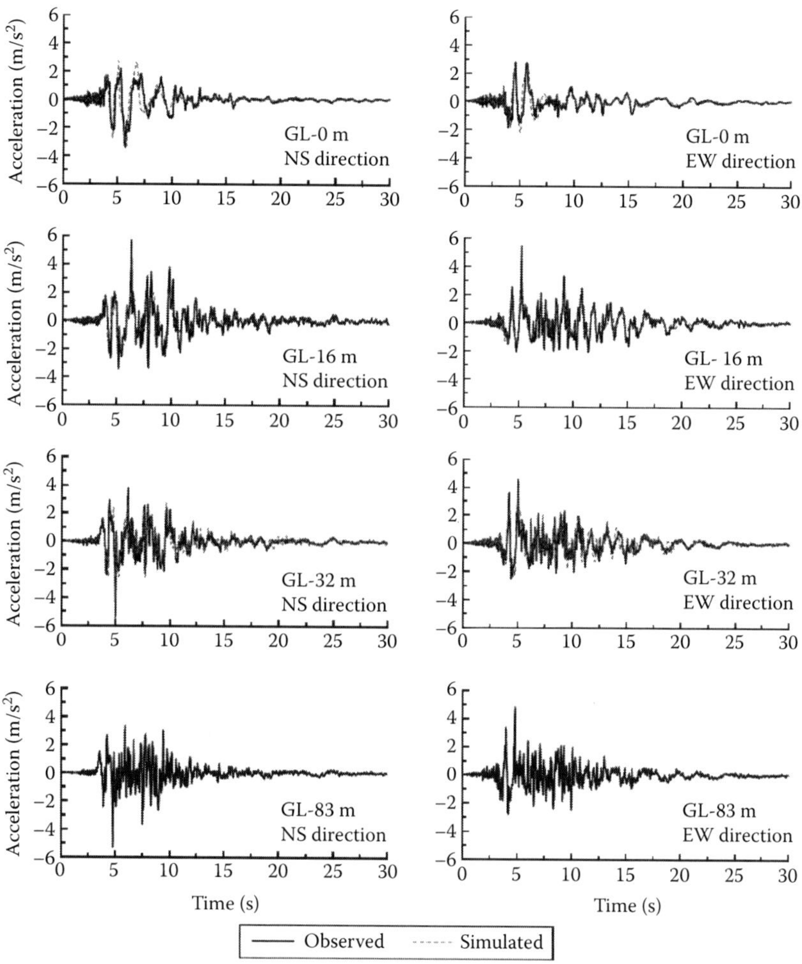

Figure 9.12 Time histories of observed and simulated horizontal acceleration.

9.6 NUMERICAL ANALYSIS OF THE DYNAMIC BEHAVIOR OF A PILE FOUNDATION CONSIDERING LIQUEFACTION

Many structures were damaged during the 1995 Hyogo-ken Nambu earthquake. It was found from the field investigations after the earthquake that not only the pile heads but also the lower parts of the piles had cracked or failed. This phenomenon indicates that both the inertia force from the upper structures and the kinematic interaction between

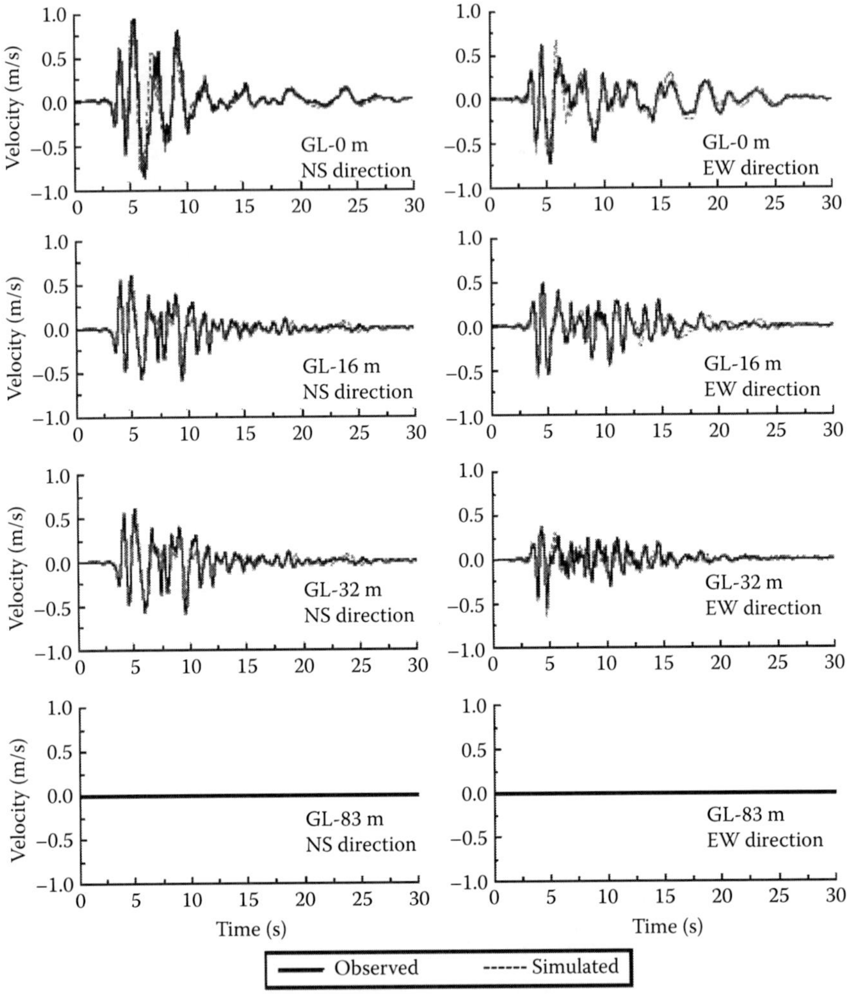

Figure 9.13 Time histories of observed and simulated relative horizontal velocity.

the piles and the ground play important roles in the mechanical behavior of piles. In particular, when the ground surrounding a structure liquefies due to seismic excitations, the behavior of the piles is more complicated. Damage related to liquefaction may involve cases in which the pile foundation is damaged due to the lateral flow of liquefied soils, or the piles fail at the boundary between two different soil layers, of which one liquefies while the other does not. In this study, a series of numerical simulations were conducted to study the dynamic behavior

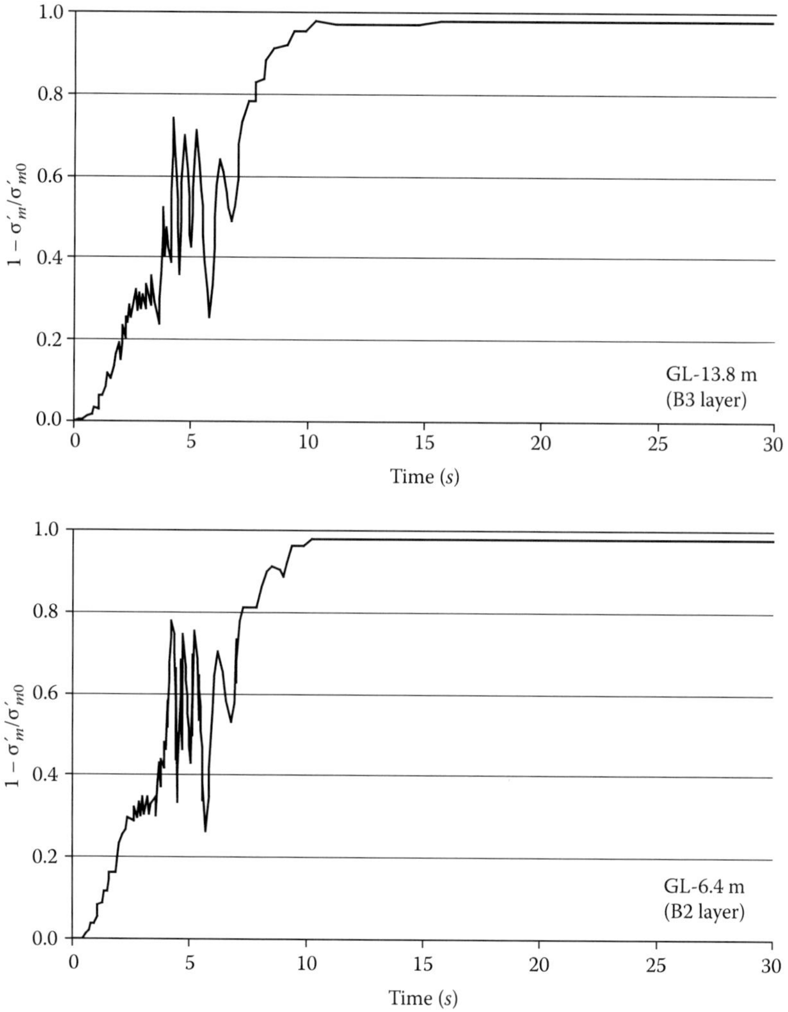

Figure 9.14 Time histories of simulated effective stress decreasing ratio.

of a single-pile foundation constructed in a two-layer ground, whose upper layer is filled with sandy soil that is dense sand, reclaimed soil, medium dense sand or loose sand, and whose lower layer is filled with clayey soil employing a three-dimensional liquefaction analysis method (code name: LIQCA3D) to clarify the mechanism of the interactions among the soil–pile–structure (Lu 2002; Oka, Lu, Uzuoka, and Zhang 2004; Lu, Oka, and Zhang 2008).

Table 9.3 Main material parameters for the soil

Soils	*Density*	M_f^*	M_m^*	G_0/σ_{m0}	B_0^*, B_1^*, C_f	D_0,n
Loose	2.0	0.80	0.70	500.0	2500,25,0	1.0,1.0
Medium	2.0	1.00	0.80	1060.0	4000,40,0	1.0,2.0
Reclaimed	2.0	1.19	0.91	2140.0	5500,55,0	1.0,4.0
Dense	2.0	1.10	0.85	1980.0	8500,85,0	1.0,2.5
Clay	1.7	1.31	1.28	300.0	500,50,0	—

9.6.1 Simulation methods

The two-layer ground considered in this chapter is a typical one near the shore of a major Japanese urban city, such as Kobe. In order to study the influence of the soil characteristics, four different sandy materials are considered for an upper sandy ground: dense sand, medium dense sand, loose sand, and reclaimed soil. Table 9.3 shows the parameters involved in the constitutive models for the different types of soil (Lu et al. 2008). The constitutive model for sand is described in detail in Section 9.2. On the other hand, an axial force-dependent (AFD) model (Zhang and Kimura 2002), in which the nonlinear behavior of steel and concrete is properly described and shown in Figure 9.15, is used to describe the dynamic behavior of the RC pile that is 1.5 m in diameter. The parameters of the RC pile are shown in Table 9.4 (Lu et al. 2008).

The governing equations for the coupling problems between the soil skeleton and the pore water pressure are obtained based on the two-phase mixture theory. The liquefaction analysis is formulated using u (displacement) – p (pore pressure). The side boundaries of the simulated system are

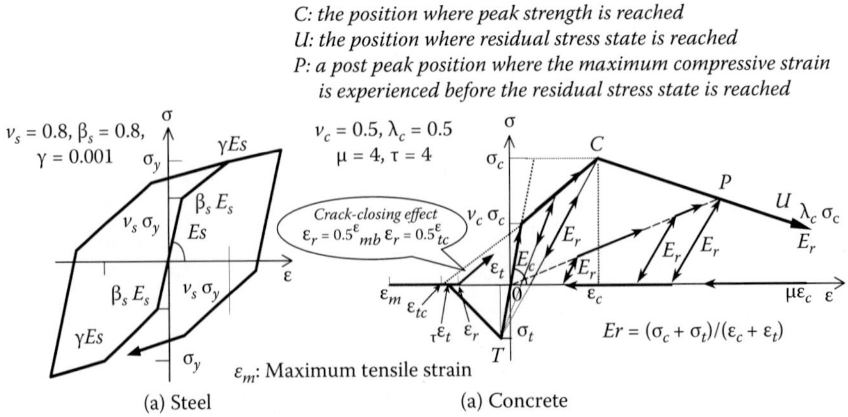

Figure 9.15 Stress–strain relations of steel and concrete adopted in the AFD model for RC material.

Table 9.4 Main material parameters for the pile

Young's modulus of concrete E_c (kN/m^2)	2.5×10^7
Diameter of pile D (m)	1.5
Compressive strength of concrete f_c (kN/m^2)	36000.00
Tensile strength of concrete f_t (kN/m^2)	3000.00
Degrading parameter of concrete β_c	0.20690
Young's Modulus of steel E (kN/m^2)	2.1×10^8
Diameter of reinforcement d (m)	0.029
Number of reinforcement N	24
Yielding strength of steel Y_s (kN/m^2)	3.8×10^5

assumed to be equal-displacement boundaries, the bottom of the system is fixed, and the boundaries are impermeable except for the surface of the ground. In this dynamic analysis, a stiffness-matrix-dependent Rayleigh type damping is adopted and the direct integration method of Newmark β is used in this dynamic analysis with a time interval of 0.01 sec. The groundwater table is 1.5 m beneath the ground surface. The mass of the superstructure is 80,000 kg and the height of pier is 8 m. Figure 9.16 shows the seismic wave used in this study, which is an NS component of the earthquake recorded on Port Island during the 1995 Hyogo-Ken Nambu earthquake, and Figure 9.17 shows the configuration of the single-pile system and the finite element mesh used in the calculation.

9.6.2 Results and discussions

Figure 9.18 shows the history of the effective stress decreasing ratio (ESDR $\equiv (\sigma'_{m0} - \sigma'_m)/\sigma'_{m0}$; σ'_{m0} is the initial mean effective stress) of the soil in the middle of different types of sandy layers. Liquefaction occurs when the

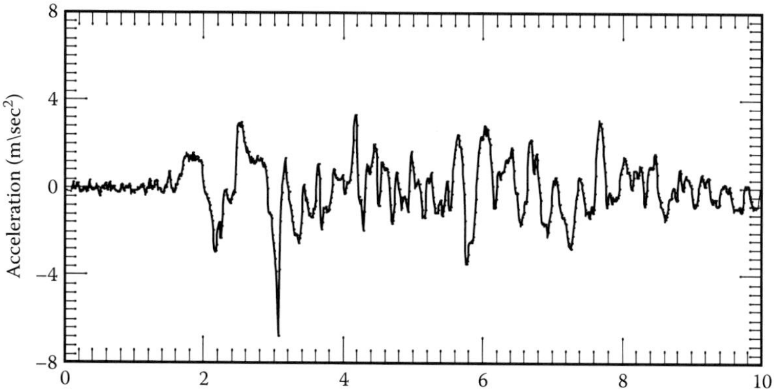

Figure 9.16 Input wave.

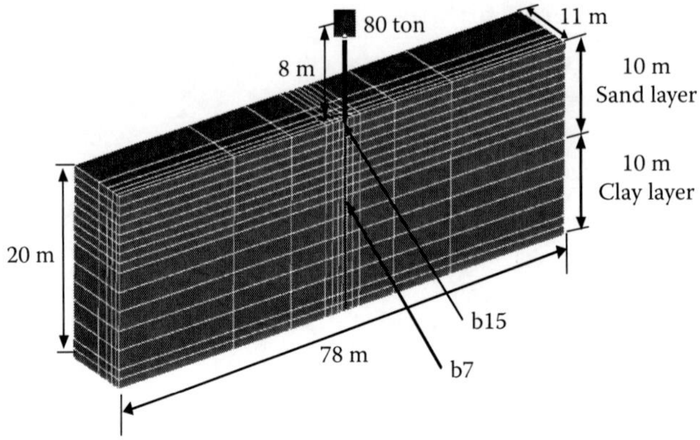

Figure 9.17 Finite element mesh.

ESDR is equal to 1. It can be seen that loose sand easily liquefies entirely, while medium sand and reclaimed soil almost liquefy at the end of the major seismic event (t = 10 sec). The effective stress of dense sand does not decrease much at all. Figure 9.19 shows the histories of the bending moments at the pile head and in the pile segment at the boundary between the soil layers. Since the dense sand layer does not liquefy at all, the earthquake wave motion does not deamplify and the largest bending moment occurs at the pile head among the cases. On the other hand, the larger bending moments occur in the pile at the boundary between the layers

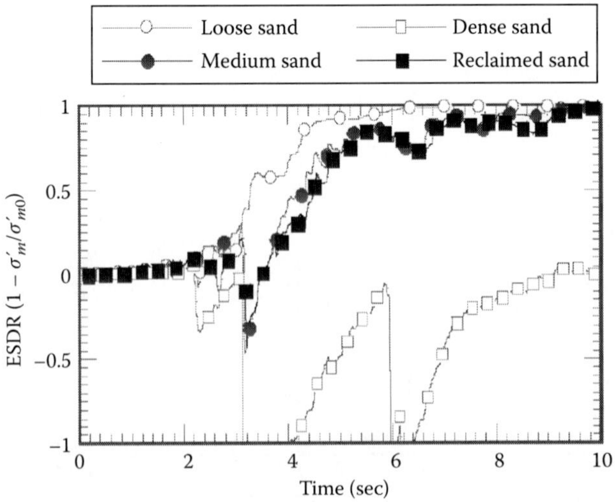

Figure 9.18 Effective stress decreasing ratio–time profile.

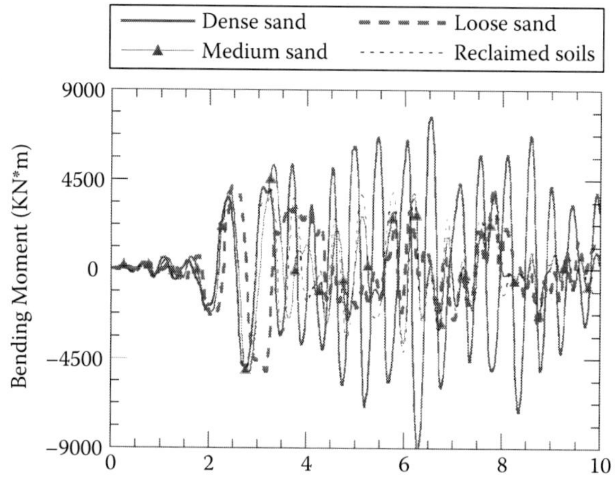

(a) At the pile head

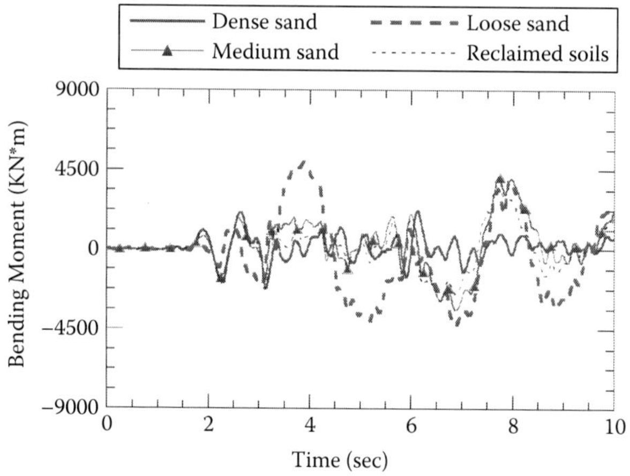

(b) In the pile segment at the soil–layer boundary

Figure 9.19 Histories of the bending moments in the pile.

at $t = 4$ sec and $t = 7$ sec for loose sand, and medium dense sand and reclaimed soil, respectively, when the effective stress of the sand layers decreases significantly. Figure 9.20 shows the distribution of the bending moments when the maximum bending moment takes place in each case and Figure 9.21 shows the distribution of the bending moments at the end of the seismic event. The figures show that although the maximum bending

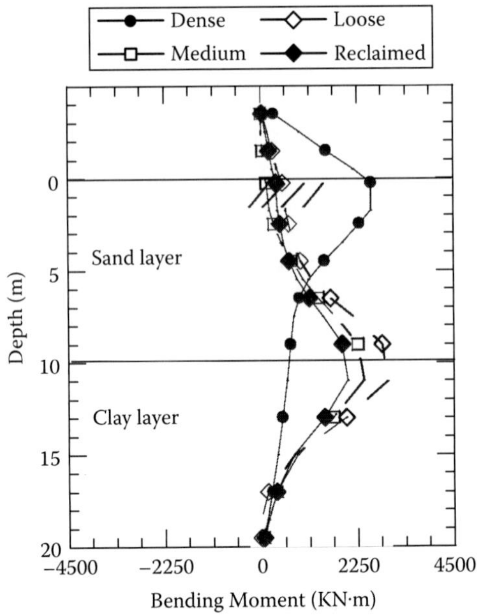

Figure 9.20 Distribution of the bending moment when the maximum bending moment at the end of the seismic event (t = 10 sec).

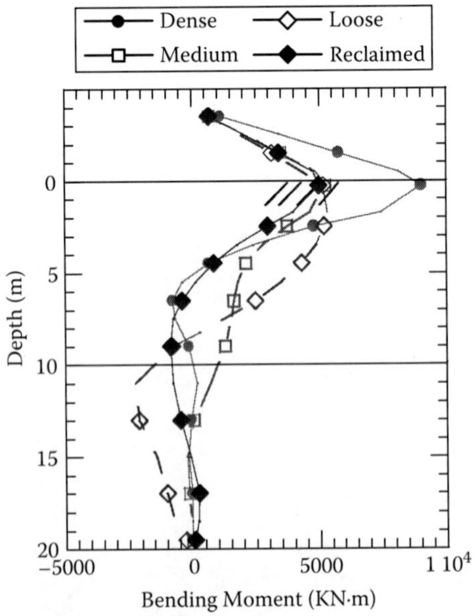

Figure 9.21 Distribution of the bending moment at the time when the maximum bending moment occurred at the bottom of the pier.

moment takes place at the pile head (b15) in every case, the development of bending moments in the ground varies due to the features of the soil.

A large bending moment takes place in the lower pile segment (b7) in the cases of liquefiable soil, but at the upper pile segment (b15) in the case of dense sand at the end of the seismic event.

REFERENCES

Adachi, T., and Oka, F. 1982. Constitutive equations for sands and overconsolidated clays and assigned works for sand. Constitutive relations for soils; Results of the International Workshop on Constitutive Relations for Soils, Grenoble, G. Gudehus, F. Darve, and I. Vardoulakis, eds., Balkema, Rotterdam, 141–157.

Armstrong, P.J., and Frederick, C.O. 1966. A mathematical representation of the multiaxial Bauschinger effect, C. E. G. B. Report RD/B/N 731.

Chaboche, J.L., and Rousselier, G. 1983. On the plastic and viscoplastic constitutive equations Part I and Part II. *J. Pressure Vessel Tech., Trans. ASME*, 105:153–164.

Desai, C.S., and Siriwardane, H.J. 1980. A concept of correction functions to account for non-associative characteristics of geologic media, *Int. J. Numer. Anal. Meth. Geomech.*, 4:377–387.

DiMaggio, F.L., and Sandler, I.S. 1971. Material model for granular soils, *J. Eng. Mech., ASCE*, 97(EM 3):935–950.

Drucker, D.C., and Prager, W. 1952. Soil mechanics and plastic analysis or limit design, *Quart. Appl. Math.*, 10:157–164.

Earthquake Engineering Committee (JSCE). 2003. Soil liquefaction under level II earthquake motion subcommittee, 2003. Report on the Soil Liquefaction under Level II Earthquake Motion (written by F. Oka, K. Furuya, and R. Uzuoka), 157–201 (In Japanese).

Ghaboussi, J., and Momen, H. 1979. Plasticity model for cyclic behaviour of sands, Proceedings of 3rd International Conference on Numerical Methods in Geomechanics, Aachen, W. Wittke, ed., Balkema, Rotterdam, 423–434.

Hashiguchi, K. 1980. Constitutive equations of elastoplastic materials with elastic-plastic transition, *J. Appl. Mech., ASME*, 47:266–272.

Hashiguchi, K., and Ueno, M. 1977. Elastoplastic constitutive law of granular materials, In Constitutive Equations of Soils, S. Murayama and A.N. Schofield, eds., Proceedings of 9th ICSMFE, Special session 9, Tokyoto, JSSMFE, 73–82.

Ishihara, K., and Kabilamany, K. 1990. Stress dilatancy and hardening laws for rigid granular model of sand, *Soil Dynamics Earthquake Eng.*, 9(2):66–77.

Iwasaki, T., and Tatsuoka, F. 1977. Effects of grain size and grading on dynamic shear moduli of sands, *Soils and Foundations*, 17(3):19–35.

Kato, M., Oka, F., Yashima, A., and Tanaka, Y. 1994. Dissipation of excess pore water pressure by gravel drain and its analysis, Tsuchi-to-Kiso, JSSMFE, 42(4):39–44 (in Japanese).

LIQCA Research and Development Group (Representative: F. Oka). 2005. Users manual for LIQCA2D04 (2005 released print), http://www.nakisuna2.kuciv.kyoto-u.ac.jp/liqca.htm.

Lu, C.W. 2002. Numerical study of soil-pile interaction during earthquakes considering liquefaction, PhD thesis, Kyoto University.

Lu, C.W., Oka, F. and Zhang, F. 2008. Analysis of soil–pile–structure interaction in a two -layer ground during earthquakes considering liquefaction, *Int. J. Numer. Anal. Meth. Geomech.*, 32:863–895.

Masing, G. 1926. Eigenspannungen und Verfestigung beim Messing.Proc. 2nd Int. Congr. Appl. Mech., Zurich, 332–335.

Naghdi, P.M., and Trapp, J.A. 1975. Restrictions on constitutive equations of finitely deformed elastic-plastic materials. *Q. J. Mech. Appl. Math.*, 28(1):25–46.

Nishi, K., and Esashi, Y. 1978. Stress-strain relationships of sand, *Proc. JSCE*, 280:111–122.

Nishi, K., and Kanatani, M. 1990. Constitutive relations for sand under cyclic loading based on elasto-plasticity theory, *Soils and Foundations*, 30(2):43–59.

Oka, F. 1982. Constitutive equations for granular materials in cyclic loadings. Proceedings of IUTAM Conference on Deformation and Failure of Granular Materials, Delft, P.A. Vermeer and H.J. Luger, eds., Balkema, Rotterdam, 297–306.

Oka, F. 1992. A cyclic elasto-viscoplastic constitutive model for clay based on the non-linear-hardening rule, Proceedings of 4th International Symposium on Numerical Models in Geomechanics, Swansea, G.N. Pande and S. Pietruszczak, eds., Balkema, Rotterdam, 1:105–114.

Oka, F., Furuya, K., and Uzuoka, R. 2004. Numerical simulation of cyclic behavior of dense sand using a cyclic elasto-plastic model, Proceedings of International Symposium on Cyclic Behavior of Soils and Liquefaction Phenomena, Th. Triantafyllidis, ed., Balkema, Rotterdam, 85–90.

Oka, F., and Kimoto, S. 2008. An elasto-viscoplastic constitutive model and its application to the sample obtained from the seabed ground at Nankai trough, Zairyo, *Japanese Soc. Mater. Sci.*, 57(3):237–242.

Oka, F., Kodaka, T., Koizumi, T., and Sunami, S. 2001. An effective stress based liquefaction analysis based on finite deformation theory, Proceedings of 10th IACMAG, Tucson Arizona, C.S. Desai, T. Kundu, S. Harpalani, D. Contractor, and J. Kemeny, eds., Balkema, Rotterdam, 1113–1116.

Oka, F., Kodaka, T., Kim, Y.-S. 2004. A cyclic viscoelastic-viscoplastic constitutive model for clay and liquefaction analysis of multi-layered ground, *Int. J. Numerical and Analytical Methods in Geomechanics*, 28(2):131–179.

Oka, F., Lu, C.-W., Uzuoka, R. and Zhang, F. 2004. Numerical study of structure-group pile foundations using an effective stress based liquefaction analysis method, Proceedings of 13th WCEE, Vancouver, Canada, Paper No. 3338.

Oka, F., Sugito, M., Yashima, A., Furumoto, Y. and Yamada, K. 2000. Time-dependent ground motion amplification at reclaimed land after the 1995 Hyogo-Ken-Nambu Earthquake, Proceedings of 12th WCEE: No. 2046.

Oka, F., Uzuoka, R. Tateishi, A., and Yashima, A. 2003. A cyclic elasto-plastic model for sand ant its application to liquefaction analysis, Constitutive Modeling of Geomaterials, selected contributions from the Frank L. DiMaggio Symposium, Inelastic Behavior Committee Engineering Mechanics Division, ASCE, CRC Press, 75–99.

Oka, F., and Washizu, H. 1981. Constitutive equations for sand and overconsolidated clays, Proceedings of International Conference on Recent Advances in Earthquake Engineering and Soil Dynamics, St. Louis, S. Prakash, ed., 1:71–74.

Oka, F., and Yashima, A. 1995. A cyclic elasto-viscoplastic model for cohesive soil, Proceedings of XI European Conference on Soil Mechanics and Foundation Engineering, Copenhagen, The Danish Geotechnical Society:6.145–6.150.

Oka, F., Yashima, A., and Kato, M. 1993. Numerical analysis of wave-induced liquefaction in seabed, Proc. 3rd Int. Offshore Polar Eng. Conf., 591–598.

Oka, F., Yashima, A., Kato, M., and Nakajima, Y. 1994. An analysis of seepage failure using an elasto-plastic constitutive equation and its application, *J. JSCE*, 493, III-27:127–135 (in Japanese).

Oka, F., Yashima, A., Kato, M., and Sekiguchi, K. 1992. A constitutive model for sand based on the non-linear kinematic hardening rule and its application. Proc. 10th WCEE:2529–2534.

Oka, F., Yashima, A., Shibata, T., Kato, M., and Uzuoka, R. 1994. FEM-FDM coupled liquefaction analysis of a porous soil using an elasto-plastic model, *Appl. Sci. Res.*, 52:209–245.

Oka, F., Yashima, A., Tateishi, A., Taguchi, Y., and Yamashita, S. 1999. A cyclic elasto-plastic constitutive model for sand considering a plastic-strain dependence of the shear modulus, *Géotechnique*, 49(5):661–680.

Pastor, M., and Zienkiewicz, O.C. 1986. A generalized plasticity, hierarchical model for sand under monotonic and cyclic loading, Proceedings of 2nd International. Symposium on Numerical Models in Geomechanics Ghent, G.N. Pande and W.F. Van Impe, eds., M. Jackson & Son, Cornwall, 131–150.

Poorooshasb, H.B., Holubec, I., and Scherbourne, A.N. 1966. Yielding and flow of sand in triaxial compression: Part I, *Can. Geotech. J.*, 3(4):179–190.

Poorooshasb, H.B., Holubec, I., and Scherbourne, A.N. 1967. Yielding and flow of sand in triaxial compression: Part II and Part III, *Can. Geotech. J.*, 4(4):376–397.

Pradhan, T.B.S., and Tatsuoka, F. 1989. On stress-dilatancy equations of sand subjected to cyclic loading, *Soils and Foundations*, 29(1):65–81.

Pradhan, T.B.S., Tatsuoka, F., and Sato, Y. 1989. Experimental stress-dilatancy relations of sand subjected to cyclic loading, *Soils and Foundations*, 29(1):45–64.

Prévost, J.H., and Keane, C.M. 1990. Multimechanism elasto-plastic model for soils. *J. Eng. Mech., ASCE*, 116(9):1924–1944.

Shibata, T., Oka, F., and Ozawa, Y. 1996. Characteristics of ground deformation due to liquefaction, Soils and Foundations, Special issue on Geotechnical Aspects of the Jan. 17 1995 Hyogo-ken-Nambu Earthquake, 65–79.

Shibata, T., Oka, F., Yashima, A., Goto, H., Goto, K., and Takezawa, K. 1992. A numerical simulation of shaking table test of coal fly ash deposit with cement mixing column, In Numerical Models in Geomechanics, G.N. Pande and S. Pietruszczak, eds., Balkema, Rotterdam, 411–420.

Sugito, M., Oka, F., Yashima, A., Furumoto, Y., and Yamada, K. 2000. Time-dependent ground motion amplification characteristics at reclaimed land after the 1995 Hyogoken Nambu Earthquake, *Eng. Geol.*, 56:137–150.

Suzuki, Y., Hatanaka, M., and Uchida, A. 1997. Drained and undrained shear strengths of a gravelly fill of weathered granite from Kobe Port Island, *J. Struct. Constr. Eng.*, AIJ, 49:867–873 (in Japanese).

Taguchi, Y., Tateishi, A., Oka, F., and Yashima, A. 1995. A cyclic elastoplastic model based on the generalized flow rule and its application, Proceedings of 5th International Symposium on Numerical Models in Geomechanics, Davos, Switzerland, G.N. Pande and S. Pietruszczak, eds., Balkema, 57–62.

Taguchi, Y., Tateishi, A., Oka, F., and Yashima, A. 1997a. Numerical simulation of soil-foundation interaction behavior during subsoil liquefaction, Proceedings of 6th International Symposium on Numerical Models in Geomechanics, S. Pietruszczak and G.N. Pande, eds., Blakema, Rotterdam, 561–566.

Taguchi, Y., Tateishi, A., Oka, F., and Yashima, A. 1997b. Three-dimensional liquefaction analysis method and array record simulation in Great Hanshin Earthquake, Proceedings of 11th World Conference on Earthquake Engineering, Acapulco, Sociedad Mexicana de Ingenieria Sismica, Balkema (CD-ROM), Rotterdam, 1042.

Uzuoka, R. 2000. Analytical study on the mechanical behavior and prediction of soil liquefaction and flow, PhD thesis, Gifu University (in Japanese).

Vermeer, P.A. 1978. A double hardening model for sand, *Géotechnique*, 28(4):413–433.

Yashima, A., Oka, F., and Kanami, H. 2000. 3-D analysis to evaluate of soil improvement on liquefaction of man-made island, Proceedings of 12th WCEE, No. 2116.

Zhang, F., and Kimura, M. 2002. Numerical prediction of the dynamic behaviors of an RC group-pile foundation, *Soils and Foundations*, 42(3):77–92.

Chapter 10

Recent advances in computational geomechanics

In this chapter, two selected subjects on recent advances in geomechanics are presented: the thermo-hydro-mechanical coupled problem and the seepage–deformation coupled analysis of unsaturated soil.

10.1 THERMO-HYDRO-MECHANICAL COUPLED FINITE ELEMENT METHOD

In this section, we develop a thermo-hydro-mechanical coupled finite element method by considering the conservation of energy in addition to the soil-water coupled finite element method. Biot's (1956) theory of porous media is adopted to give the governing equations for the soil–water coupling problem presented in Chapter 2.

Thermal consolidation problems have been studied by many researchers (e.g., Campanella and Mitchell 1968; Baldi et al. 1988; Akagi and Komiya 1995; Delage et al. 2000; Sultan et al. 2002). Boudali et al. (1994) and Leroueil and Marques (1996) demonstrated the temperature dependency of clay through experiments on natural clay, and Cekerevac and Laloui (2004) reported the thermal effects on the mechanical behavior of clay in experiments. Thermo-hydro-mechanical (THM) coupling problems are very important in the field of geomechanics. Thermo-hydro-mechanics has been applied to several problems such as nuclear waste disposal, ground heating, thermal consolidation, cleanup techniques for contaminated grounds, rapid landslides, and earthquake faulting (e.g., Komine and Ogata 2002; Cleall et al. 2006; Laloui et al. 2006; Vardoulakis 2002; Sulem and Famin 2009). Vardoulakis (2002) studied the effect of thermal softening on catastrophic landslides. Kimoto et al. (2010) developed a chemo-thermo-mechanical coupled numerical method to analyze the behavior of ground during the dissociation of methane hydrate.

The temperature-dependent viscoplastic behavior was incorporated into the elastoviscoplastic model by Yashima et al. (1998) based on the experimental results by Boudali et al. (1994). They showed that the viscoplastic parameter, m', in Equation (5.19) is not dependent on the temperature, but the viscoplastic parameter C in Equation (5.19) is dependent on the temperature. The model was then incorporated into the finite element program and successfully applied to the thermal consolidation analysis using the finite element method with the energy balance law (Oka et al. 2005). The other application of the temperature-dependent model is the analysis of the ground deformation due to the dissociation of a methane hydrate (Kimoto et al. 2010).

10.1.1 Temperature-dependent viscoplastic parameter

Yashima et al. (1998) introduced the temperature dependency of the viscoplastic parameter based on the experimental data by Boudali et al. (1994). Figure 10.1 shows the relations between the logarithm of the strain rate and the preconsolidation pressure. Parameter m' seems to be independent of the temperature. This means that viscoplastic parameter, C, may depend on the temperature. As a result, Yashima et al. (1998) showed the following relation between consolidation yield stress, σ'_p, and temperature, θ (see Figure 10.1):

$$\frac{\sigma'_p}{\sigma'_{pr}} = \left[\frac{\theta_r}{\theta}\right]^{\alpha} \tag{10.1}$$

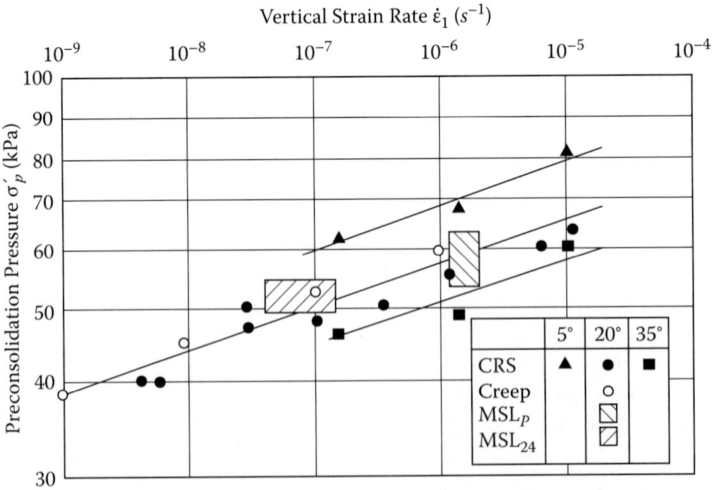

Figure 10.1 Relation between logarithm of strain rate and consolidation yield stress. (After Leroueil, S., and Marques, M.E.S. 1996, Geotechnical Special Publication No. 61 ASCE, T.C. Sheahan and V. Kaliakin, eds., 1–60.)

in which σ'_{pr} is the value of σ'_p at referential temperature θ_r and α is the gradient of line $\log\sigma'_p - \log\theta$.

If the stress ratio during one-dimensional compression is assumed to be constant and the initial hardening parameter $\sigma'^{(s)}_{myi}$ in Equation (5.60) corresponds to consolidation yield stress, σ'_p, namely,

$$\frac{\sigma'_{mai}}{\sigma'^{(s)}_{myi}} = \frac{\sigma'_p}{\sigma'_0} \tag{10.2}$$

where σ'_0 is a consolidation stress. Temperature-dependent viscoplastic parameter $C(\theta)$ is rewritten from Equation (5.60) as

$$C(\theta) = C'\exp\left\{m'\tilde{M}^*\left(-\ln\left[\frac{\sigma'_p}{\sigma'_0}\right]\right)\right\} \tag{10.3}$$

Substituting Equation (10.1) into Equation (10.3) yields

$$C(\theta) = C(\theta_r)\exp\left\{m'\tilde{M}^*\left(-\ln\left[\frac{\theta_r}{\theta}\right]^{\alpha}\right)\right\} \tag{10.4}$$

$$C(\theta_r) = C'\exp\left\{m'\tilde{M}^*\left(-\ln\left[\frac{\sigma'_{pr}}{\sigma'_0}\right]\right)\right\} \tag{10.5}$$

From Equations (10.4) and (10.5), the temperature dependency of viscoplastic parameter C is obtained as follows:

$$\frac{C(\theta)}{C(\theta_r)} = \left[\frac{\theta}{\theta_r}\right]^{\bar{\beta}}, \quad \bar{\beta} = \alpha m'\tilde{M}^* \tag{10.6}$$

in which the coefficient of dilatancy, $\tilde{M}^*$, is a function of the stress ratio in the overconsolidated (OC) region (see Equation 5.51). However, the thermoviscoplastic parameter $\bar{\beta}$ is independently determined as a material constant. Adopting the temperature-dependent viscoplastic parameter, Equation (10.6), into the viscoplastic flow rule, Equation (5.56), an elasto-thermo-viscoplastic model is derived.

Figure 10.2 shows simulated results for constant rate-of-strain (CRS) odometer tests using the above-mentioned model (Yashima et al. 1998) with the experimental results at strain rates of $1\times10^{-5}\,s^{-1}$ and $1.6\times10^{-7}\,s^{-1}$ and temperatures of 5°C and 35°C from tests performed by Boudali et al. (1994). It is seen that the simulated results reproduce all the compression curves well.

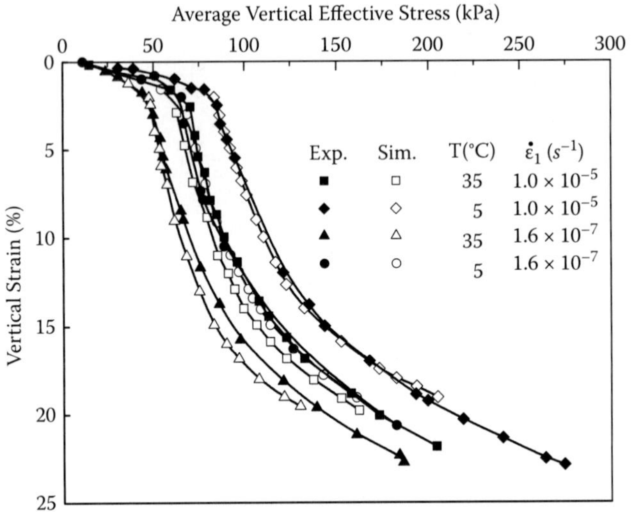

(a)

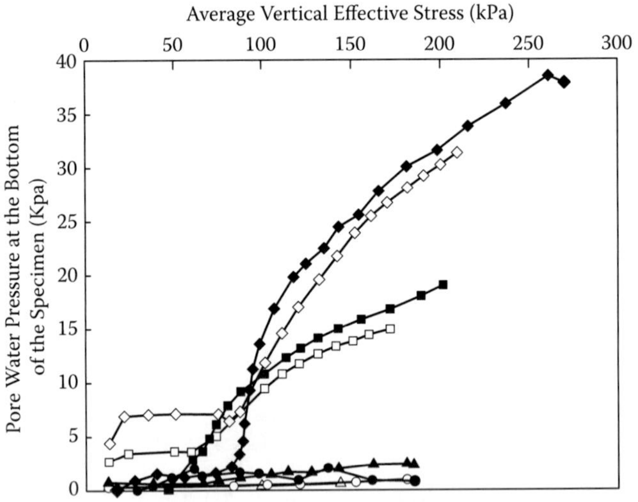

(b)

Figure 10.2 (a) Simulated and experimental results of CRS tests with different temperatures. (b) Simulated and experimental results of CRS tests with different temperatures. (From Yashima, A., Leroueil, S., Oka, F., and Guntoro, I., 1998, *Soils and Foundations*, 38(2):63–73.)

10.1.2 Elastic and temperature-dependent stretching

An additive decomposition of the total stretching tensor, D_{ij}, into elastic stretching D_{ij}^{e} and viscoplastic stretching D_{ij}^{vp} is assumed such that

$$D_{ij} = D_{ij}^{e} + D_{ij}^{vp} \tag{10.7}$$

Elastic stretching, D_{ij}^{e}, is given by a generalized Hooke's law, namely,

$$D_{ij}^{e} = \frac{1}{2G}\hat{\dot{S}}_{ij} + \frac{\kappa}{3(1+e)\sigma'_m}\dot{\sigma}'_m\delta_{ij} \tag{10.8}$$

where $\hat{\dot{S}}_{ij}$ is the Jaumman rate of the deviatoric Cauchy stress tensor, σ'_m is the mean effective stress, G is the elastic shear modulus, e is the void ratio, κ is the swelling index, and the superimposed dot denotes the time differentiation.

In the thermo-hydro-mechanical coupled analysis, we added a new term to the constitutive equations, which is contributed by the change in temperature. Total stretching tensor, D_{ij}, is assumed as

$$D_{ij} = D_{ij}^{e} + D_{ij}^{vp} + \frac{1}{3}\beta_\theta\dot{\theta}\delta_{ij} \tag{10.9}$$

where β_θ is the coefficient of thermal expansion.

The relation between the elastic stretching tensor, D_{ij}^{e}, and the Jaumann rate of Cauchy stress tensor, $\hat{T}'_{ij}$, can be obtained as

$$\hat{T}'_{ij} = C_{ijkl}^{e} D_{kl}^{e} \tag{10.10}$$

which leads to the relation

$$\hat{T}'_{ij} = C_{ijkl}^{e}(D_{kl} - D_{kl}^{vp} - \frac{1}{3}\beta_\theta\dot{\theta}\delta_{kl}) \tag{10.11}$$

From the flow rule, the viscoplastic stretching tensor, D_{ij}^{vp}, which corresponds to Equation (5.58), can be obtained as

$$D_{ij}^{vp} = \gamma < \Phi_1(f_y) > \frac{\partial f_p}{\partial T'_{ij}} \tag{10.12}$$

In the analysis, the tangent modulus method (Peirce et al. 1984) is adopted in order to evaluate the viscoplastic stretching tensor, D_{ij}^{vp}. Next, the relation between the rate of effective stress and the stretching tensor can be written in matrix form, as shown in the following equation:

$$\{\hat{T}'\} = [C]\{D\} - \{Q\} - \frac{1}{3}\beta_\theta \dot{\theta}[C]\{I\} \tag{10.13}$$

Substituting Equation (6.195) into Equation (10.13) yields

$$\{\dot{T}'\} = [C]\{D\} - \{Q\} + \{W^*\} - \frac{1}{3\beta_\theta}\dot{\theta}[C]\{I\} \tag{10.14}$$

where $[C]$ is the tangential stiffness matrix, $\{Q\}$ is the relaxation stress vector, and $\{W^*\} = \{WT - TW\}$ is the vector of the spin tensor.

10.1.3 Weak form of the equilibrium equation for water–soil mixture

We will use the following weak form of the equilibrium equation from Equation (6.194):

$$\int_V \dot{T}'_{ij}\,\delta D_{ij}\,dv - \int_V T'_{jp}L_{pi}\,\delta L_{ij}\,dv + \int_V T'_{ij}D_{kk}\,\delta L_{ij}\,dv + \int_V \dot{u}_w\,\delta D_{kk}\,dv + \int_V U_{ji}\,\delta L_{ij}\,dv = \int_{\Gamma_t} \dot{s}_i\,\delta v_i\,d\Gamma \tag{10.15}$$

in which tension is positive for the pore water pressure, u_w, as seen by Equation (6.175) in Chapter 6.

By all the matrix and vector relations obtained previously, by discretizing Equation (10.15) based on the theory of virtual work, and by considering the arbitrariness of the unconstrained virtual nodal velocity, we obtain

$$[K]\{v^*\} - \int_V [B]^T\{Q\}dV + \int_V [B]^T\{W^*\}dV + [K_L]\{v^*\} + [K_v]\{\dot{u}_w^*\} + [K_\theta]\{\dot{\theta}^*\} = \{\dot{F}\} \tag{10.16}$$

in which

$$[K] = \int_V [B]^T[C][B]dV \tag{10.17}$$

$$[K_L] = \int_V [B_M]^T[D'_s][B_M]dV + \int_V [B_M]^T[U][B_M]dV + \int_V [B_M]^T\{T'\}\{B_v\}^T dV \tag{10.18}$$

$$[K_v] = \int_V \{B_v\}^T \{N_h\} dV \tag{10.19}$$

$$\{\dot{F}\} = \int_{\Gamma_t} [N]^T \{\dot{s}\} d\Gamma \tag{10.20}$$

$$[K_\theta] = \int_V [C]\{B_v\}^T \left(-\frac{1}{3}\beta_\theta\right)[N_\theta] dV \tag{10.21}$$

where $\{\dot{s}\}$ is the rate of the surface force vector, $[B]$ is the matrix that transforms the nodal velocity vector to the stretching vector, $\{v^*\}$ is the nodal velocity vector, $[N]$ is a shape function of the eight-node quadrilateral element, $[B_M]$ is the matrix that transforms the nodal velocity vector to the velocity gradient vector, $\{B_v\}$ is the vector that transforms the nodal velocity into the trace of D_{ij}, $\{\dot{u}_w^*\}$ and $\{\dot{\theta}^*\}$ are the nodal pore pressure rate vector and the nodal temperature rate vector, respectively, and $[N_h]$ and $[N_\theta]$ represent the four-node quadrilateral element shape functions for pore water pressure and temperature, respectively.

The relation between the nodal velocity vector and the nodal displacement vector can be obtained using Euler's approximation as

$$\{v^*\} = \frac{\{\Delta u^*\}}{\Delta t} \quad \{\Delta u^*\}\text{: nodal displacement increment vector} \tag{10.22}$$

Similarly, the pore water pressure rate and the temperature rate can be obtained as

$$\{\dot{u}_w^*\} = \frac{\{u_w^*\}_{t+\Delta t} - \{u_w^*\}_t}{\Delta t} \tag{10.23}$$

$$\{\dot{\theta}^*\} = \frac{\{\theta^*\}_{t+\Delta t} - \{\theta^*\}_t}{\Delta t} \tag{10.24}$$

Substituting Equations (10.22) and (10.23) into Equation (10.16) yields

$$[K]\frac{\{\Delta u^*\}}{\Delta t} + [K_L]\frac{\{\Delta u^*\}}{\Delta t} + [K_v]\frac{\{u_w^*\}_{t+\Delta t} - \{u_w^*\}_t}{\Delta t} + [K_\theta]\frac{\{\theta^*\}_{t+\Delta t} - \{\theta^*\}_t}{\Delta t}$$
$$- \int_V [B]^T \{Q\} dv + \int_V [B]^T \{W^*\} dv = \{\dot{F}\} \tag{10.25}$$

Finally, by transforming the preceding equation, the weak form of the equilibrium equation is obtained:

$$[[K]+[K_L]]\{\Delta u^*\}+[K_v]\{u_w^*\}_{t+\Delta t}+[K_\theta]\{\theta^*\}_{t+\Delta t}=\Delta t\{\dot{F}\}+[K_v]\{u_w^*\}_t+[K_\theta]\{\theta^*\}_t$$
$$+\Delta t\int_V [B]^T\{Q\}dv-\Delta t\int_V [B]^T\{W^*\}dv \tag{10.26}$$

10.1.4 Continuity equation

For the behavior of the pore water pressure, the boundary surface Γ of a closed domain V can be broken into two components, namely,

$$\Gamma=\Gamma_p+\Gamma_q \tag{10.27}$$

where Γ_p is the pore pressure boundary and Γ_q is the pore water flow boundary.

$$\text{On boundary } \Gamma_p,\ u_w=\bar{u}_w \tag{10.28}$$

$$\text{On boundary } \Gamma_q,\ q_w=\bar{q}_w \tag{10.29}$$

Considering the change in temperature, the continuity equation of the fluid phase can be written as

$$\frac{\partial}{\partial x_i}\left(\frac{k}{\gamma_w}\left(\frac{\partial u_w}{\partial x_i}+\rho^f b_i\right)\right)+D_{ii}-\frac{n}{K^f}\dot{u}_w-\alpha_w\dot{\theta}=0 \tag{10.30}$$

where k is the coefficient of permeability, γ_w is the pore water density, K_f is the volumetric modulus of pore water, α_w is the thermal expansion coefficient of the pore water, the soil particle is incompressible, and tension is positive for pore water pressure.

If the body force and the mass density of the pore water are constant, Equation (10.30) becomes

$$\frac{\partial}{\partial x_i}\left(\frac{k}{\gamma_w}\frac{\partial u_w}{\partial x_i}\right)+D_{ii}-\frac{n}{K^f}\dot{u}_w-\alpha_w\dot{\theta}=0 \tag{10.31}$$

Considering a test function of $\hat{u}_w$, which is an arbitrary function, except that $\hat{u}_w=0$ at the pore pressure prescribed boundary, a weak form of the continuity equation can be obtained as

$$\int_V \hat{u}_w\frac{\partial}{\partial x_i}\left(\frac{k}{\gamma_w}\frac{\partial u_w}{\partial x_i}\right)dv+\int_V \hat{u}_w D_{ii}\,dv-\int_V \hat{u}_w\frac{n}{K^f}\dot{u}_w\,dv-\int_V \hat{u}_w\alpha_w\dot{\theta}\,dv$$
$$-\int_{\Gamma_q}\hat{u}_w(q_{wi}-\bar{q}_{wi})\,d\Gamma=0 \tag{10.32}$$

By applying Gauss's theorem to the previous equation and disregarding the spatial dependency of the permeability, the following equation is obtained:

$$\frac{k}{\gamma_w}\int_{\Gamma_p}\hat{u}_w\frac{\partial u_w}{\partial x_i}n_i\,d\Gamma-\frac{k}{\gamma_w}\int_V\hat{u}_{w,i}\frac{\partial u_w}{\partial x_i}dv+\int_V\hat{u}_w D_{ii}\,dv-\int_V\frac{n}{K^f}\dot{u}_w\hat{u}_w dv$$

$$-\int_V\alpha_w\dot{\theta}\hat{u}_w\,dv+\int_{\Gamma q}\hat{u}_w\bar{q}_{wi}n_i d\Gamma=0 \tag{10.33}$$

Since the first term can be omitted, because $\hat{u}_w=0$ on Γ_p or the trial function satisfies the pore pressure prescribed boundary Γ_p, Equation (10.33) becomes

$$-\frac{k}{\gamma_w}\int_V\hat{u}_{w,i}\frac{\partial u_w}{\partial x_i}dV+\int_V\hat{u}_w D_{ii}dV-\int_V\frac{n}{K^f}\dot{u}_w\hat{u}_w dV-\int_V\alpha_w\dot{\theta}\hat{u}_w dV$$

$$+\int_{\Gamma q}\hat{u}_w\bar{q}_{wi}n_i d\Gamma=0 \tag{10.34}$$

Equation (10.34) is discretized as

$$-\frac{k}{\gamma_w}\int_V\{\hat{u}_w^*\}^T[B_h]^T[B_h]\{u_w^*\}dV+\int_V\{\hat{u}_w^*\}^T[N_h]^T\{B_v\}\{v^*\}dV$$

$$-\int_V\{\hat{u}_w^*\}^T\frac{n}{K^f}[N_h]^T[N_h]\{\dot{u}_w^*\}dV-\int_V\{\hat{u}_w^*\}^T\alpha_w[N_h]^T[N_\theta]\{\dot{\theta}^*\}dV \tag{10.35}$$

$$+\int_{\Gamma q}\{\hat{u}_w^*\}^T[N_h]^T\bar{q}_{wi}n_i d\Gamma=0$$

where

$$u_w=[N_h]\{u_w^*\} \tag{10.36}$$

$$\{u_{w,i}\}=[N_{h,i}]\{u_w^*\}=[B_h]\{u_w^*\} \tag{10.37}$$

$$[N_{h,i}]=\nabla\{N_h\} \tag{10.38}$$

$$D_{ii}=trD=\{B_v\}^T\{v^*\} \tag{10.39}$$

$$\theta=[N_\theta]\{\theta^*\} \tag{10.40}$$

By taking the arbitration of the test function, considering the arbitrariness of the test function, unconstrained nodal pore pressure $\{\hat{u}_w^*\}^T$, we have

$$-\frac{k}{\gamma_w}\int_V [B_h]^T[B_h]dV\left\{u_w^*\right\}+\int_V [N_h]^T\{B_v\}\; dV\{v^*\}$$

$$-\int_V \frac{n}{K^f}[N_h]^T\;[N_h]\{\dot{u}_w^*\}dV-\int_V \alpha_w[N_h]^T\;[N_\theta]\{\dot{\theta}^*\}dV-\int_{\Gamma q}[N_h]^T\bar{q}_i^f n_i d\Gamma=0 \tag{10.41}$$

From Equation (10.40), we define thc following terms:

$$[K_h]=\frac{k}{\gamma_w}\int_V [B_h]^T[B_h]dV \tag{10.42}$$

$$[K_v]^T=\int_V [N_h]^T[B_v]dV \tag{10.43}$$

$$[K_c]=-\int_V \frac{n}{K^f}[N_h]^T[N_h]dV \tag{10.44}$$

$$[K_{\theta w}]=-\int_V \alpha_w\,[N_h]^T[N_\theta]dV \tag{10.45}$$

$$\{V\}=\int_{\Gamma_q}[N_h]^T\bar{q}_{wi}n_i d\Gamma \tag{10.46}$$

Substituting these matrices into Equation (10.41) yields

$$[K_v]^T\{v^*\}-[K_h]\left\{u_w^*\right\}+[K_c]\{\dot{u}_w^*\}+[K_{\theta w}]\{\dot{\theta}^*\}=\{V\} \tag{10.47}$$

By using Euler's approximation, the following relation between the nodal velocity vector and the nodal displacement vector can be obtained:

$$\{v^*\}=\frac{\{\Delta u^*\}}{\Delta t} \tag{10.48}$$

where $\{\Delta u^*\}$ is the nodal displacement.

For the pore water pressure rates at the nodes, the temperature rates at the nodes are approximated by the finite difference scheme, as shown by Equations (10.23) and (10.24).

Substituting Equations (10.48), (10.23), and (10.24) into Equation (10.47) gives

$$[K_v]^T\left\{\Delta u^*\right\}-\Delta t[K_h]\left\{u_w^*\right\}_{t+\Delta t}+[K_{\theta w}]\{\theta^*\}_{t+\Delta t}+[K_c]\{u_w^*\}_{t+\Delta t}$$

$$=[K_c]\{u_w^*\}_t+[K_{\theta w}]\{\theta^*\}_t+\Delta t[V] \tag{10.49}$$

10.1.5 Balance of energy

The first law of thermodynamics presented in Chapter 1, namely, the conservation of energy (Equation 1.138), is given as

$$\rho\dot{e}=T'_{ij}D_{ij}-h_{i,i} \tag{10.50}$$

where $\dot{e}$ is the rate of internal energy, T'_{ij} is the effective Cauchy stress tensor, and h_i is the heat flux vector.

As for the formulation of h_i, we can use the Cattaneo equation (Müller and Ruggeri 1998), namely,

$$h_i+\tau\dot{h}_i=-k_\theta\theta_{,i} \tag{10.51}$$

in which k_θ is the coefficient of heat conductivity, θ is the temperature, and τ is the heat flux parameter regarded as the relaxation time for the stationary state. If we use the Cattaneo equation we have a hyperbolic type of heat conduction equation instead of a parabolic equation. When τ is equal to zero, Equation (10.51) becomes Fourier's law.

The constitutive relation between the elastic stretching tensor, D_{ij}^e, and the rate of internal energy, $\dot{e}$, is assumed as follows:

$$\rho\dot{e}=\rho c\dot{\theta}+T'_{ij}D_{ij}^e \tag{10.52}$$

where ρ is the density of soil and c is the specific heat capacity.

By taking Equations (10.50) and (10.52), the conservation of energy is written using the viscoplastic stretching tensor, D_{ij}^{vp}, as

$$\rho c\dot{\theta}=T'_{ij}D_{ij}^{vp}-h_{i,i} \tag{10.53}$$

Using the boundary condition for the heat flow, the boundary can be divided into two parts, namely, Γ_θ and Γ_T as $\Gamma=\Gamma_\theta+\Gamma_T$.

On boundary Γ_θ $\theta = \overline{\theta}$

On boundary Γ_T $h_i = \overline{h}_i$

where the upper bar denotes the prescribed values.

Adopting the boundary conditions, the weak form of the conservation of energy can be written as follows:

$$\int_V (\rho c\dot{\theta} - T'_{ij}D_{ij}^{vp} + h_{i,i})\hat{\theta}\,dv + \lambda \int_{\Gamma_T} (h_i - \overline{h}_i)\hat{\theta} n_i\, d\Gamma = 0 \tag{10.54}$$

in which $\hat{\theta}$ is the test function that is zero on Γ_θ and $\overline{h}_i$ is the specified value of the heat flux. Since λ is arbitrary, we can take $\lambda = -1$. Using Fourier's law in Equation (10.51) and adopting the Gauss theorem into Equation (10.54) provides

$$\int_V (\rho c\dot{\theta} - T'_{ij}D_{ij}^{vp})\hat{\theta}\,dv + \int_V k_\theta \theta_{,i}\hat{\theta}_{,i}\,dv + \int_{\Gamma_\theta} \overline{h}_i \hat{\theta} n_i\, d\Gamma = 0 \tag{10.55}$$

For the discretization, the shape function $[N_\theta]$ is used

$$\{\theta\} = [N_\theta]\{\theta^*\}, \quad \{\theta_{,i}\} = [N_{\theta,i}]\{\theta^*\} = [B_\theta]\{\theta^*\} \tag{10.56}$$

where superscript asterisk (*) indicates the nodal value.

Euler's approximation is used for the rate of temperature, $\{\dot{\theta}\}$, and the rate of heat flux, $\{\dot{h}_i\}$, and considering arbitrariness, $\{\hat{\theta}^*\}^T$, we have from Equation (10.55)

$$([K_{T1}] + [K_{T2}])\{\theta^*\}_{t+\Delta t} = \{F_T\} - \{H\} \tag{10.57}$$

$$[K_{T1}] = \rho c \int_V [N_\theta]^T [N_\theta]\,dv \ , \quad [K_{T2}] = \Delta t \int_V k_\theta [B_\theta]^T [B_\theta]\,dv \tag{10.58}$$

$$\{F_T\} = [K_{T1}]\{\theta^*\}_t + \Delta t \int_V [N_\theta]^T T'_{ij} D_{ij}^{vp}\,dv \tag{10.59}$$

$$\{H\} = \Delta t \int_{\Gamma_T} [N_\theta]^T \{\overline{h}_i\}\{n_i\}\,d\Gamma \tag{10.60}$$

in which subscript $t + \Delta t$ denotes an unknown value, and subscript t indicates the latest known value.

The element for the temperature is a four-node quadrilateral isoparametric element with four Gaussian integration points. Combining Equations

(10.26), (10.49), and (10.58) gives the governing equation for the finite element formulation as

$$\begin{bmatrix} [K]+[K_L] & [K_v] & [K_\theta] \\ [K_v]^T & -\Delta t[K_h]+[K_c] & [K_{\theta w}] \\ 0 & 0 & [K_{T1}]+[K_{T2}] \end{bmatrix} \begin{bmatrix} \{\Delta u^*\} \\ \{u_w^*\}_{t+\Delta t} \\ \{\theta^*\}_{t+\Delta t} \end{bmatrix}$$

$$= \begin{bmatrix} \Delta t\{\dot{F}\}+[K_v]\{u_w^*\}_t+[K_\theta]\{\theta^*\}_t+\Delta t\int_V [B]^T\{Q\}dV-\Delta t\int_V [B]^T\{W^*\}dV \\ [K_c]\{u_w^*\}_t+[K_{\theta w}]\{\theta^*\}_t+\Delta t\{V\} \\ \{F_T\}-\{H\} \end{bmatrix} \tag{10.61}$$

10.1.6 Simulation of thermal consolidation

Heating water-saturated soil induces volume expansion of the soil particles and the pore water. Even if the heated soil is under drained conditions, excess pore water pressure is generated. Dissipation of the heat-generated pore pressure induces consolidation, which is called thermal consolidation (Campanella and Mitchell 1968; Baldi et al. 1988; Delage et al. 2000; Sultan et al. 2002). Figure 10.3 shows the experimental results by Campanella and Mitchell (1968) during the increase and decrease in temperature under a constant confining pressure. It can be seen that the consolidation and swelling occur during the increase and decrease in temperature. In this section, we conducted a simulation of thermal consolidation due to the dissipation of pore water pressure induced by heat and viscoplastic thermal softening.

For simplified and practical formulations, both the grain particles and the fluid are assumed to be mechanically incompressible. Based on the finite deformation theory presented in Chapter 6, an updated Lagrangian method with the objective Jaumann rate of Cauchy stress is used for the weak form of the rate type of equilibrium equations for the whole soil–water mixture. As for the element type, an eight-node quadrilateral isoparametric element with a reduced Gaussian two-point Gauss quadrature rule for integration is used for the displacement, while the pore water pressure and the temperature are defined by a four-node quadrilateral isoparametric element.

The heat-generated pore pressure is simulated using the model by Campanella and Mitchell (1968), which is given as

$$\Delta u_w = \frac{n\Delta\theta(\alpha_s-\alpha_w)+\alpha_{st}\Delta\theta}{m_v} \tag{10.62}$$

$$m_v = \frac{\kappa}{(1+e)\sigma'_m} \tag{10.63}$$

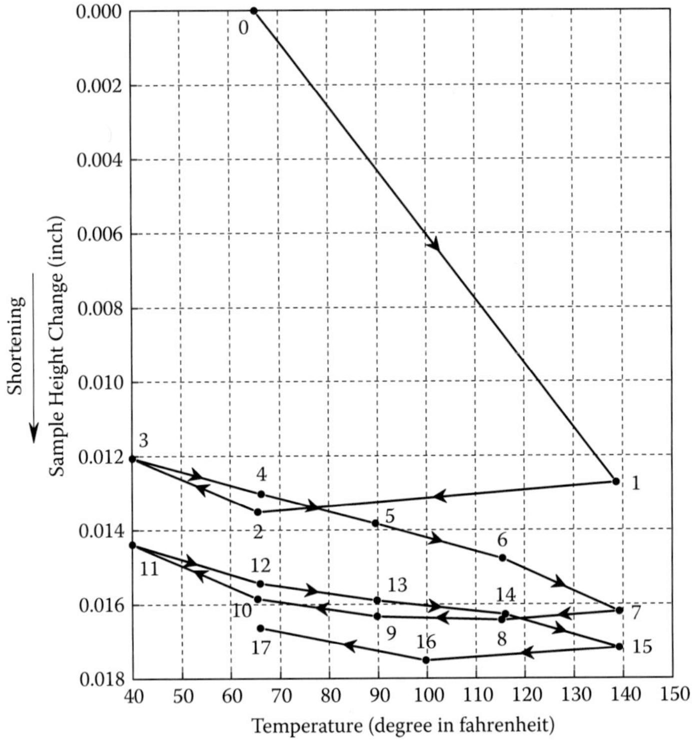

Figure 10.3 Effect of temperature variations on height and volume change. (After Campanella, R.G., and Mitchell, J.K., 1968, J. SMFE, ASCE, 94, 3:709–734.)

in which Δu_w is the heat-generated pore pressure, $\Delta\theta$ is the change in temperature, n is the porosity, m_v is the coefficient of volume compressibility, α_s is the thermal coefficient of expansion for the soil particles, and α_{st} is a physicochemical coefficient of the change in structural volume. In this analysis, $\alpha_s = 0.35 \times 10^{-4}$ (1/°C) and $\alpha_{st} = -0.50 \times 10^{-4}$(1/°C), which are the same values used by Campanella and Mitchell (1968). As for the thermal coefficient of expansion for water, α_w, we employed an empirical formula by Baldi et al. (1988):

$$\alpha_w = (139.0 + 6.1 \times \theta) \times 10^{-6} \tag{10.64}$$

In addition, the temperature dependency of the coefficient of permeability, k, is also considered, namely,

$$k = \frac{K\gamma_w(\theta)}{\mu(\theta)} \tag{10.65}$$

Table 10.1 Material parameters

Compression index	λ	0.172
Swelling index	κ	0.054
Initial void ratio	e_0	0.72
Initial mean effective stress (kPa)	σ'_{me}	392
Compression yield stress (kPa)	σ'_{mbi}	392
Coefficient of earth pressure at rest	K_0	1.0
Viscoplastic parameter	m'	21.5
Viscoplastic parameter (1/s)	$C(\theta_r)$	4.5×10^{-8}
Stress ratio at critical state	M^*_m	1.05
Initial elastic shear modulus (kPa)	G	5500
Coefficient of permeability (m/s)	k	1.54×10^{-8}
Referential temperature (°C)	θ_r	20.0
Soil density (t/m^3)	ρ	1.96
Specific heat capacity (J/kg·°C)	c	0.87
Coefficient of heat conductivity (J/kg·°C)	k_θ	0.14
Heat flux parameter (s)	τ	0.0
Thermo-viscoplastic parameter	a	0.15
Structural parameter (kPa)	σ'_{maf}	350
Structural parameter	β	10

where K is the intrinsic permeability and $\gamma_w(\theta)$ is the unit weight of water, in which the temperature dependency of $\gamma_w(\theta)$ is assumed to be negligible. As for the temperature dependency of the viscosity of water, μ, an empirical formula, $\mu = -0.00046575ln(\theta)+0.00239138$ $(Pa{\cdot}s)$, is used (Delage et al. 2000).

Other material parameters that were used in the simulation of thermal consolidation are listed in Table 10.1. Figure 10.4 depicts the boundary conditions and the size of the specimen for the simulation of thermal consolidation. An isothermal boundary is imposed at the left side of the specimen as the installation of a heat source, for which temperature increases gradually from its initial value of 20°C at $t = 0$ (hr) to 60°C at $t = 4$ (hr) (see Figure 10.5). The upper boundary is also an isothermal boundary of the atmospheric temperature of 20°C.

The settlement at node A is shown in Figure 10.6. We can observe around 1.8 cm of settlement at $t = 150$ (hr) and it is seen that the deformation does not converge.

The distribution of temperature, the pore water pressure, the mean effective stress, and the accumulated viscoplastic volumetric strain are demonstrated in Figure 10.7. It can be observed from these figures that thermal consolidation due to the dissipation of the pore water pressure induced by

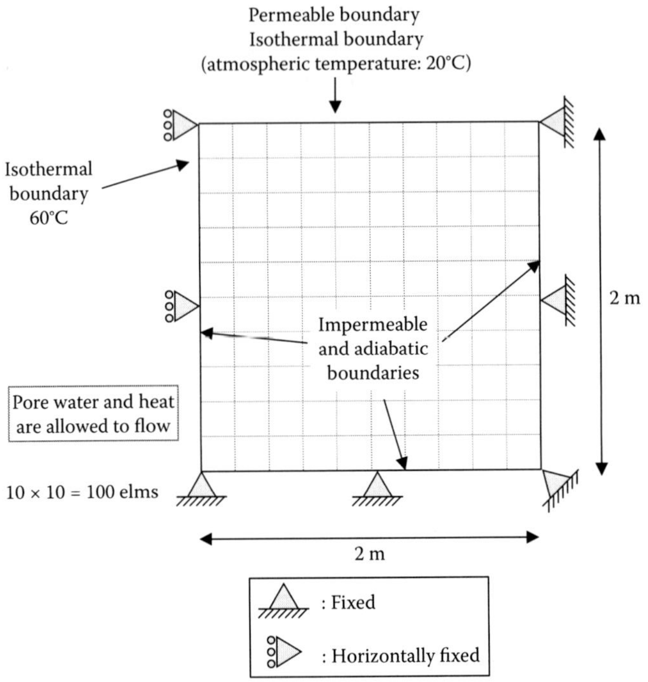

Figure 10.4 Finite element mesh and boundary conditions for pore water pressure and temperature.

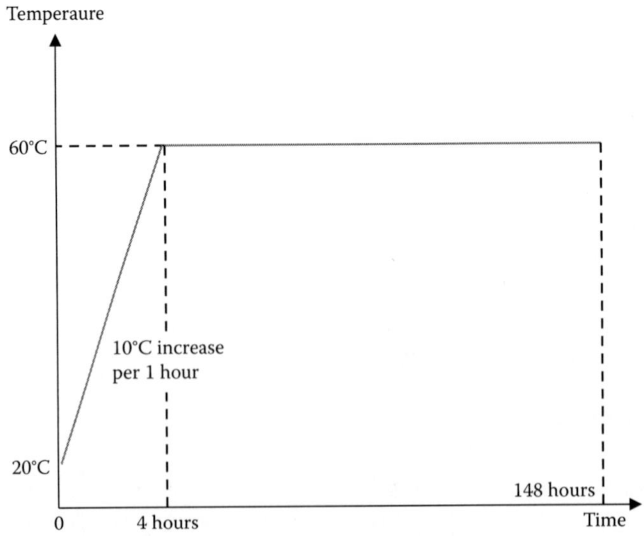

Figure 10.5 Increase in temperature at the heat source.

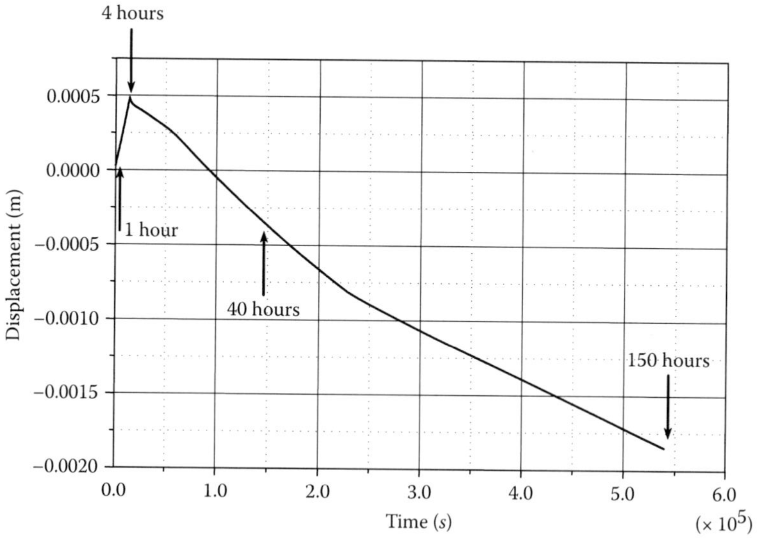

Figure 10.6 Displacement–time profile at point A.

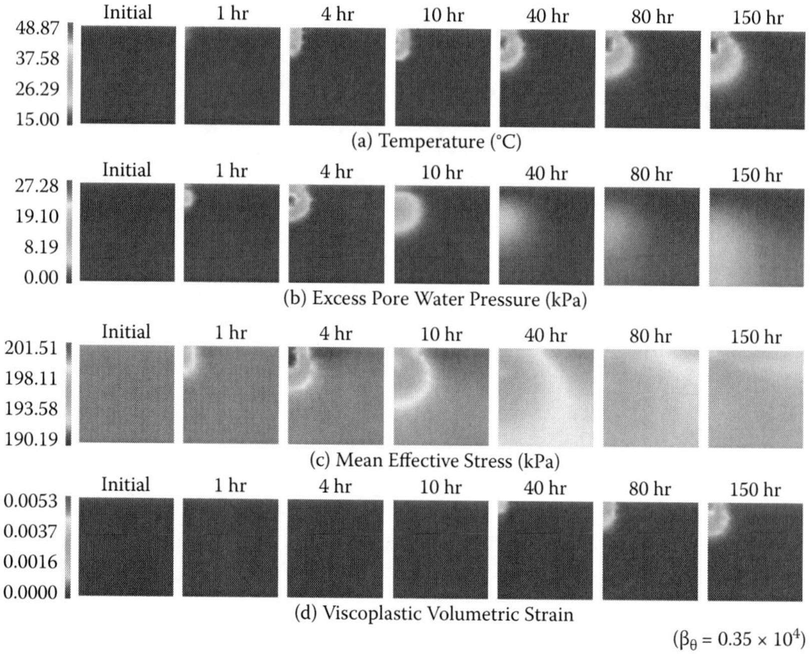

Figure 10.7 Distributions of temperature, excess pore water pressure, mean effective stress, and viscoplastic volumetric strain.

the change in temperature is reproduced. At the beginning of heating, the temperature gradually increases near the heat source. Corresponding to the temperature elevation, pore water pressure is generated near the heat source. The maximum displacement in Figure 10.4 corresponds to the development of large pore water pressure.

Since the viscosity of clay is dependent on temperature and the plastic stress power changes the temperature of clay, it is necessary for the analysis method of clay to consider the variation in temperature. In this chapter, to address the temperature dependency of saturated clay and to deal with the change in temperature induced by the plastic stress power and the heat conduction, we proposed an elasto-thermo-viscoplastic constitutive model for clay and developed a thermo-hydro-mechanical coupled finite element method with the constitutive model through the finite element formulation of the energy balance equation. Using the newly developed analysis method, simulations of thermal consolidation were conducted. We can simulate the thermal consolidation phenomena by use of the proposed method.

10.2 SEEPAGE–DEFORMATION COUPLED ANALYSIS OF UNSATURATED RIVER EMBANKMENT USING MULTIPHASE ELASTOVISCOPLASTIC THEORY

10.2.1 Introduction

In recent years, many natural disasters due to floods associated with torrential rains, typhoons, hurricanes, and so forth have occurred around the world. In many cases, river embankments have failed due to seepage and overflow. A multiphase deformation analysis of a river embankment has recently been carried out using an air–soil–water coupled finite element method considering the unsaturated seepage flow. A numerical model for unsaturated soil has been constructed based on the porous media theory and the elastoviscoplastic constitutive model shown in Chapters 2 and 5. As for the stress variables in the formulation of unsaturated soil, we use the skeleton stress and suction simultaneously. The skeleton stress has been validated by experimental results (Oka, Kodaka, et al. 2010; Kimoto et al. 2011). The skeleton stress is used in the constitutive model instead of Terzaghi's effective stress for saturated soil, and suction is incorporated through the constitutive parameters of the model. An air–soil–water coupled finite element method is developed using the governing equations for the three-phase soil based on the nonlinear finite deformation theory, that is, the updated Lagrangian method. Two-dimensional numerical analyses of the river embankment under seepage conditions have been conducted for the high river water level.

10.2.2 Governing equations and analysis method

For the governing equations, such as the equilibrium equation and the mass balance equation for unsaturated soil, we use the method presented in Chapter 2. In the numerical analysis, we use the updated Lagrangian method described in Chapter 6.

10.2.3 Constitutive model for unsaturated soil

In the analysis, the saturated elastoviscoplastic model for the overstress-type of viscoplasticity with soil structure degradation presented in Chapter 5 has been extended to unsaturated soil using the skeleton stress and including the effects of suction (Oka, Kimoto, et al. 2010; Kimoto et al. 2011). The collapse behavior of unsaturated soil is macroscopic evidence of the structural instability of the soil skeleton and it is independent of the chosen stress variables (Oka et al. 2008). In the present model, the collapse behavior is described by the shrinkage of both the overconsolidated boundary surface and the static yield surface due to the decrease in suction.

It is assumed that the strain rate tensor consists of an elastic stretching tensor, D^e_{ij}, and a viscoplastic stretching tensor, D^{vp}_{ij}, as

$$D_{ij} = D^e_{ij} + D^{vp}_{ij} \tag{10.66}$$

The elastic stretching tensor is given by a generalized Hooke's law, namely,

$$D^e_{ij} = \frac{1}{2G}\hat{\dot{S}}_{ij} + \frac{\kappa}{3(1+e)}\frac{\dot{\sigma}'_m}{\sigma'_m}\delta_{ij} \tag{10.67}$$

where $\hat{\dot{S}}_{ij}$ is the Jaumann rate of the deviatoric Cauchy stress tensor, σ'_m is the mean skeleton stress, G is the elastic shear modulus, e is the initial void ratio, κ is the swelling index, and the superimposed dot denotes the time differentiation.

10.2.3.1 Overconsolidation boundary surface

The overconsolidation boundary surface separates the normal consolidated (NC) region, $f_b \geq 0$, from the overconsolidated region, $f_b < 0$, as follows:

$$f_b = \bar{\eta}^*_{(0)} + M^*_m \ln\frac{\sigma'_m}{\sigma'_{mb}} = 0 \tag{10.68}$$

$$\bar{\eta}^*_{(0)} = \left\{\left(\eta^*_{ij} - \eta^*_{ij(0)}\right)\left(\eta^*_{ij} - \eta^*_{ij(0)}\right)\right\}^{\frac{1}{2}} \tag{10.69}$$

where η^*_{ij} is the stress ratio tensor ($\eta^*_{ij} = s_{ij}/\sigma'_m$) and (0) denotes the state at the end of the consolidation, in other words, the initial state before the shear test. M^*_m is the value of $\eta^* = \sqrt{\eta^*_{ij}\eta^*_{ij}}$ when the volumetric strain increment changes from negative to positive dilatancy, which is equal to ratio M_f at the critical state. σ'_{mb} is the strain-hardening parameter, which controls the size of the boundary surface.

The suction effect is introduced into the value of σ'_{mb} as

$$\sigma'_{mb} = \sigma'_{ma} \exp\left(\frac{1+e}{\lambda-\kappa}\varepsilon^{vp}_{kk}\right)\left[1+S_I \exp\left\{-s_d\left(\frac{P^c_i}{P^c}-1\right)\right\}\right] \tag{10.70}$$

where ε^{vp}_{kk} is the viscoplastic volumetric strain, P^c is the present suction value, P^c_i is a reference suction, and S_I denotes the increase in the yield stress when the suction increases from zero to reference value P^c_i. S_d controls the rate of increase or decrease in σ_{mb} with suction and σ_{ma} is a strain-softening parameter used to describe the degradation caused by structural changes, namely,

$$\sigma'_{ma} = \sigma'_{maf} + (\sigma'_{mai} - \sigma'_{maf})\exp(-\beta z) \tag{10.71}$$

$$z = \int_0^t \dot{z}\,dt \quad \text{with} \quad \dot{z} = \sqrt{\dot{\varepsilon}^{vp}_{ij}\dot{\varepsilon}^{vp}_{ij}} \tag{10.72}$$

in which σ'_{mai} and σ'_{maf} are the initial and the final values of σ'_{ma}, respectively, while β controls the rate of degradation with viscoplastic strain and $\dot{\varepsilon}^{vp}_{ij}$ is the viscoplastic strain rate.

10.2.3.2 Static yield function

To describe the mechanical behavior of the soil at its static equilibrium state, a Cam-clay type of static yield function is assumed as

$$f_y = \bar{\eta}^*_{(0)} + \tilde{M}^* \ln\frac{\sigma'_m}{\sigma'^{(s)}_{my}} = 0 \tag{10.73}$$

where $\tilde{M}^*$ is assumed to be constant in the NC region, but varies with the current stress in the OC region (see Chapter 5).

The static strain-hardening parameter $\sigma'^{(s)}_{my}$ controls the size of the static yield surface. In the same way as for the overconsolidation boundary surface, parameter $\sigma'^{(s)}_{my}$ varies with the changes in suction as well as with the changes in viscoplastic volumetric strain and structural degradation as

$$\sigma'^{(s)}_{my} = \frac{\sigma'^{(s)}_{myi}}{\sigma'_{mai}} \sigma'_{ma} \exp\left(\frac{1+e}{\lambda-\kappa}\varepsilon^{vp}_{kk}\right)\left[1+s_I \exp\left\{-s_d\left(\frac{P^c_i}{P^c}-1\right)\right\}\right] \tag{10.74}$$

where $\sigma'^{(s)}_{myi}$ is the initial value of $\sigma'^{(s)}_{my}$.

10.2.3.3 Viscoplastic potential function

The viscoplastic potential function is given by

$$f_p = \bar{\eta}^*_{(0)} + \tilde{M}^* \ln \frac{\sigma'_m}{\sigma'_{mp}} = 0 \tag{10.75}$$

where σ'_{mp} denotes the mean skeleton stress at the intersection of the viscoplastic potential function surface and the σ'_m axis.

10.2.3.4 Viscoplastic flow rule

The viscoplastic stretching tensor is an extention of Perzyna's viscoplastic theory and is given as

$$D^{vp}_{ij} = \langle \Phi_{ijkl}(f_y) \rangle \frac{\partial f_p}{\partial \sigma'_{kl}} \tag{10.76}$$

where the symbol $\langle\rangle$ is defined as

$$\langle \Phi_{ijkl}(f_y) \rangle = \begin{cases} \Phi_{ijkl}(f_y) & ; \quad f_y > 0 \\ 0 & ; \quad f_y \le 0 \end{cases} \tag{10.77}$$

in which Φ_{ijkl} denotes a material function for rate sensitivity. Herein, the value of f_y is assumed to be positive for any stress state in this model; in other words, the stress state always exists outside the static yield function, so that viscoplastic deformation always occurs. Based on the experimental results of constant strain-rate triaxial tests, the material function Φ_{ijkl} is defined by an exponential function (Kimoto and Oka 2005):

$$\Phi_{ijkl}(f_y) = C_{ijkl}\sigma'_m \exp\left\{ m' \left(\bar{\eta}^*_{(0)} + \tilde{M}^* \ln \frac{\sigma'_m}{\sigma'^{(s)}_{my}} \right) \right\} \tag{10.78}$$

where m' is the viscoplastic parameter that controls the rate sensitivity and viscoplastic parameter C_{ijkl} is a fourth-rank isotropic tensor given by

$$C_{ijkl} = a\delta_{ij}\delta_{kl} + b(\delta_{ik}\delta_{jl} + \delta_{il}\delta_{jk}),$$

$$C_1 = 2b, C_2 = 3a + 2b, \tag{10.79}$$

where a and b are material parameters, which are related to the deviatoric component C_1 and volumetric component C_2 of the viscoplastic parameter.

10.2.3.5 Constitutive model for pore water: soil–water characteristic curve

Since saturation is a function of the suction, that is, the pressure head, the time rate for saturation is given by

$$n\dot{S}_r = n\frac{dS_r}{d\theta}\frac{d\theta}{d\psi}\frac{d\psi}{dp^c}\dot{p}^c = \frac{1}{\gamma_w}\frac{d\theta}{d\psi}\dot{p}^c \tag{10.80}$$

where $\theta = \frac{V_w}{V}$ is the volumetric water content, p^c is the matrix suction $p^c = (p^a - p^f)$, $\psi = p^c/\gamma_w$ is the pressure head for the suction, and $C = \left|\frac{d\theta}{d\psi}\right|$ is the specific water content. It is worth noting that the saturation is a decreasing function of the suction.

The soil–water characteristic model proposed by Van Genuchten (1980) is used to describe the unsaturated seepage characteristics for which effective saturation, S_e, is adopted as

$$S_e = \frac{\theta - \theta_r}{\theta_s - \theta_r} = \frac{nS_r - \theta_r}{\theta_s - \theta_r} \tag{10.81}$$

where θ is the volumetric water content, θ_s is the volumetric water content in the saturated state, which is equal to porosity n, and θ_r is the residual volumetric water content retained by the soil at a large value of suction head, which is a disconnected pendular water meniscus. For relatively large and uniform sand particles, such as those of Toyoura sand, θ_r becomes zero, which is equal to common saturation.

In order to determine the soil–water characteristics, effective saturation, S_e, can be related to negative pressure head, ψ, through the following relation:

$$S_e = \left(1 + |\alpha\psi|^{n'}\right)^{-m} \tag{10.82}$$

where α is a scaling parameter that has the dimensions of the inverse of ψ, and n' and m determine the shape of the soil–water characteristic curve. The relation between n' and m leads to an S-shaped type of soil–water characteristic curve, namely,

$$m = 1 - \frac{1}{n'} \tag{10.83}$$

In the present analysis, we follow the Guide for Structural Investigations of River Embankments (Japan Institute of Construction Engineering 2002), and set $\alpha = 2$ and $n' = 4$.

Specific water content, C, used in Equation (10.80), can be calculated as

$$C(\equiv \left|\frac{d\theta}{d\psi}\right|) = \alpha(n'-1)(\theta_s - \theta_r)S_e^{1/m}(1 - S_e^{1/m})^m \tag{10.84}$$

The specific permeability coefficient, k_r, which is a ratio of the unsaturated to the saturated permeability, is defined by

$$k_r = S_e^{1/2}\{\,1-(1-S_e^{1/m})^m\,\}^2 \tag{10.85}$$

Applying the aforementioned relations, we can describe the unsaturated seepage characteristics. In the analysis, the unsaturated region is treated in the following manner. In the embankment, the initial suction, that is, the initial negative pore water pressure is assumed to be constant. Below the water level, the pore water pressure is given by the hydrostatic pressure. In the transition region between the water level and the suction constant region, we assume that the distribution of the pore water pressure is linearly interpolated as in Figure 10.7. When the pressure head is negative, the increase in the soil modulus due to suction is considered. The effective saturation and the saturation are calculated with Equations (10.81) and (10.82) using the negative pressure head, that is, suction. Applying the obtained effective saturation, specific water content, C, is then calculated by Equation (10.84) and specific permeability, k_r, is calculated by Equation (10.85).

10.2.4 Simulation of the behavior of unsaturated soil by elastoviscoplastic model

For unsaturated soil, we have applied the viscoplastic model to the experimental results of drained triaxial tests on DL clay (silt). Figure 10.8 indicates the simulated results and the material parameters used in the simulation are

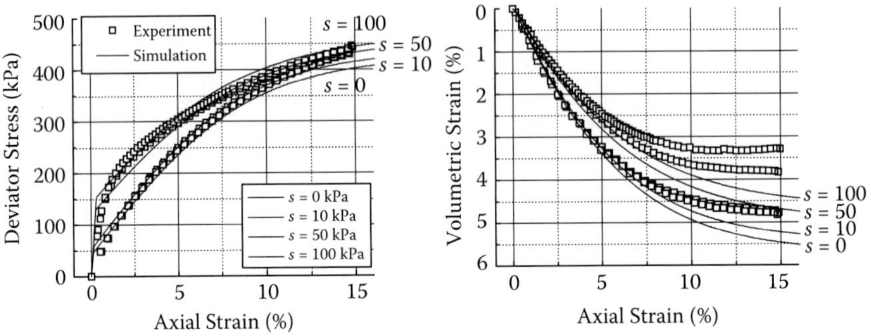

Figure 10.8 Stress–strain relations and volume change characteristics.

listed in Table 10.2. From this figure, the extended model can describe the effect of suction on the triaxial behavior.

10.2.5 Numerical analysis of seepage–deformation behavior of a levee

10.2.5.1 Analysis method

Weak forms of the governing equations are discretized in space and solved by the finite element method. In this formulation, an updated Lagrangian method with the objective Jaumann rate of Cauchy stress is adopted, as presented in Chapter 6. The independent variables are the pore water pressure,

Table 10.2 Material parameters

Initial suction s (kPa)	0	10	50	100
Initial void ratio, e	1.01	1.03	1.07	1.04
Compression index, λ	0.095	0.095	0.114	0.114
Swelling index, κ	0.0086	0.0086	0.0102	0.0102
Initial mean skeleton stress, σ_{m0}(kPa)	200	205	217	222
Viscoplastic parameter, m'		52.0		
Viscoplastic parameter, C_1(1/s)		1.0×10^{-11}		
Viscoplastic parameter, C_2 (1/s)		1.5×10^{-11}		
Stress ratio at critical state, M^*_m	1.00	1.00	1.00	1.00
Elastic shear modulus, G(kPa)	34800	36600	45100	46800
Compression yield stress, σ'_{mbi}(kPa)	200	205	217	222
Suction effect parameter, S_I		0.5		
Suction effect parameter, s_d		0.25		
Slope of stress path, $\Delta q/\Delta\sigma'_m$	2.99	2.98	3.01	3.01

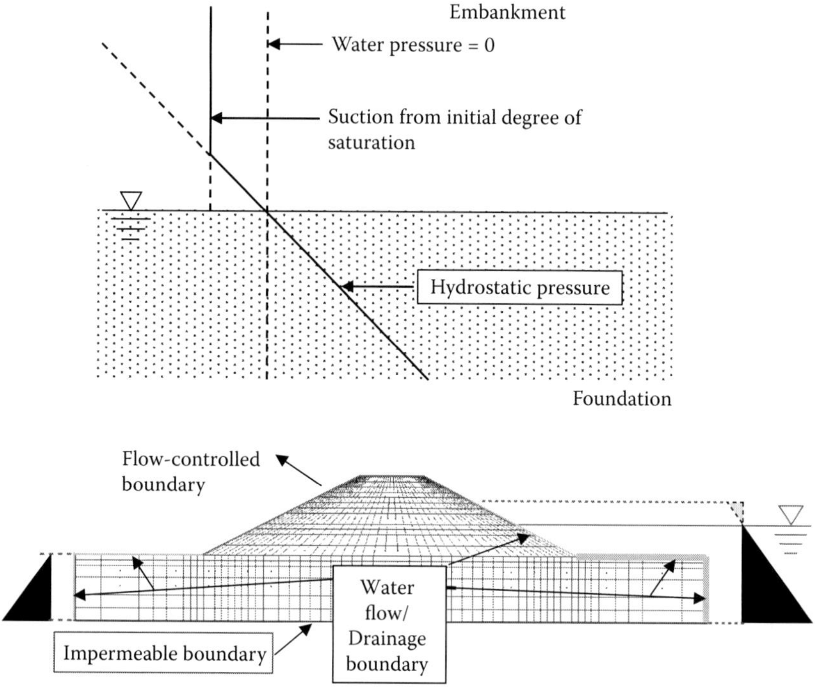

Figure 10.9 Initial pressure head and finite element mesh and boundary conditions.

the pore air pressure, and the nodal velocity. In the finite element formulation, an eight-node quadrilateral element with a reduced Gaussian integration is used for the displacement, and four nodes are used for the pore water pressure and the pore air pressure. The backward finite difference method is used for the time discretization.

10.2.5.2 Deformation during the seepage flow

Figure 10.9 shows the model of the river embankment and the finite element mesh used in the analysis with boundary conditions. The boundary of the embankment is defined as follows. The boundary that contacts with water or air is a drainage boundary if the pore fluid pressure is positive, but the boundary between the embankment and the air is assumed to be a no-water-flow boundary if the pore fluid pressure is negative to avoid the fictitious water flow-in. The soil parameters used in the simulation are listed on Tables 10.3 and 10.4. The parameters are determined for DL clay obtained by Oka, Kodaka, et al. (2010). The water level of the river has been increased for 18 hours and then remains constant, as shown in

Table 10.3 Soil parameters for DL clay used in the analysis

Compression index, λ	0.136
Swelling index, κ	0.0175
Initial void ratio, e_0	1.05
Initial elastic shear modulus, G_0	32400 (kPa)
Initial mean skeleton stress, σ'_{mi}	205 (kPa)
Stress ratio at critical state, M^*_m	1.01
Viscoplastic parameter, m'	23.0
Viscoplastic parameter, C_1	1.0×10^{-11} (1/s)
Viscoplastic parameter, C_2	1.5×10^{-11} (1/s)
Structural parameter, β	0.0
Suction parameter, S_I	0.2
Suction parameter, s_d	0.25
Water permeability for full saturation, k_s^W	1.0×10^{-5} (m/s)
Air Permeability for full saturation, k_s^G	1.0×10^{-3} (m/s)

Figure 10.10. Figure 10.11 shows the distribution of the saturation with different times. The phreatic surface proceeds from the river side to the land side of the embankment. Figures 10.12 and 10.13 indicate the distributions of the pore water pressure and the mean skeleton stress during the seepage flow. The increases in pore water pressure associated with the advancement of the seepage surface, and correspondingly with the mean skeleton stress, decrease with the advancing seepage. Figure 10.14 indicates the development of the air pressure with time. The magnitude of the air pressure is relatively small but well simulated in the analysis.

Table 10.4 Soil parameters for hydraulic properties

Parameter	*Value*
Shape parameter for water permeability, a	3.0
Shape parameter for air permeability, b	2.3
Maximum saturation, $S_{r\,max}$	1.0
Minimum saturation, $S_{r\,min}$	0.0
Van Genuchten parameter, α	0.2(1/kPa)
Van Genuchten parameter, n	0.0
Parameter for initial stress analysis	Value
Young's modulus, E	7900 (kPa)
Poisson ratio, ν	0.33
Internal friction angle, ϕ	30.0
Cohesion, c	0.0

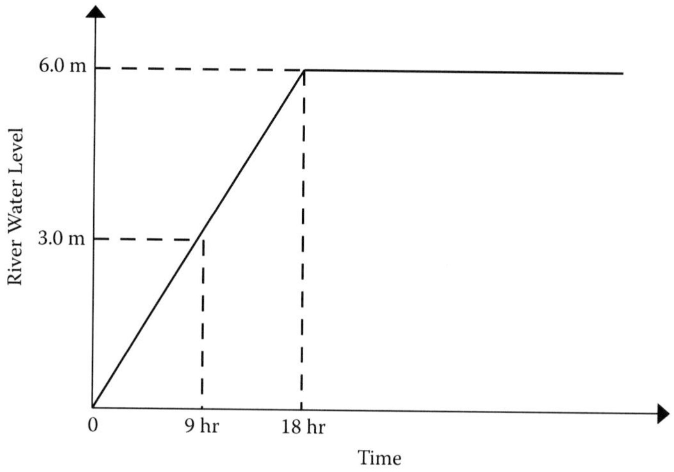

Figure 10.10 River water level–time profile.

The development of the accumulated plastic shear strain $\gamma^p = \int d\gamma^p$, $d\gamma^p = (de_{ij}^{vp} de_{ij}^{vp})^{1/2}, de_{ij}^{vp}$ (viscoplastic deviatoric strain increment) during the seepage flow is presented in Figure 10.15. The accumulated viscoplastic shear strain is an indicator of the inelastic deformation history. From this figure, in an early stage of seepage flow, the strain develops near the surface of the slope on the river side of the embankment, then a larger strain of more than 1% occurs at the toe of the embankment on the land side of the levee.

Figure 10.16 shows the distribution of the horizontal local hydraulic gradient. In this simulation, the maximum value of 1.26 has been reached inside the embankment on the river side after 18 hours. After 250 hours, the maximum value of 0.66 of the horizontal gradient is observed at the toe of the embankment.

Two-dimensional numerical deformation analyses of a river embankment under seepage conditions have been conducted based on the mixture theory. In modeling the unsaturated soil, we have adopted the skeleton stress and the suction as stress variables. The use of the skeleton stress is consistent with the definition of partial stresses in the mixture theory. In the analysis, we have adopted the finite element method with an updated Lagrangian scheme. For the constitutive model for soil, we used the viscoplastic model that can take the degradation into account. We have numerically analyzed the seepage–deformation coupled behavior of the river embankment with an increase in the river water level. Using the proposed seepage–deformation coupled three-phase analysis method, we can predict the behavior of levees during the increase in the river water level. The numerical results indicate that the large value of the horizontal hydraulic gradient corresponds to the large deformation at the toe of the embankment.

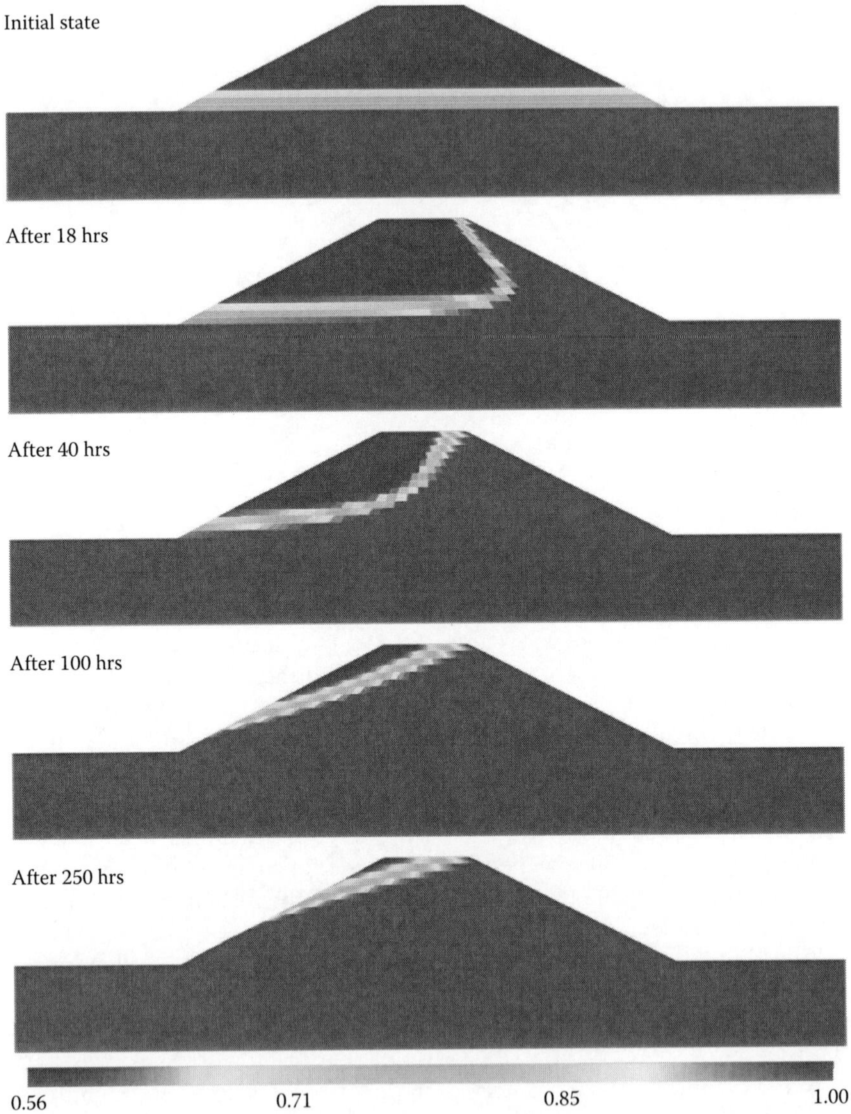

Figure 10.11 Distribution of the saturation with different times.

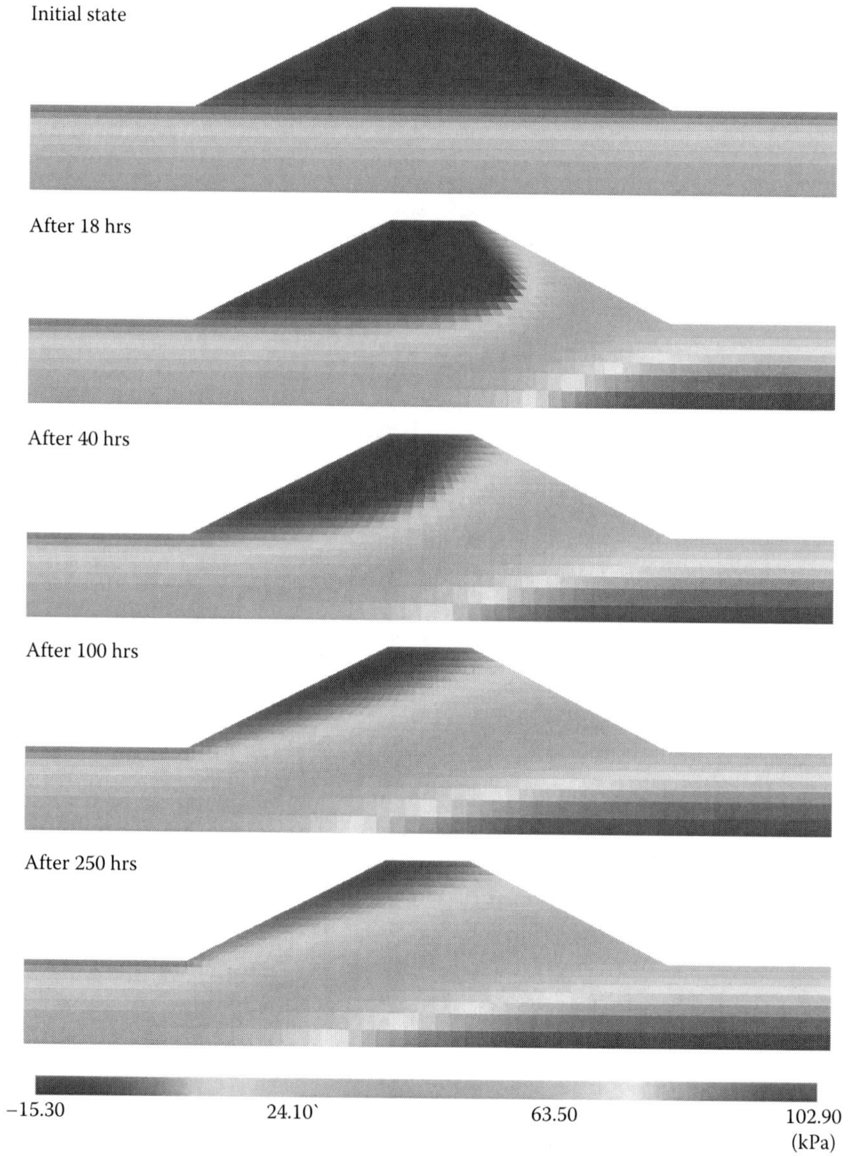

Figure 10.12 Distribution of the pore water pressure during the seepage flow.

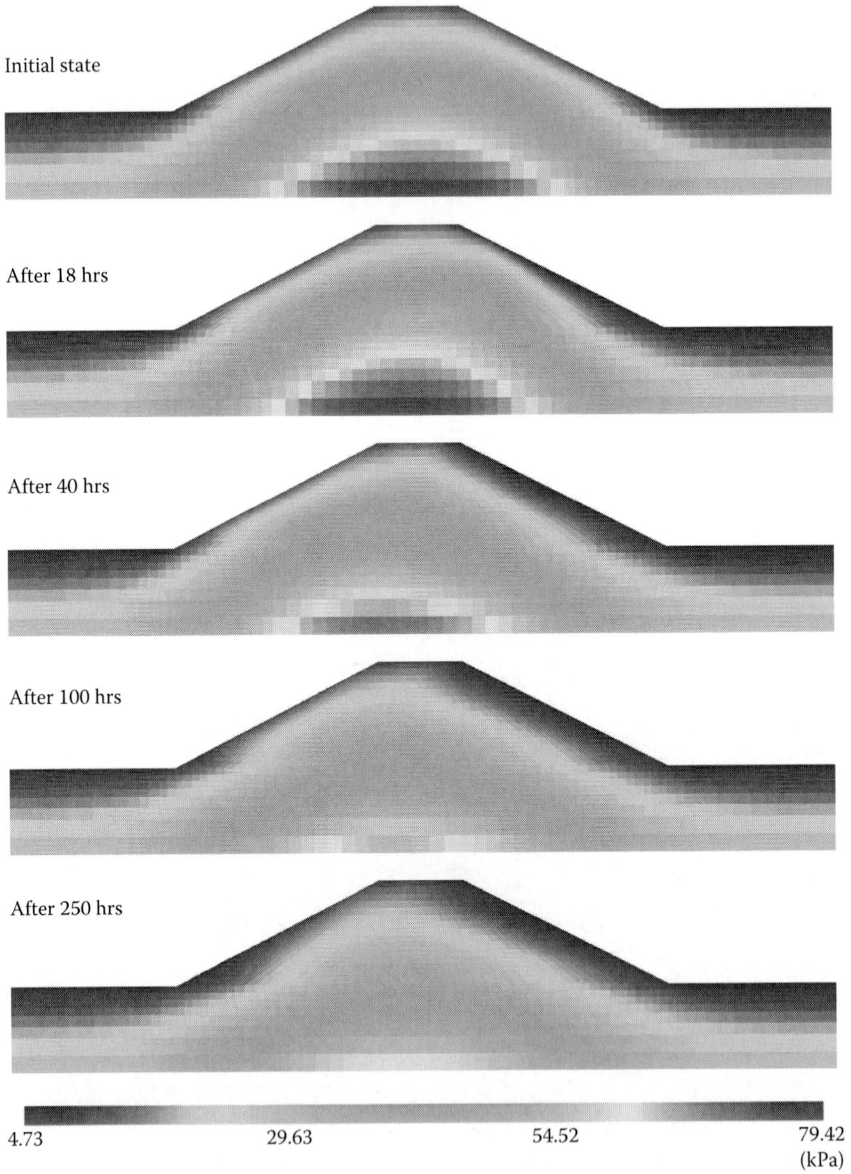

Figure 10.13 Distribution of the mean skeleton stress during the seepage flow.

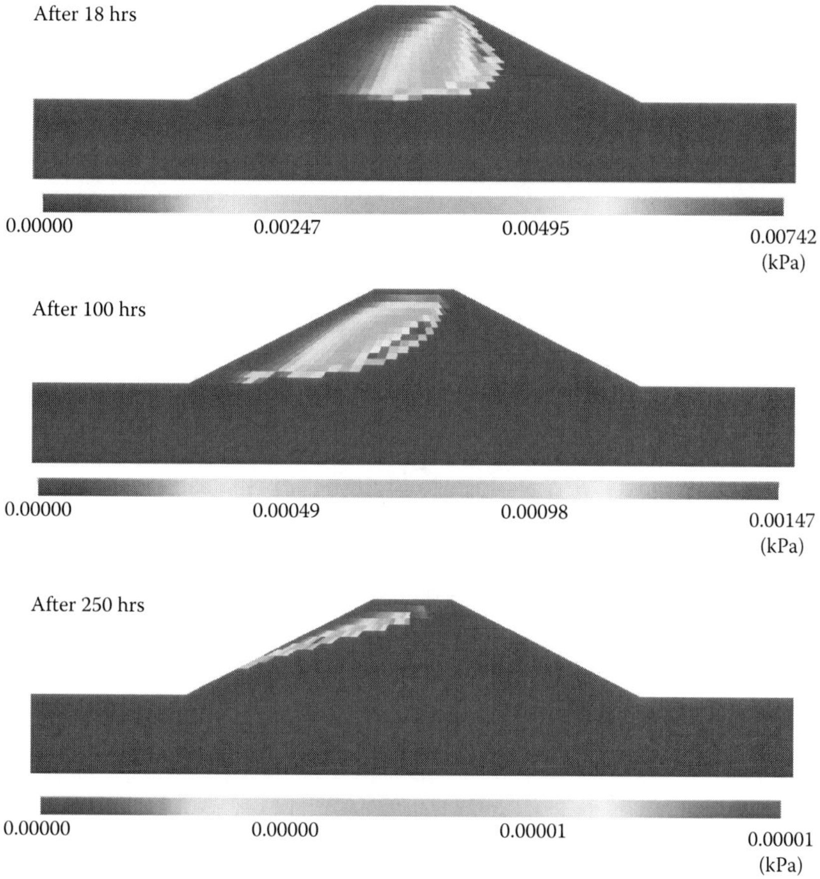

Figure 10.14 Distribution of air pressure with time.

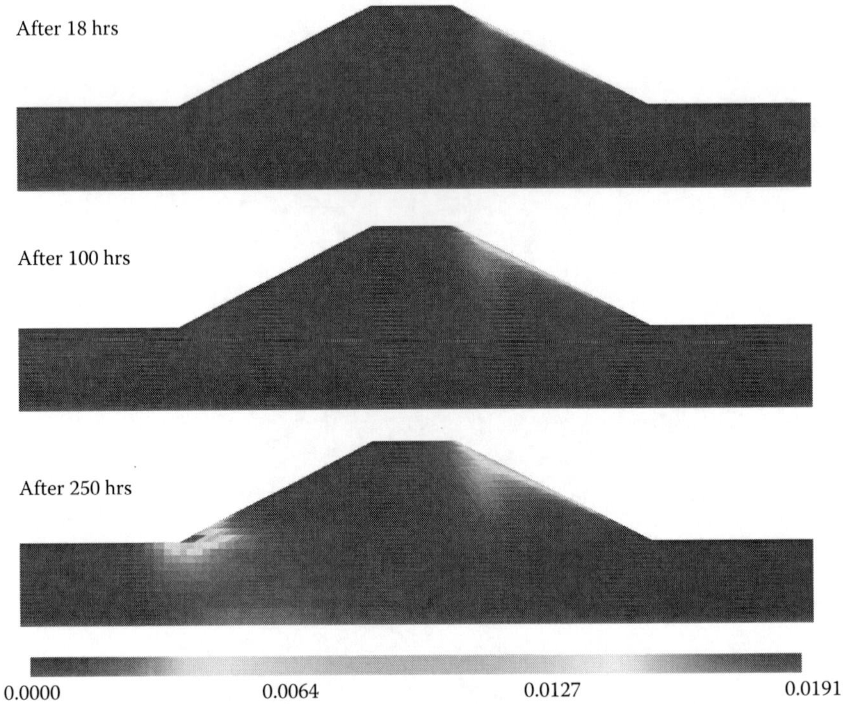

Figure 10.15 Distribution of the accumulated viscoplastic deviatoric strain.

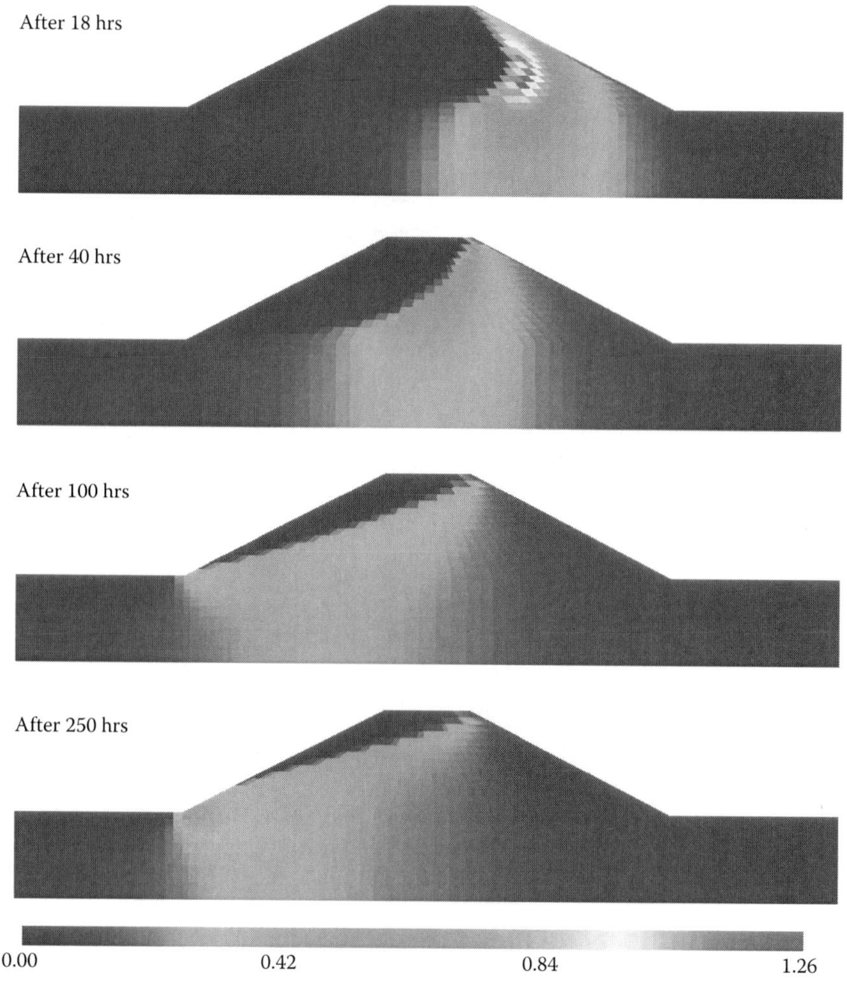

Figure 10.16 Distribution of the horizontal local hydraulic gradient.

REFERENCES

Adachi, T., and Oka, F. 1982. Constitutive equations for normally consolidated clay based on elasto-viscoplasticity, *Soils and Foundations*, 22(4):57–70.

Akagi, H., and Komiya, K. 1995. Constant rate of strain consolidation properties of clayey soil at high temperature, Proc. Int. Symp. Compression Consolidation of Clayey Soils, Hiroshima, H. Yoshikuni and O. Kusakabe eds., Balkema, Rotterdam, 1:3–8.

Baldi, G., Hueckel, T., and Pellegrini, R. 1988. Thermal volume change of the mineral-water system in low-porosity clay soils, *Can. Geotech. J.*, 25:807–825.

Biot, M. A. 1956. Theory of propagation of elastic waves in a fluid-saturated porous solid, *J. Acoust. Soc. Am.*, 28(2):168–178.

Boudali, M., Leroueil, S., and Srinivasa Murthy, B. R. 1994. Viscous behaviour of natural clays, Proc. 13th Int. Conf. on SMFE, New Delhi, India, 411–416.

Campanella, R.G., and Mitchell, J.K. 1968. Influence of temperature variations on soil behavior, *J. SMFE, ASCE*, 94, 3:709–734.

Cattaneo, C. 1958. A form of heat conduction equation which eliminates the paradox of instantaneous propagation, *Comp. Rend*, 247:431–433.

Cekerevac, C., and Laloui, L. 2004. Experimental study of thermal effects on the mechanical behaviour of a clay, *Int. J. Numer. Anal. Meth. Geomech.*, 28:209–228.

Cleall, P.J., Melhuish, T.A. and Thomas, H.R. 2006. Modelling the three-dimensional behaviour of a prototype nuclear waste repository, *Eng. Geo.*, 85:212–220.

Delage, P., Sultan, N., and Cui, Y.J. 2000. On the thermal consolidation of Boom clay, *Can. Geotech. J.*, 37:343–354.

Ehlers, W., Graf, T., and Ammann, M. 2004. Deformation and localization analysis of partially saturated soil, *Comp. Meth. Appl. Mech. Eng.*, 193:2885–2910.

Higo, Y. 2003. Instability and strain localization analysis of water-saturated clay by elasto-viscoplastic constitutive models, PhD thesis, Kyoto University, Japan.

Higo, Y., Oka, F., Kodaka, T., and Kimoto, S. 2006. Three-dimensional strain localization of water-saturated clay and numerical simulation using an elasto-viscoplastic model, *Philos. Mag.*, 86(21–22):3205–3240.

Japan Institute of Construction Engineering. 2002. Guide for structure investigations of river embankments (in Japanese).

Jommi, C. 2000. Remarks on the constitutive modelling of unsaturated soils, In Experimental Evidence and Theoretical Approaches in Unsaturated Soils, A.Tarantio and C. Mancuso, eds., Balkema, Rotterdam, 139–153.

Kim, Young Seok. 2004. Elasto-viscoplastic modeling and analysis for cohesive soil considering suction and temperature effects, PhD thesis, Kyoto University.

Kimoto, S., and Oka, F. 2003. An elasto-viscoplastic model for clay considering destructuralization and prediction of compaction bands, Proceedings of International Workshop on Prediction and Simulation Methods in Geomechanics, F. Oka, I. Vardoulakis, A. Murakami, and T. Kodaka, eds., Athens, Greece, TC34 of ISSMGE, 65–68.

Kimoto, S., and Oka, F. 2005. An elasto-viscoplastic model for clay considering destructuralization and consolidation analysis of unstable behavior, *Soils and Foundations*, 45(2):29–42.

Kimoto, S., Oka, F., Fukutani, J., Yabuki, T., and Nakashima, K. 2011. Monotonic and cyclic behavior of unsaturated sandy soil under drained and fully undrained conditions, *Soils and Foundations*, 51(4):663–681.

Kimoto, S., Oka, F. and Fushita, T. 2010. A chemo-thermo-mechanically coupled analysis of ground deformation induced by gas hydrate dissociation, *Int. J. Mech. Sci.*, 52:365–376.

Kimoto, S., Oka, F., Fushita, T., and Fujiwaki, M. 2007. A chemo-thermo-mechanically coupled numerical simulation of the subsurface ground deformations due to methane hydrate dissociation, *Comput. Geotech*, 34(4):216–228.

Kimoto, S., Oka, F., and Higo, Y. 2004. Strain localization analysis of elasto-viscoplastic soil considering structural degradation, *Comput. Meth. Appl. Mech. Eng.*, 193:2845–2866.

Kimoto, S., Oka, F., Fukutani, J., Yubuki, T. and Nakashima, K. 2011. Monotonic and cyclic behavior of unsaturated sandy soil under drained and fully undrained conditions, *Soils and Foundations*, 51(4):663–681.

Komine, H., and Ogata, N. 2002. Hydraulic properties of bentonite buffer and backfill materials and simplified evaluation, J. Geotech. Eng., JSCE, 708/III-59:133–144 (in Japanese).

Laloui, L., Nuth, M., and Vulliet, L. 2006. Experimental and numerical investigations of the behaviour of a heat exchanger pile, *Int. J. Num. Anal. Meth. Geomech*, 30(8):763–781.

Leroueil, S., Kabbaj, M., Tavenas, F., and Bouchard, R. 1985. Stress-strain-strain rate relation for the compressibility of sensitive natural clays, *Géotechnique*, 35(2):159–180.

Leroueil, S., and Marques, M.E.S. 1996. State of the art: Importance of strain rate and temperature effects in geotechnical engineering, Geotechnical Special Publication No. 61 ASCE, T.C. Sheahan and V. Kaliakin, eds., 1–60.

Müller, I., and Ruggeri, T. 1998. *Rational Extended Thermodynamics, 2nd ed.*, Springer, New York, 9–14.

Oka, F., Feng, H., Kimoto, S., Kodaka, T., and Suzuki, H. 2008. A numerical simulation of triaxial tests of unsaturated soil at constant water and constant air content by using an elasto-viscoplastic model, Proceedings of 1st European Conference on Unsaturated Soils, D. Toll and S.J. Wheeler, eds., Taylor & Francis/CRC Press, 735–741.

Oka, F., Higo, Y., and Kimoto, S. 2002. Effect of dilatancy on the strain localization of water-saturated elasto-viscoplastic soil, *Int. J. Solids Struct.*, 39:3625–3647.

Oka, F., Kimoto, S., Takada, N., Gotoh, H., and Higo, Y. 2010. A seepage-deformation coupled analysis of an unsaturated river embankment using a multiphase elasto-viscoplastic theory, *Soils and Foundations*, 50(4):483–494.

Oka, F., Kodaka, T., Kimoto, S., Kato, R., and Sunami, S. 2007. Hydro-mechanical coupled analysis of unsaturated river embankment due to seepage flow, *Key Eng. Mater.*, 340–341:1223–1230.

Oka, F., Kodaka, T., Suzuki, H., Kim, Y., Nishimatsu, N., and Kimoto, S. 2010. Experimental study on the behavior of unsaturated compacted silt under triaxial compression, *Soils and Foundations*, 50(1):27–44.

Oka, F., Yashima, A., and Leroueil, S. 1997. An elasto-thermo-viscoplastic model for natural clay and its application, Proceedings of 6th International Symposium on Numerical Models in Geomechanics, S. Pietruszczak and G. Pande, eds., Balkema, Rotterdam, 105–110.

Oka, F., Yashima, A., Sawada, K., and Aifantis, E.C. 2000. Instability of gradient-dependent elastoviscoplastic model for clay and strain localization, *Comput. Meth. Appl. Mech. Eng.*, 183:67–86.

Peirce, D., Shih, F., and Needleman, A. 1984. A tangent modulus method for rate dependent solids, *Comput. Struct.*, 18(5):845–887.

Schrefler, B. A. 1995. F.E. in environmental engineering-coupled thermo-hydro-mechanical processes in porous media including pollutant transport, *Arch. Comput. Meth. Eng.*, 2(3):1–54.

Suga, G., Hashimoto, H., and Ishikawa, T. 1984. Final report on the research of overflow levee (Interpretation edition), Report of the Civil Engineering Research Institute, 2074 (in Japanese).

Sulem, J., and Famin, V. 2009. Thermal decomposition of carbonates in fault zone:slip-weaking and temperature-limiting effects, *J. Geophys. Res.*, 114, B03309, doi10.1029/2008JB006004.

Sultan, N., Delage, P., and Cui, Y.J. 2002. Temperature effects on the volume change behaviour of boom clay, *Eng. Geol.*, 64:135–145.

Van Genuchten, M. T. 1980. A closed-form equation for predicting the hydraulic conductivity of unsaturated soils, *Soil Sci. Soc. Am. J.*, 44:892–898.

Vardoulakis, I. 2002. Dynamic thermo-poro-mechanical analysis of catastrophic landslides, *Géotechnique*, 52(3):157–171.

Yashima, A., Leroueil, S., Oka, F., and Guntoro, I. 1998. Modelling temperature and strain rate dependent behavior of clays: One-dimensional consolidation, *Soils and Foundations*, 38(2):63–73.

Index